BAINTE DEN STOC
WITHDRAWN FROM
DÚN LAOGHAIRE-RATHDOWN COUNTY
LIBRARY STOCK

THE WORLD BOOK ENCYCLOPEDIA OF
PEOPLE AND PLACES

THE WORLD BOOK ENCYCLOPEDIA OF

PEOPLE AND PLACES

2
D-H

a Scott Fetzer company
Chicago
www.worldbookonline.com

COVER ACKNOWLEDGMENTS

View toward Chora from Ellás on Serifos Island, Cyclades, Greece
© Simeone Giovanni, SIME/4Corners Images

View of Seine River from Île Saint Louis toward Notre Dame, Paris, France
© Simeone Giovanni, SIME/4Corners Images

Leabharlanna Dhún Laoghaire · Ráth An Dúin

For information about other World Book publications, visit our Web site at http://www.worldbookonline.com or call 1-800-WORLDBK (1-800-967-5325).

For information about sales to schools and libraries, call 1-800-975-3250 (United States); 1-800-837-5365 (Canada).

© 2008 World Book, Inc. All rights reserved. This volume may not be reproduced in whole or in part in any form without prior written permission from the publisher.

WORLD BOOK and the GLOBE DEVICE are registered trademarks or trademarks of World Book, Inc.

Previous editions © 2007, 2005 World Book, Inc.; © 2004, 2003, 2002, 2000, 1998, 1996, 1995, 1994, 1993, 1992 World Book, Inc., and Bertelsmann GmbH. All rights reserved.

Library of Congress Cataloging-in-Publication Data

The World Book encyclopedia of people and places.
p. cm.
Summary: "Alphabetically arranged set presents profile of individual nations and other political/geographical units including overview of history, geography, economy, people, culture, and government for each. Includes cumulative index and web site resources"--Provided by publisher.
Includes bibliographical references and index.
ISBN 978-0-7166-3757-8
1. Encyclopedias and dictionaries. 2. Geography--Encyclopedias. I. World Book, Inc.
AE5 .W563 2008
031--dc22
2007039279

Printed in Singapore.

18 19 20 21 22 12 11 10 09 08

STAFF

Vice President and Editor in Chief, World Book, Inc.
Paul A. Kobasa

EDITORIAL

Associate Director, Supplementary Publications
Scott Thomas

Managing Editor, Supplementary Publications
Barbara A. Mayes

Project Editor
Shawn Brennan

Senior Editor
Kristina A. Vaicikonis

Editor
Micah Savaglio
John Stowe

Statistics Editor
William M. Harrod

Researchers
Madolynn Cronk
Lynn Durbin
Cheryl Graham
Karen McCormack

Writers
Pamela Bliss
Kathy Klein
Susan Messer
Rita Vander Meulen

Consultant
Kempton Webb
Prof. Emeritus of Geography, Columbia University

Administrative Assistant
Ethel Matthews

PUBLICATIONS OPERATIONS

Director
Tony Tills

Manager
Loranne K. Shields

Editorial Administrator
Janet T. Peterson

Indexing Services Manager
David Pofelski

Information Services
Stephanie Kitchen

GRAPHICS AND DESIGN

Associate Director
Sandra M. Dyrlund

Associate Manager, Design
Brenda B. Tropinski

Senior Designer
Isaiah W. Sheppard, Jr.

Associate Manager, Photography
Tom Evans

Contributing Photographs Editor
Carol Parden

Senior Manager, Cartographic Services
H. George Stoll

Manager, Cartographic Services
Wayne K. Pichler

Senior Cartographer
John M. Rejba

Coordinator
John Whitney

PRODUCTION

Director, Manufacturing and Pre-Press
Carma Fazio

Manager, Manufacturing
Steven Hueppchen

Production/Technology Manager
Anne Fritzinger

Proofreader
Emilie Schrage

MARKETING

Chief Marketing Officer
Patricia Ginnis

Director, Direct Marketing
Mark Willy

Associate Director, School and Library Marketing
Jennifer Parello

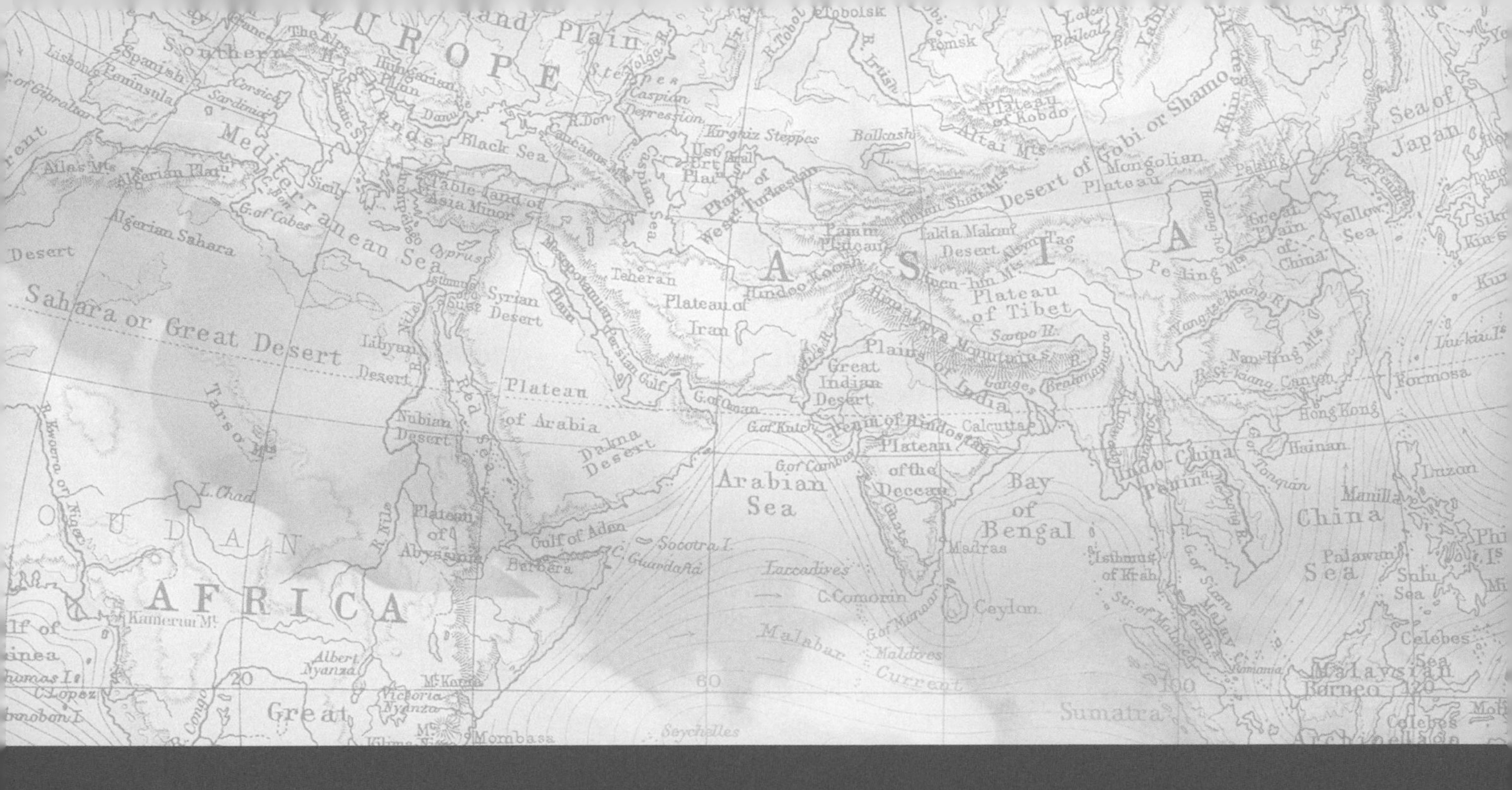
EUROPE
ASIA
AFRICA
Mediterranean Sea
Black Sea
Caspian Sea
Sahara or Great Desert
Plateau of Arabia
Arabian Sea
Bay of Bengal
Plateau of Tibet
Desert of Gobi or Shamo
Mongolian Plateau
Sea of Japan
Yellow Sea
China Sea
Great Indian Desert
Plains of India
Himalaya Mountains
Plateau of the Deccan
Ceylon
Madras
Calcutta
Teheran
Plateau of Iran
Syrian Desert
Libyan Desert
Nubian Desert
Plateau of Abyssinia
Gulf of Aden
Socotra I.
Laccadives
Maldives
Malabar Current
Sumatra
Borneo
Hainan
Formosa
Manilla
Luzon
Hong Kong
Canton
Indo-China Peninsula
Seychelles
Mombasa
Tobolsk
Tomsk
Kirghiz Steppes
Balkash
Altai Mts
Plateau of Kobdo
Plateau of Asia Minor
Algerian Sahara
Atlas Mts
Spanish Peninsula
L. Chad
Kamerun Mt
Albert Nyanza
Victoria Nyanza
Palawan
Sulu Sea
Celebes Sea
Celebes
Malaysian

CONTENTS

C. Mendocino
San Francisco
AMERICA
Californian Penin^a
Cancer
C. San Lucas
Mexican Plateau
Rio Grande
Southern Plain
Western Plain
Mississippi
New York
Gulf Stream
C. Hatteras
Bermudas
Atlantic
Florida Penin^a
Bahama Is
Gulf of Mexico
West Indian Archipelago
Hayti
Cuba
Yucatan Penin.
Jamaica
Caribbean Sea
C. Gallinas
North Equ
Popocatepetl
Isthmus of Tehuantepec
Isthmus of Panama
Orinoco
Plateau of Guiana
Guiana
Current
Equator
120
100
Galapagos
Rio Negro
Yapura
Selvas
Marañon
Amazon
C. Blanco

Political World Map

The world has 194 independent countries and about 40 dependencies. An independent country controls its own affairs. Dependencies are controlled in some way by independent countries. In most cases, an independent country is responsible for the dependency's foreign relations and defense, and some of the dependency's local affairs. However, many dependencies have complete control of their local affairs.

By 2007, the world's population surpassed 6.75 billion. Almost all of the world's people live in independent countries. Only about 12 million people live in dependencies.

Some regions of the world, including Antarctica and certain desert areas, have no permanent population. The most densely populated regions of the world are in Europe and in southern and eastern Asia. The world's largest country in terms of population is China, which has more than 1.3 billion people. The independent country with the smallest population is Vatican City, with only about 1,000 people. Vatican City, covering only 1/6 square mile (0.4 square kilometer), is also the smallest in terms of size. The world's largest nation in terms of area is Russia, which covers 6,592,850 square miles (17,075,400 square kilometers).

Every nation depends on other nations in some ways. The interdependence of the entire world and its peoples is called *globalism.* Nations trade with one another to earn money and to obtain manufactured goods or the natural resources that they lack. Nations with similar interests and political beliefs may pledge to support one another in case of war. Developed countries provide developing nations with financial aid and technical assistance. Such aid strengthens trade as well as defense ties.

Nations of the World

Name	Map key
Afghanistan	D 13
Albania	C 11
Algeria	D 10
Andorra	C 10‡
Angola	F 10
Antigua and Barbuda	E 6
Argentina	G 6
Armenia	D 12
Australia	G 16
Austria	C 10
Azerbaijan	D 12
Bahamas	D 6
Bahrain	D 12
Bangladesh	D 14
Barbados	E 7
Belarus	C 11
Belgium	C 10
Belize	E 5
Benin	E 10
Bhutan	D 14
Bolivia	F 6
Bosnia-Herzegovina	C 10
Botswana	G 11
Brazil	F 7
Brunei	E 15

Name	Map key
Bulgaria	C 11
Burkina Faso	E 9
Burundi	F 11
Cambodia	E 15
Cameroon	E 10
Canada	C 4
Cape Verde	E 8
Central African Republic	E 10
Chad	E 10
Chile	G 6
China	D 14
Colombia	E 6
Comoros	F 12
Congo (Brazzaville)	F 10
Congo (Kinshasa)	F 11
Costa Rica	E 5
Côte d'Ivoire	E 9
Croatia	C 10
Cuba	D 5
Cyprus	D 11
Czech Republic	C 10
Denmark	C 10
Djibouti	E 12
Dominica	E 6

Name	Map key
Dominican Republic	E 6
East Timor	F 16
Ecuador	F 6
Egypt	D 11
El Salvador	E 5
Equatorial Guinea	E 10
Eritrea	E 12
Estonia	C 11
Ethiopia	E 11
Federated States of Micronesia	E 17
Fiji	F 1
Finland	B 11
France	C 10
Gabon	F 10
Gambia	E 9
Georgia	C 12
Germany	C 10
Ghana	E 9
Great Britain	C 9
Greece	D 11
Grenada	E 6
Guatemala	E 5
Guinea	E 9
Guinea-Bissau	E 9

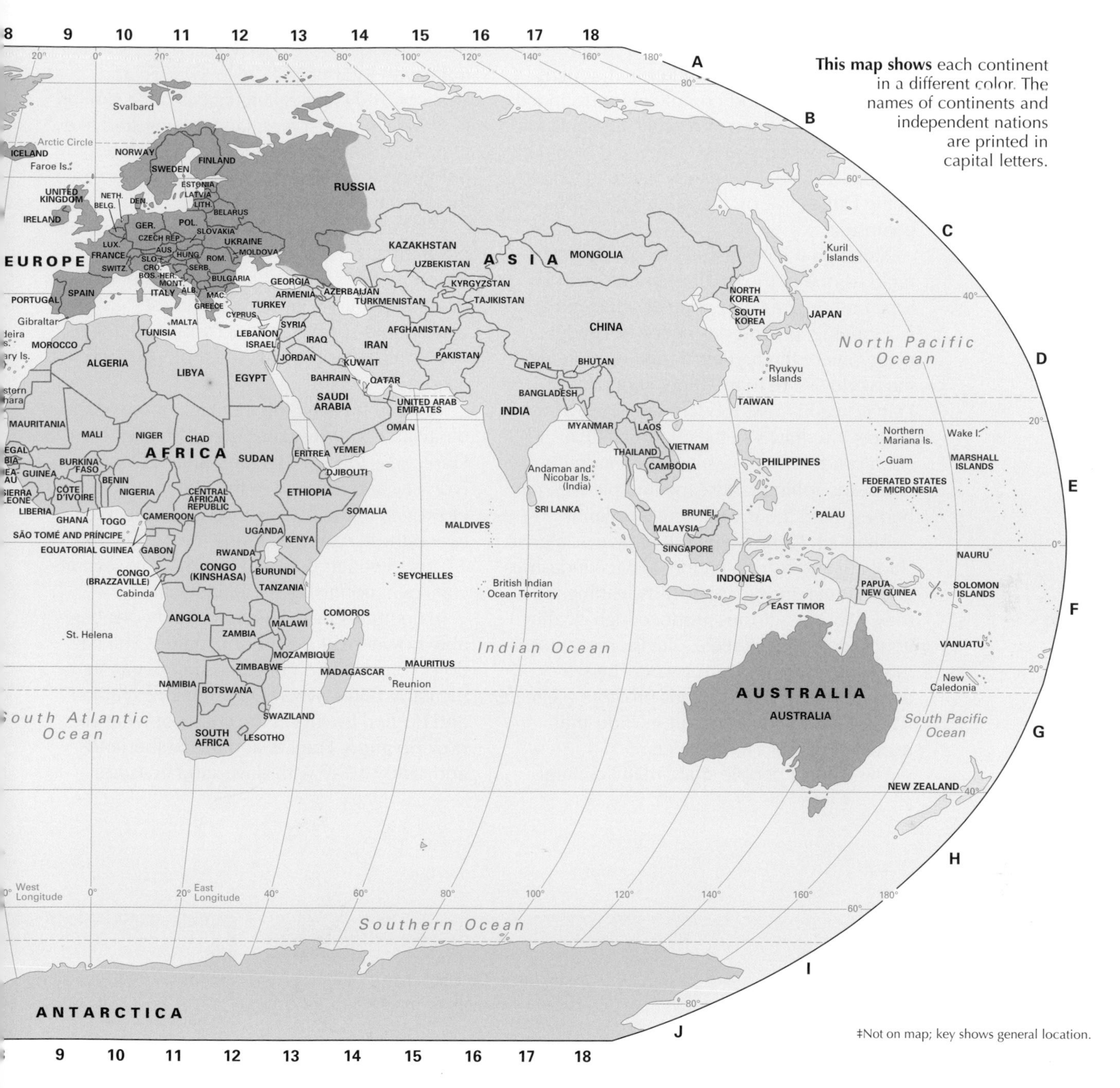

This map shows each continent in a different color. The names of continents and independent nations are printed in capital letters.

‡Not on map; key shows general location.

Name	Map key
Guyana	E 7
Haiti	E 6
Honduras	E 5
Hungary	C 10
Iceland	B 9
India	D 13
Indonesia	F 16
Iran	D 12
Iraq	D 12
Ireland	C 9
Israel	D 11
Italy	C 10
Jamaica	E 6
Japan	D 16
Jordan	D 11
Kazakhstan	C 13
Kenya	E 11
Kiribati	F 1
Korea, North	C 16
Korea, South	D 16
Kuwait	D 12
Kyrgyzstan	C 13
Laos	E 15
Latvia	C 11

Name	Map key
Lebanon	D 11
Lesotho	G 11
Liberia	E 9
Libya	D 10
Liechtenstein	C 10‡
Lithuania	C 11
Luxembourg	C 10
Macedonia	C 11
Madagascar	F 12
Malawi	F 11
Malaysia	E 15
Maldives	E 13
Mali	E 9
Malta	D 10
Marshall Islands	E 18
Mauritania	D 9
Mauritius	G 12
Mexico	D 4
Moldova	C 11
Monaco	C 10‡
Mongolia	C 15
Montenegro	C 10
Morocco	D 9
Mozambique	F 11
Myanmar	D 14

Name	Map key
Namibia	G 10
Nauru	F 18
Nepal	D 14
Netherlands	C 10
New Zealand	G 18
Nicaragua	E 5
Niger	E 10
Nigeria	E 10
Norway	B 10
Oman	E 12
Pakistan	D 13
Palau	E 16
Panama	E 5
Papua New Guinea	F 17
Paraguay	G 7
Peru	F 6
Philippines	E 16
Poland	C 10
Portugal	D 9
Qatar	D 12
Romania	C 11
Russia	C 13
Rwanda	F 11
St. Kitts and Nevis	E 6‡

Name	Map key
St. Lucia	E 6
St. Vincent and the Grenadines	E 6
Samoa	F 1
San Marino	C 10‡
São Tomé and Príncipe	E 10
Saudi Arabia	D 12
Senegal	E 9
Serbia	C 10
Seychelles	F 12
Sierra Leone	E 9
Singapore	E 15
Slovakia	C 11
Slovenia	C 11
Solomon Islands	F 18
Somalia	E 12
South Africa	G 11
Spain	C 9
Sri Lanka	E 14
Sudan	E 11
Suriname	E 7
Swaziland	G 11
Sweden	B 10
Switzerland	C 10
Syria	D 11

Name	Map key
Taiwan	D 16
Tajikistan	D 14
Tanzania	F 11
Thailand	E 15
Togo	E 9
Tonga	F 1
Trinidad and Tobago	E 6
Tunisia	D 10
Turkey	D 11
Turkmenistan	D 13
Tuvalu	F 1
Uganda	E 11
Ukraine	C 11
United Arab Emirates	D 12
United States	C 4
Uruguay	G 7
Uzbekistan	D 14
Vanuatu	F 18
Vatican City	C 10‡
Venezuela	E 6
Vietnam	E 15
Yemen	E 12
Zambia	F 11
Zimbabwe	G 11

Denmark

Denmark is a small kingdom in northern Europe that is almost completely surrounded by water. The Jutland Peninsula accounts for almost 70 per cent of the land, and Denmark also has 482 nearby islands. More than half of the Danes live on the islands near the peninsula.

Although it lacks natural resources and ranks as one of the smallest countries in Europe both in area and population, Denmark is a thriving, prosperous nation. Its people enjoy one of the highest standards of living in the world. More than 70 per cent of the Danes live in busy, modern cities. However, visitors are drawn to the charming Danish countryside, where castles and windmills rise above the rolling landscape and picturesque houses stand amid well-kept farmlands.

Although Denmark is still known as a great shipping and fishing nation, service industries and manufacturing are now its leading industries. Danish factories produce high-quality goods including stereos, television sets, furniture, porcelain, and silverware. The sea around the country's islands provides a cheap way for Denmark to import its industrial needs and export its products.

The Danish people share many cultural traits with their neighbors in Sweden and Norway. The Danish language is quite similar to Swedish and Norwegian tongues, and like their Scandinavian neighbors, most Danes belong to the Evangelical Lutheran Church, the official church of Denmark.

The nation's highly developed welfare program also resembles the systems of Sweden and Norway. Denmark was one of the first countries in the world to establish widespread social services by introducing public relief for the sick, unemployed, and aged.

The common Nordic heritage of the Scandinavian countries dates from 1397, when Denmark's Queen Margaret united Denmark, Norway, and Sweden in the Union of Kalmar. However, many battles have taken place between these nations. During the 1600's and 1700's, Sweden defeated Denmark in several wars over control of the Baltic Sea.

In centuries past, Denmark held considerable power over neighboring lands. The nation's long tradition of expansion began in the early 800's, when Danish Vikings raided and burned towns on the coasts of what are now Belgium, France, and the Netherlands, and sailed away with slaves and treasure.

FACT BOX

COUNTRY

Official name: Kongeriget Danmark (Kingdom of Denmark)
Capital: Copenhagen
Terrain: Low and flat to gently rolling plains
Area: 16,639 sq. mi. (43,094 km^2)
Climate: Temperate; humid and overcast; mild, windy winters and cool summers
Main rivers: Guden, Skjern
Highest elevation: Yding Skovhoj, 568 ft. (173 m)
Lowest elevation: Sea level along the coasts

GOVERNMENT

Form of government: Constitutional monarchy
Head of state: Monarch
Head of government: Prime minister
Administrative areas: 14 amter (counties), 2 kommunes (municipalities); Greenland is a province of Denmark; the Faroe Islands are a self-governing administrative division
Legislature: Folketing (Parliament) with 179 members serving four-year terms
Court system: Supreme Court
Armed forces: 30,000 troops

PEOPLE

Estimated 2008 population: 5,463,000
Population density: 328 persons per sq. mi. (127 per km^2)
Population distribution: 86% urban, 14% rural
Life expectancy in years: Male: 75, Female: 79
Doctors per 1,000 people: 3.7
Percentage of age-appropriate population enrolled in the following educational levels: Primary: 105*, Secondary: 129*, Further: 63
Languages spoken: Danish, Faroese

Denmark consists of the Jutland Peninsula and hundreds of islands to the east. The peninsula projects north from mainland Europe, toward the other Scandinavian nations. The islands lie between the Kattegat, a narrow extension of the North Sea, and the Baltic Sea.

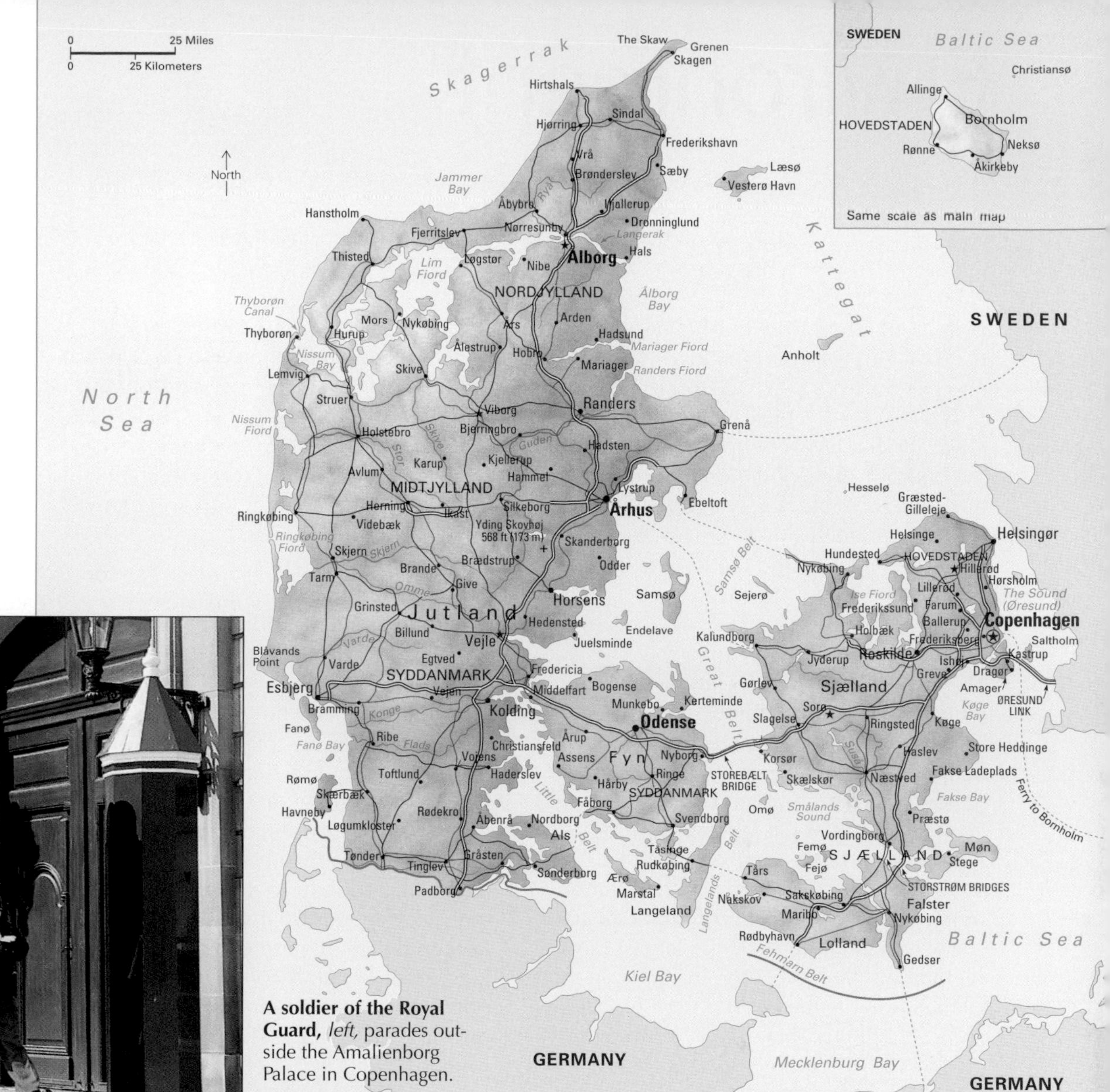

A soldier of the Royal Guard, *left,* parades outside the Amalienborg Palace in Copenhagen.

Greenlandic (an Inuit dialect)
German (small minority)
English is the predominant second language

Religions: Evangelical Lutheran 97%
other Protestant and Roman Catholic

*Enrollment ratios compare the number of students enrolled to the population which, by age, should be enrolled. A ratio higher than 100 indicates that students older or younger than the typical age range are also enrolled.

TECHNOLOGY

Radios per 1,000 people: 1,400

Televisions per 1,000 people: 859

Computers per 1,000 people: 576.8

ECONOMY

Currency: Danish krone

Gross domestic product (GDP) in 2004: $212.2 billion U.S.

Real annual growth rate (2003–2004): 2.1%

GDP per capita (2004): $32,200 U.S.

Goods exported: Machinery and instruments, meat and meat products, fuels, dairy products, ships, fish, chemicals

Goods imported: Machinery and equipment, petroleum, chemicals, grain and foodstuffs, textiles, paper

Trading partners: European Union, Norway, United States

In 865, the Danes invaded England and settled in the eastern half of the country.

Denmark chose to remain neutral in World War I (1914–1918). Early in World War II (1939-1945), German troops invaded Denmark, and the Danes surrendered after a few hours of fighting. After the war, U.S. aid helped the Danes rebuild industries that had been damaged.

Today, despite an economic recession during the 1970's and 1980's that brought inflation and unemployment, Denmark's economy remains strong by worldwide standards. However, the Danish economy is greatly affected by international trends and developments, since Denmark must sell its products to other countries to pay for the fuels and metals it imports. Denmark belongs to the European Union, but reserves the right to abstain from a common currency or defense policy.

Environment

The landscape of Denmark owes most of its features to the Ice Age, when glaciers moved slowly over the region from about 2 million to 10,000 years ago. The advancing glaciers moved rocks and boulders with them, and many were crushed almost to powder. This debris formed *moraines* (banks and ridges) where the ice stopped.

These glacial deposits almost completely cover the flat layers of limestone that form the bedrock of the Jutland Peninsula and the islands. White, finely grained limestone formations are visible in only a few places, such as Lim Fiord and the islands of Møn and Bornholm.

Land regions

The smooth curves of Denmark's Western Dune Coast consist chiefly of great sandy beaches that close off many fiords once connected to the sea. Inland, the Western Sand Plains were formed when melting glaciers flowed over the land, depositing quantities of sand.

The East-Central Hills, which include much of Jutland and almost all the nearby islands, make up Denmark's largest land region, and the deep moraine soils on these islands provide the best farmland in Denmark. Here, Danish farmers grow such crops as barley, potatoes, sugar beets, and a leafy herb known as *canola.*

Most of the crops grown on these islands are used for animal feed, since livestock production is the major activity on most Danish farms. About two-thirds of the country's agricultural products are exported, and Denmark is world famous for its meat and dairy products, including cheeses, butter, bacon, and ham.

The Lim Fiord winds through northern Jutland for 112 miles (180 kilometers), forming an inland lagoon 15 miles (24 kilometers) wide. A beach on the Western Dune Coast closes the fiord's outlet to the North Sea, so small vessels use the Thyborøn Canal to travel between the fiord and the sea.

Sjælland, the nation's largest island and the most thickly populated part of Denmark, lies in the East-Central Hills. Most of Copenhagen, Denmark's capital and largest city, stands on this island, and nearly half of the nation's industries are located in the Copenhagen area.

Located on the northeast coast of Jutland, the Northern Flat Plains were once part of the seabed. The region rose up from the water when the weight of the ancient glaciers was lifted as they melted. Like the East-Central Hills, the Northern Flat Plains have deep, rich moraine soil and many farms.

Natural resources

Unlike the other Scandinavian nations, only about 10 per cent of Denmark is forested, and these woodlands supply less than half the nation's annual timber requirement.

Denmark has few other natural resources. Although the country gets some natural gas and petroleum from wells in the North Sea, it must still import petroleum products. The only other minerals mined are chalk and industrial clay, and Danish industry depends on imports for its coal, iron, and most other metals. Because the land is flat or gently rolling, the rivers cannot be used to generate hydroelectric power. However, the sea, which almost surrounds Denmark, is rich in such fish as cod, herring, Norway pout, sand lances, sprat, and whiting.

Climate

Mainly because it is almost surrounded by water, Denmark has a mild, humid climate that is affected by sea winds throughout the year. In winter, the sea is warmer than the land, and in summer, it is cooler. As a result, west winds from the sea warm Denmark in winter and cool the land in summer.

The climate varies little throughout the country because Denmark has no mountains to block the sea winds. However, the moisture-bearing sea winds reach western Denmark first, so that area receives more rainfall than eastern Denmark. Snow falls from 20 to 30 days a year, but usually melts quickly.

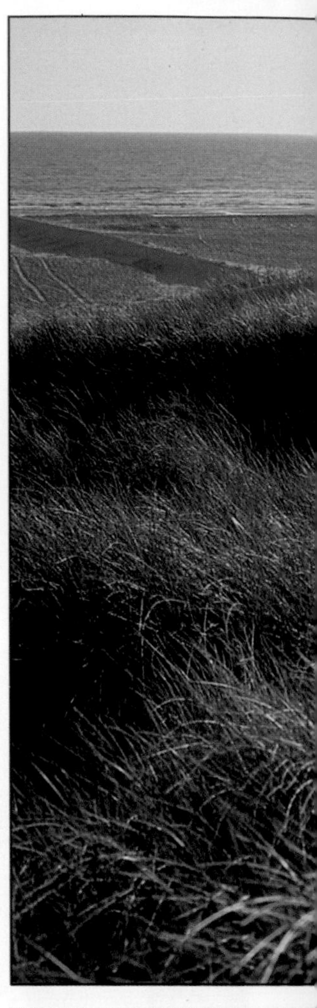

Fishing boats are pulled up for the night onto the smooth, dune-lined beaches of western Jutland, *left.*

Except for the extreme southeast of the island of Bornholm, Denmark, *map right,* consists of a glacial deposit over a limestone base. The landscape is made up of small hills, moors, ridges, hilly islands, and raised seabed.

Denmark's coastal waters are rich fishing grounds that provide a bountiful catch.

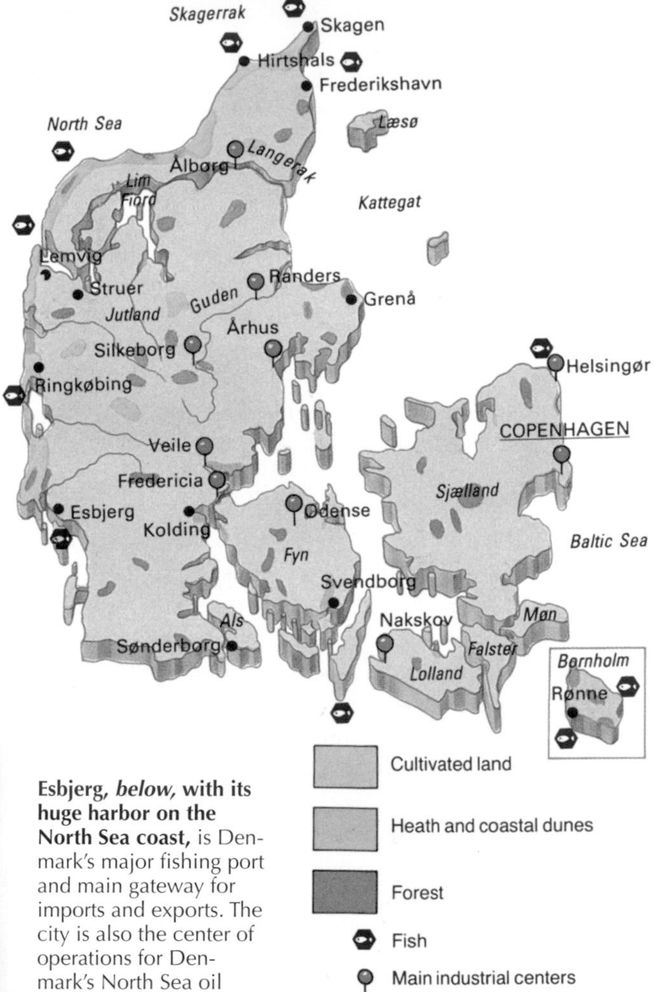

Esbjerg, *below,* with its huge harbor on the North Sea coast, is Denmark's major fishing port and main gateway for imports and exports. The city is also the center of operations for Denmark's North Sea oil exploration.

The sheer cliffs carved by the sea from cream-colored chalk (finely grained limestone) are a famous landmark on the island of Møn, *far left.* Known as Møns Klint, the cliffs tower 400 feet (120 meters) above a narrow beach that can be reached only by steep steps.

A whitewashed farmhouse stands in lush green fields on the island of Aero, *left.* The farmland that makes up about 75 per cent of Denmark's total area gets its fertile soil from deposits of sand and fine clay left behind by the glaciers after the last Ice Age.

Copenhagen

Copenhagen, Denmark's capital and largest city, traces its origins to the mid-1000's, when it was a small fishing village. Today, the city is Denmark's major port and also serves as the center of the nation's economic, political, and cultural activity. Most of the city lies on the east coast of the island of Sjælland, while other sections are on Amager, an island just east of Sjælland.

A historic city

Until the mid-1100's, Copenhagen—whose well-sheltered port provides immediate access to the sea—was continually raided by pirates. In 1167, Archbishop Absalon of Roskilde built a castle to protect the harbor, which encouraged the growth and development of Copenhagen as a trade center. In 1254, Copenhagen was granted a town charter.

As the town continued to grow in size and importance within the castle's protective walls, it attracted the jealous wrath of German Hanseatic merchants, who destroyed the castle in 1369. The settlement was rebuilt and in 1443, it became the capital of Denmark.

During the next 400 years, Copenhagen suffered through terrible times. In the 1600's, the town was attacked by Swedish forces. In the 1700's, much of it was destroyed by fires, while epidemics killed many of its people. The town had barely recovered before being bombarded by the British fleet in 1801 and 1807, during the Napoleonic Wars.

Copenhagen bounced back each time from the devastation, and continued to grow as an economic, military, and political center. During the 1850's, the city expanded to the north and west, and in the late 1800's, it also experienced rapid economic growth and began to develop industries.

Modern Copenhagen

Considered by many to be one of the most delightful capitals in the world, present-day Copenhagen boasts a wealth of beautiful, historic buildings, many of them built in the 1600's, during the reign of King Christian IV.

Town Hall Square, at the center of Copenhagen, is surrounded by open-air cafes and historic buildings, including the Town Hall, which was built in the late 1800's. The Dragon Fountain, in the center of the square, is dedicated to Hans Christian Andersen, Denmark's most famous writer. Andersen's fairy tales have enchanted generations of the world's children.

The Little Mermaid, *top,* a bronze statue of one of Hans Christian Andersen's most beloved characters, watches over Copenhagen's harbor.

The Tivoli Gardens amusement park, in the center of the city, is a popular meeting place for Danes of all ages. The Chinese pagoda, lit by hundreds of tiny lights, is one of the park's many popular attractions.

Christian IV became known as Denmark's greatest builder. He sponsored the construction of some of Copenhagen's most elegant buildings and even helped design some of them, including the Rosenborg Palace (1606) and the Stock Exchange (1624).

The Rosenborg Palace, once the king's spring and autumn residence, is now a historical museum containing a breathtaking display of all the Danish crown jewels, as well as the personal effects of Danish kings from the time of Christian IV. The Stock Exchange, the oldest building of its kind still used for its original purpose, is remarkable for its spire adorned with intertwined dragons' tails.

Another distinctive feature of Copenhagen is the world-famous Tivoli Gardens amusement park, built in 1843. About 4 million visitors come to Tivoli every summer to enjoy its various forms of entertainment, including shows, games, rides, restaurants, museums, and vast flower gardens. Tivoli also boasts a concert hall and puppet theater. On certain nights, the park closes with a magnificent fireworks display.

East of the Tivoli Gardens stands Christiansborg Castle, the seat of the Danish Parliament (Folketing) and Supreme Court. The castle stands on the same site as the castle built by Archbishop Absalon, and the ruins of the original castle can be seen under the present building. Farther east stands Amalienborg Palace, a square of four mansions that has served as the royal residence since 1794.

In addition to its historic buildings, Copenhagen boasts a wealth of fascinating museums. The National Museum, built in the 1740's, houses the world's finest collection of Stone Age tools, as well as runic stones dating from Viking times. The Museum of Decorative Arts displays a collection of ceramics, silverware, and Flemish tapestries from the Middle Ages.

Picturesque houses built by wealthy Danish merchants during the 1700's and 1800's, *left,* line the Nyhavn Canal. The quays along Nyhavn are filled with old sailing ships, including an 1893 lightship that is now a floating museum.

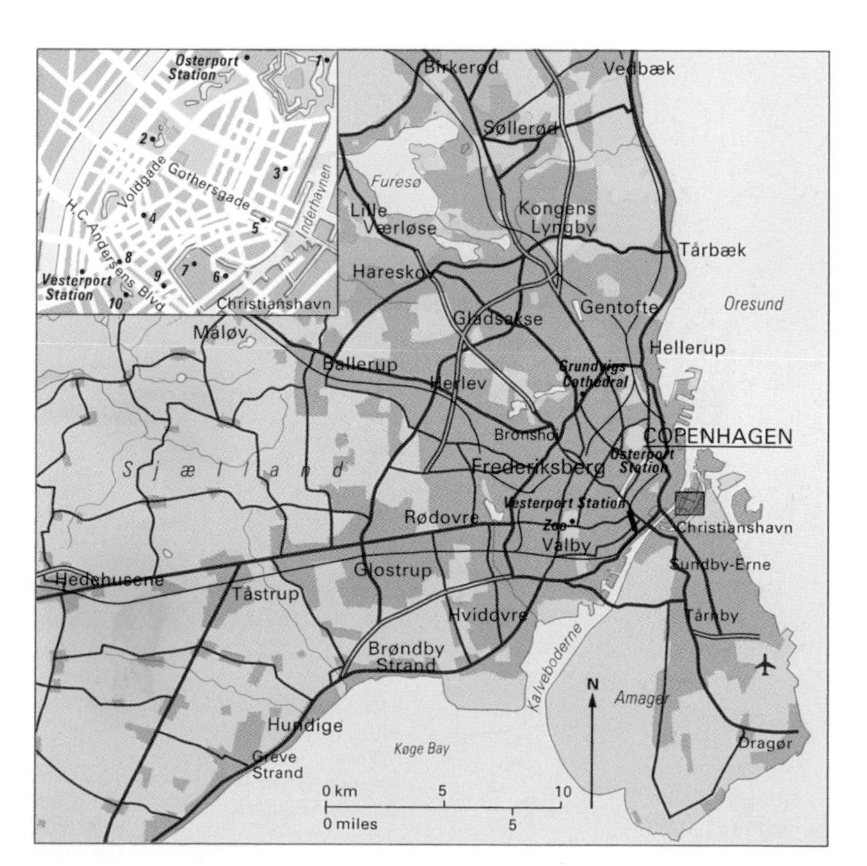

Copenhagen features many cultural and historic sites, including the statue *The Little Mermaid* (1), the Botanical Gardens (2), the Amalienborg Palace (3), the University of Copenhagen (4), the Nyhavn Canal (5), the Stock Exchange (6), the Christiansborg Castle (7), the Town Hall Square (8), the National Museum (9), and Tivoli Gardens (10).

Djibouti

Djibouti is a small country on the eastern coast of Africa, where the continent nearly touches the Arabian Peninsula. Its hot, dry climate, near-barren land, and lack of resources have left it poor and underdeveloped. The country's entire manufacturing industry consists of two soft-drink plants. The only agricultural activity is livestock herding.

Djibouti's sole advantage is its location, which has helped make its capital city—also named Djibouti—a major port. A railroad linking Djibouti with Ethiopia also makes the capital a center for trade. The economy of the entire nation depends almost totally on this sea and rail trade.

In addition, Djibouti's location is important in world politics. If a powerful nation ever gained control of this area, it could interfere with the many ships that now move freely past Djibouti's coast traveling through the Suez Canal and the Red Sea between the Mediterranean Sea and the Indian Ocean.

Despite Djibouti's desert climate and desolate land, people have lived in the area since prehistoric times. During the A.D. 800's, Muslims from Arabia converted the Afars, a nomadic group living in the region, to Islam. The Afars established several Islamic states and fought wars with the Christians in neighboring Ethiopia from the 1200's through the early 1600's.

Workers cut blocks of salt from the surface of Lake Assal. Lying on the edge of Djibouti's barren coastal plain, Lake Assal is the lowest point in Africa. A rugged plateau stretches beyond a high mountain range farther inland. Plant life is sparse in this hot, dry land.

By the 1800's, however, a group of Somali nomads from the south called the Issas had taken over a large part of the Afars' grazing lands. The two groups grew hostile toward each other.

In 1862, France bought a port in the region and established a coaling station for its ships. The French eventually gained control of more land and turned it into a territory named French Somaliland. In 1967, the name was changed to the French Territory of the Afars and Issas.

Opposition to French rule grew in the 1970's, especially from the Issas. In 1977, the people voted overwhelmingly for independence in May, and on June 27, the territory became the independent nation of Djibouti.

The Afars and the Issas still make up the two main ethnic groups in Djibouti. The Afars live

FACT BOX

COUNTRY

Official name: Republic of Djibouti
Capital: Djibouti
Terrain: Coastal plain and plateau separated by central mountains
Area: 8,958 sq. mi. (23,200 km^2)
Climate: Desert; torrid, dry
Highest elevation: Mousaalli, 6,768 ft. (2,063 m)
Lowest elevation: Lake Assal, 509 ft. (155 m) below sea level

GOVERNMENT

Form of government: Republic
Head of state: President
Head of government: Prime minister
Administrative areas: 5 cercles (districts)
Legislature: Chambre des Deputes (Chamber of Deputies) with 65 members serving five-year terms
Court system: Cour Supreme (Supreme Court)
Armed forces: 9,850 troops

PEOPLE

Estimated 2008 population: 838,000
Population density: 94 persons per sq. mi. (36 per km^2)
Population distribution: 85% urban, 15% rural
Life expectancy in years: Male: 45 Female: 48
Doctors per 1,000 people: N/A
Percentage of age-appropriate population enrolled in the following educational levels: Primary: N/A Secondary: N/A Further: N/A

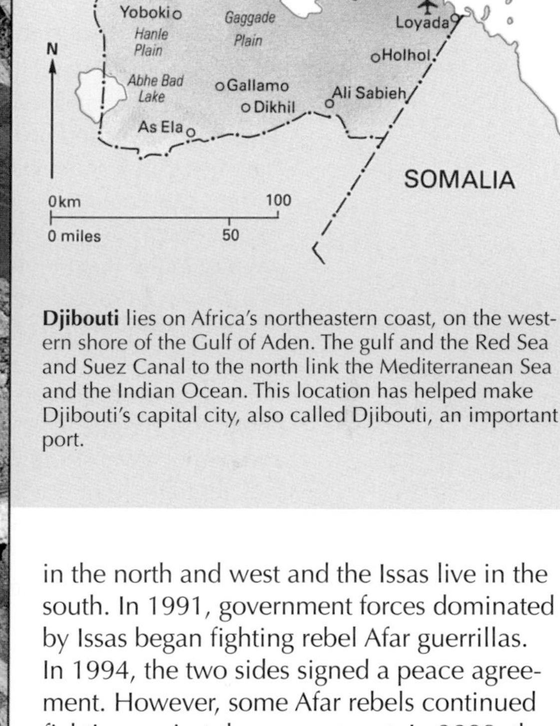

Djibouti lies on Africa's northeastern coast, on the western shore of the Gulf of Aden. The gulf and the Red Sea and Suez Canal to the north link the Mediterranean Sea and the Indian Ocean. This location has helped make Djibouti's capital city, also called Djibouti, an important port.

in the north and west and the Issas live in the south. In 1991, government forces dominated by Issas began fighting rebel Afar guerrillas. In 1994, the two sides signed a peace agreement. However, some Afar rebels continued fighting against the government. In 2000, the last remaining rebels and the government signed a peace agreement.

Many Afars and Issas speak Afar or Somali rather than Djibouti's official language of Arabic and follow the nomadic way of life of their ancestors. They wander over the desolate land with their herds of animals. Scorching heat, little water, and scarce grazing land make life difficult for the nomads, so almost 450,000 Afars and Issas have moved to the capital city of Djibouti in search of a better life. However, poverty plagues these urban dwellers, and unemployment is as high as 80 per cent.

Educational opportunities are also limited, and less than half of the population can read and write. Large numbers of Djibouti workers spend almost half their income on *khat,* a leaf that produces a feeling of well-being when it is chewed.

Languages spoken:
French (official)
Arabic (official)
Somali
Afar

Religions:
Muslim 94%
Christian 6%

TECHNOLOGY

Radios per 1,000 people: N/A

Televisions per 1,000 people: N/A

Computers per 1,000 people: N/A

ECONOMY

Currency: Djiboutian franc

Gross domestic product (GDP) in 2002: $619 million U.S.

Real annual growth rate (2001–2002): 3.5%

GDP per capita (2002): $1,300 U.S.

Goods exported: Re-exports, hides and skins, coffee (in transit)

Goods imported: Foods, beverages, transport equipment, chemicals, petroleum products

Trading partners: Somalia, Yemen, Ethiopia, France, Italy

Dominica

A small island republic in the Lesser Antilles, Dominica lies 320 miles (515 kilometers) north of Venezuela. This mountainous, tree-covered island, which was formed by volcanic eruptions, is one of the most unspoiled islands in the Caribbean. The Dominicans call their home the "Island of Adventure."

An unspoiled wilderness

Traveling around Dominica can be difficult because some roads are so poor and others have been damaged by heavy rainfall and avalanches. But visitors willing to brave the bumpy, desolate roads are rewarded with some of the most magnificent, unspoiled scenery in the West Indies. A ride across the island is an extraordinary journey through dense rain forests and over high, volcanic mountains.

The rain forests of the interior support about 200 varieties of trees, including cedar and mahogany. In addition, the forests contain abundant bird life. Flowering plants, such as hibiscus, lilies, and orchids, also blanket the land.

Rain water tumbles down the mountains in huge torrents and merges with swift-flowing rivers. The rivers, which are too rough to be used by boats other than canoes, flow into coastal bays lined by black volcanic sands.

Dominica's interior also features hundreds of waterfalls and crystal-clear lakes. Boiling Lake, with its bubbling sulfur springs, is the second largest such lake in the world.

History and government

Arawak Indians, the first people to settle in Dominica, arrived there about 2,000 years ago. About 1,000 years later, Carib Indians took over the island. Christopher Columbus sighted the island on Sunday, Nov. 3, 1493. He named the island *Dominica,* which is the Latin word for *Sunday.*

French and British settlers began to arrive in Dominica in the 1600's, and the French, British, and Carib fought for control of the island for many years. It was not until 1763 that the British gained possession of Dominica. They established large plantations and brought in black African slaves to work on them. The slaves were freed in 1834.

Volcanic mountains form the backdrop for a coastal scene on the island of Dominica. Some mountains in the north and south reach heights of more than 4,000 feet (1,200 meters).

FACT BOX

COUNTRY

Official name: Commonwealth of Dominica
Capital: Roseau
Terrain: Rugged mountains of volcanic origin
Area: 290 sq. mi. (751 km^2)
Climate: Tropical; moderated by northeast trade winds; heavy rainfall
Main river: Layou
Highest elevation: Morne Diablotin, 4,747 ft. (1,447 m)
Lowest elevation: Caribbean Sea, sea level

GOVERNMENT

Form of government: Republic
Head of state: President
Head of government: Prime minister
Administrative areas: 10 parishes
Legislature: House of Assembly with 24 members serving five-year terms
Court system: Eastern Caribbean Supreme Court
Armed forces: N/A

PEOPLE

Estimated 2008 population: 80,000
Population density: 276 persons per sq. mi. (107 per km^2)
Population distribution: 75% urban, 25% rural
Life expectancy in years:
Male: 71
Female: 77
Doctors per 1,000 people: N/A
Percentage of age-appropriate population enrolled in the following educational levels:
Primary: N/A
Secondary: N/A
Further: N/A

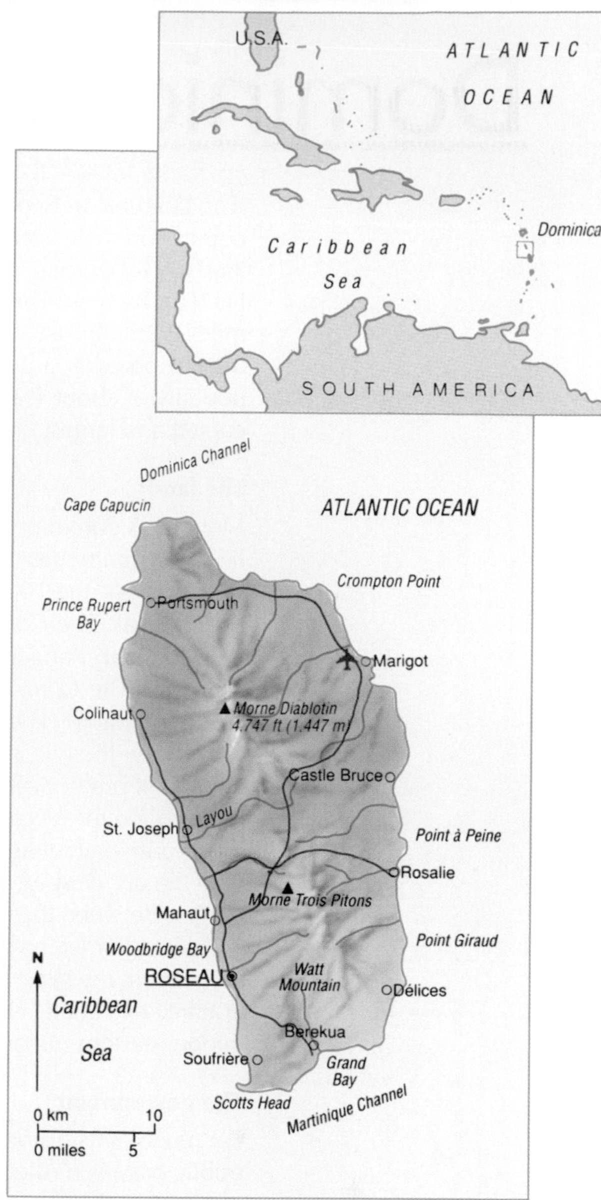

Languages spoken:
: English (official)
: French patois

Religions:
: Roman Catholic 80%
: Other Protestant

TECHNOLOGY

Radios per 1,000 people: N/A

Televisions per 1,000 people: N/A

Computers per 1,000 people: N/A

ECONOMY

Currency: East Caribbean dollar

Gross domestic product (GDP) in 2003: $384 million U.S.

Real annual growth rate (2003): -1%

GDP per capita (2003): $5,500 U.S.

Goods exported:
Mostly: bananas
Also: soap, bay oil, vegetables, grapefruit, oranges

Goods imported:
Manufactured goods, machinery and equipment, food, chemicals

Trading partners:
Caricom countries, United Kingdom, United States

Dominica, *above right,* has been an independent republic since 1978. Most of its people have African or mixed African, British, and French ancestry. About 70 per cent of the people live in urban areas while about 30 per cent live in rural villages. English is spoken in the cities, while the villagers generally speak *French patois*—a mixture of African languages and French.

Between the 1930's and the 1970's, the United Kingdom gradually gave the island control over its internal affairs. Dominica became an independent republic on Nov. 3, 1978—exactly 485 years after Columbus first sighted the island from the deck of his ship.

Today, Dominica is one of the poorest nations in the West Indies, and its many problems include poverty, economic mismanagement, and political corruption. A rugged landscape, heavy rainfall, avalanches, and hurricanes have added to the nation's difficulties. In spite of their country's problems, however, the people of Dominica welcome visitors with warmth and friendliness. The Dominicans' pride in their "Island of Adventure" far outweighs the daily struggle to make a living.

Dominican Republic

The Dominican Republic is a mountainous country on the eastern two-thirds of the island of Hispaniola. It shares the island with Haiti to the west. The country lies about 575 miles (925 kilometers) southeast of Florida in the Caribbean Sea. Santo Domingo, a busy port city of about 1-1/3 million people, is its capital and largest city.

The land

Mountains dominate the Dominican Republic. The country has the highest point in the West Indies—the 10,417-foot (3,175-meter) Duarte Peak. The *Cordillera Central,* or Central Mountain Range, runs northwest to southeast. The *Cordillera Septentrional,* or Northerly Range, is in the far north.

Between the two cordilleras lies the Cibao, an area of pine-covered slopes and a fertile plain called the *Vega Real,* or Royal Plain—the country's chief agricultural area.

In the dry west lie two other mountain ranges. Between them lies Lake Enriquillo, at 151 feet (46 meters) below sea level, the lowest point in the West Indies. Most of the nation's sugar cane is grown in the less mountainous eastern region.

The government

For much of its history, the Dominican Republic has been ruled by dictators or by other countries. Today, a president heads the country. The president is elected by the people for a four-year term and appoints a Cabinet to help run the government and governors to head the provinces.

The people also elect the national legislature. It consists of a 32-member Senate and a 150-member Chamber of Deputies. Members of both houses are elected to four-year terms.

History

Christopher Columbus landed on the island of Hispaniola in 1492. Thousands of Spanish colonists soon followed, many seeking gold. In 1496, they founded La Nueva Isabela, now Santo Domingo, the first city in the Western Hemisphere founded by Europeans.

After the Spaniards nearly wiped out the Indians who were living on the island, many colonists left Hispaniola for more prosperous settlements. The king of Spain then ordered the remaining colonists to move to the Santo Domingo area. This action cost Spain its control of the northern and western sections of the island. France eventually took over the western end, now Haiti.

When black slaves in Haiti revolted, they gained control of the whole island. From 1822 to 1844, Haiti controlled all of Hispaniola. In 1844, Dominican heroes Juan Pablo Duarte, Francisco del Rosario Sánchez, and

FACT BOX

COUNTRY

Official name: Republica Dominicana (Dominican Republic)
Capital: Santo Domingo
Terrain: Rugged highlands and mountains with fertile valleys interspersed
Area: 18,815 sq. mi. (48,730 km^2)
Climate: Tropical maritime; little seasonal temperature variation; seasonal variation in rainfall
Main rivers: Yaque del Norte, Yaque del Sur, Yuna
Highest elevation: Pico Duarte, 10,417 ft. (3,175 m)
Lowest elevation: Lago Enriquillo, 150 ft. (46 m) below sea level

GOVERNMENT

Form of government: Republic democracy
Head of state: President
Head of government: President
Administrative areas: 31 provincias (provinces), 1 distrito (district)
Legislature: Congreso Nacional (National Congress) consisting of the Senado (Senate) with 32 members serving four-year terms and the Camara de Diputados (Chamber of Deputies) with 150 members serving four-year terms
Court system: Corte Suprema (Supreme Court)
Armed forces: 24,500 troops

PEOPLE

Estimated 2008 population: 9,290,000
Population density: 494 persons per sq. mi. (191 per km^2)
Population distribution: 64% urban, 36% rural
Life expectancy in years:
Male: 67
Female: 70
Doctors per 1,000 people: 1.9
Percentage of age-appropriate population enrolled in the following educational levels:
Primary: 124*
Secondary: 59
Further: 34

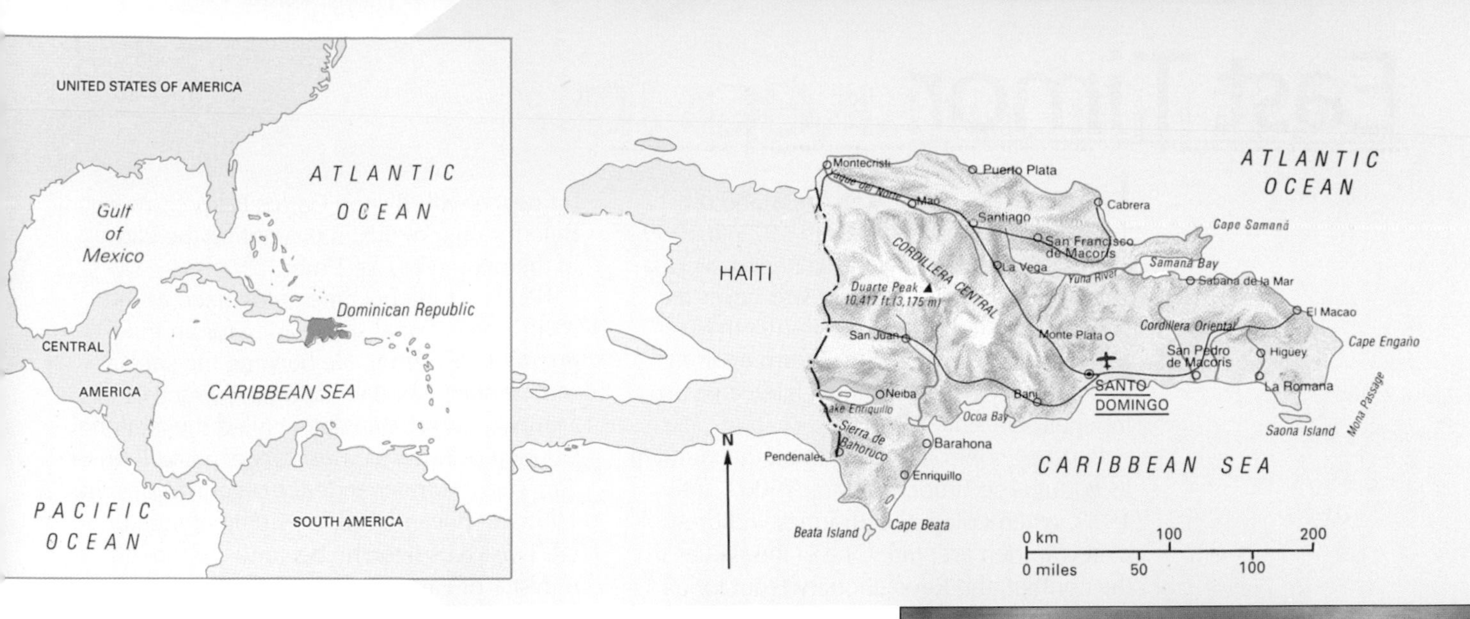

Ramón Mella led a successful revolt against the Haitians.

Spain governed the Dominican Republic from 1861 to 1865 to protect it from Haiti. Dictator Ulises (Lilis) Heureaux ruled the country from 1882 to 1899 and left it in debt. The United States took control of collecting customs duties and repaying the nation's debts from 1905 to 1941. It also sent U.S. Marines to keep order in the Dominican Republic from 1916 to 1924.

Rafael Leonidas Trujillo Molina seized power in a military revolt in 1930. Trujillo ruled harshly for 30 years, allowing little freedom and imprisoning or killing his opponents. He ruled efficiently, and the country prospered economically. But the people did not benefit because profits went to the Trujillo family.

Conspirators shot and killed Trujillo in 1961. A power struggle then began among the military, the upper class, the people who wanted a democracy, and Communists. In 1965, the United States again sent troops to maintain order, to protect Americans there, and to keep Communists from taking over. Other countries sent troops too.

The last foreign troops left the Dominican Republic in 1966. Since that time, the country has had regular elections for its presidency.

In May 2004, flash floods from torrential rains caused widespread destruction in the Dominican Republic. About 680 people were killed or reported missing. Extensive deforestation contributed to the flooding.

The rugged landscape of the Dominican Republic is dominated by the Cordillera Central, a high mountain range.

The Dominican Republic, a land of rich valleys and high, forested mountains, occupies the eastern two-thirds of the island of Hispaniola in the Caribbean Sea.

Language spoken: Spanish

Religion: Roman Catholic 95%

*Enrollment ratios compare the number of students enrolled to the population which, by age, should be enrolled. A ratio higher than 100 indicates that students older or younger than the typical age range are also enrolled.

TECHNOLOGY

Radios per 1,000 people: 181

Televisions per 1,000 people: 97

Computers per 1,000 people: N/A

ECONOMY

Currency: Dominican peso

Gross domestic product (GDP) in 2004: $55.68 billion U.S.

Real annual growth rate (2003–2004): 1.7%

GDP per capita (2004): $6,300 U.S.

Goods exported: Ferronickel, sugar, gold, silver, coffee, cocoa, tobacco, meats

Goods imported: Foodstuffs, petroleum, cotton and fabrics, chemicals and pharmaceuticals

Trading partners: United States, Venezuela, Belgium, Mexico, Asia

East Timor

East Timor is a country on the island of Timor in Southeast Asia. From 1975 to 1999, Indonesia claimed the region known as East Timor, but its authority there was never recognized by the United Nations (UN). In 1999, East Timor began transforming itself into an independent country. It became an independent state in 2002.

Portugal controlled East Timor, then known as Portuguese Timor, from the 1500's until 1975, when colonial authorities withdrew. A civil war then erupted. One of the parties in the conflict, the Revolutionary Front for an Independent East Timor (Fretilin), declared East Timor's independence in November 1975. In December of that year, Indonesian forces invaded. In July 1976, Indonesia annexed East Timor as its 27th province.

Indonesia spent large sums of money in East Timor, but many of the people continued to resist Indonesian occupation. During the 1990's , the United States and other nations joined nongovernmental organizations in accusing Indonesia of serious human rights violations in East Timor.

Peace efforts

Jose Ramos-Horta, an East Timorese politician and diplomat, shared the 1996 Nobel Prize for peace with Bishop Carlos Felipe Ximenes Belo for their efforts in promoting the cause of the people of East Timor.

Ramos-Horta became recognized as the main international spokesman for an independent East Timor. He became the special representative of the National Council of Maubere Resistance, later called the National Council of Timorese Resistance, a coalition of Timorese proindependence organizations. In 1998, he became vice president of the council. Two years later, he became the foreign minister of East Timor's transitional government.

The man with whom he shared the Nobel Prize, Carlos Felipe Ximenes Belo, was born in a remote village in East Timor, where his family had lived for many generations. The Vatican consecrated him as bishop in 1988. As the leader of the church in East Timor, Belo became an active defender of the human rights of his people against the military forces of Indonesia.

Independence from Indonesia

In an UN-sponsored referendum held in August 1999, the people of East Timor voted overwhelmingly for independence from Indonesia. Following the vote, armed

FACT BOX

COUNTRY

Official name: The Democratic Republic of East Timor
Capital: Dili
Terrain: coastal plains, mountainous interior
Area: 5,743 sq. mi. (14,874 km²)
Climate: Tropical; dry season offset by heavy monsoon rains
Main river(s): none; many unnavigable streams
Highest elevation: Tata Mai Lau, 9,721 ft. (2,963 m)
Lowest elevation: sea level

GOVERNMENT

Form of government: Republic
Head of state: President
Head of government: Prime minister
Administrative areas: 13 districts
Legislature: National Assembly of 88 members [to be reduced to between 52 and 65 members after 2006 election]
Court system: Supreme Court of Justice
Armed forces: 1,250 troops

PEOPLE

Estimated 2008 population: 995,000
Population density: 173 persons per sq. mi. (67 per km²)
Population distribution: 78% rural, 22% urban
Life expectancy in years:
Male: 48
Female: 49
Doctors per 1,000 people: N/A
Percentage of age-appropriate population enrolled in the following educational levels:
Primary: N/A
Secondary: N/A
Further: N/A

pro-Indonesian militias, backed by some elements of the Indonesian military, began attacking and killing East Timorese citizens. Thousands of people were driven from their homes, and much of the East Timor capital of Dili was burned.

In mid-September, a UN-sanctioned multinational force began arriving in East Timor to try to restore peace to the region. In October, Indonesia's highest governmental body voted to accept the results of the referendum and to end Indonesia's claim to East Timor. The UN then set up an interim administration in East Timor to help prepare the region for full independence. In 2001, the people of East Timor elected an Assembly to create a constitution.

In April 2002, Xanana Gusmao was the winner of East Timor's first presidential election, with nearly 83 per cent of the more than 378,500 votes cast. Under the country's new Constitution, the president has primarily a symbolic role. The president can, however, veto some legislation passed by the Assembly. Parliament and the prime minister are the real sources of power in the country. In May 2002, East Timor was internationally recognized as an independent nation.

A mother and daughter wade in a rice paddy near Baukau on the northern coast of East Timor. Agriculture and fishing are important economic activities in East Timor.

Languages spoken:
Portuguese, Tetum, and other indigenous languages

Religions:
Roman Catholic 90%
other Protestant and traditional religions

TECHNOLOGY

Radios per 1,000 people: N/A

Televisions per 1,000 people: N/A

Computers per 1,000 people: N/A

ECONOMY

Currency: U.S. dollar

Gross domestic product in 2004: $370 million

Real annual growth rate (2003–2004): 1%

GDP per capita (2004): $400

Goods exported: coffee, forest products, fish

Goods imported: rice, petroleum products, construction materials

Trading partners: Indonesia

Thailand
ASIA
Vietnam
Philippines
Pacific Ocean
Malaysia
Equator
Indonesia
EAST TIMOR
Indian Ocean
Australia

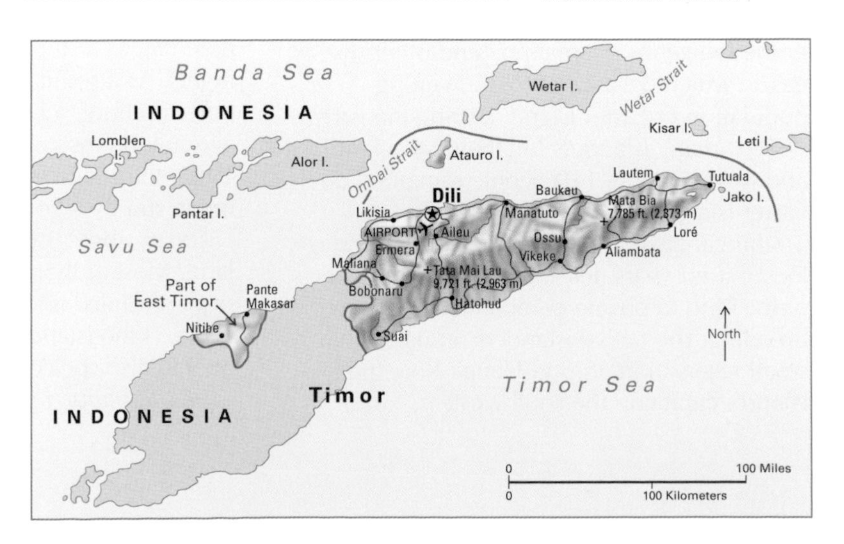

The terrain of East Timor consists of coastal plains and a mountainous interior.

Easter Island

Remote Easter Island lies about 2,300 miles (3,700 kilometers) west of Chile. It is the easternmost island in Polynesia, more than 4,000 miles (6,400 kilometers) east of New Zealand. The stony, volcanic island covers 64 square miles (166 square kilometers). Its only fresh water comes from wells, pools, and crater lakes in the island's three extinct volcanoes.

History and people

Scientists believe that Easter Island was settled about A.D. 400, but no one is sure who the first inhabitants were. Some authorities say the earliest settlers were American Indians who came from Peru. Others believe they were Polynesians.

The early islanders created Easter Island's famous stone statues, which are called *moai* (pronounced *MOH eye*). A bloody war between two groups of Easter Islanders broke out about 1680. Over the next 150 years, the victors in that war and their descendants toppled the moai from their platforms—in most cases breaking the necks of the statues. Today, about 15 moai have been restored to their original positions.

Jacob Roggeveen, a Dutch explorer, was the first European to see Easter Island. He discovered it on Easter Sunday, 1722, and gave the island its name. In 1862, slave ships from Peru arrived. Their crews kidnapped about 1,400 Easter Islanders and brought them to Peru to work on plantations. All but 100 of these islanders died in Peru, and the survivors were taken back to Easter Island in 1863.

During the voyage home, an additional 85 islanders died. The 15 survivors carried home the germs of smallpox and other diseases, which devastated the remaining population of Easter Island. During the early 1870's, many islanders left their homeland, and in 1877, only 110 people remained on Easter Island.

Chile annexed the island in 1888, but neglected it for years, leasing about 90 per cent of the land to private companies for sheep breeding. The native islanders retreated to a small reservation around Hanga Roa, the island's capital in the southwest.

In the 1950's, Chile discovered the island's potential as a stopping point for international flights. As a result, Chile built a military airport, a school, a hospital, shops, and an agricultural institute on Easter Island. In 1965, the island was made a Chilean department, run by a governor and a native mayor.

Today, about 3,000 people live on the island. Most of them are Polynesians, and the rest are Chileans. Spanish, the language of Chile, is the island's official language, but the people also speak a Polynesian language called *Rapanui,* for the island's Polynesian

Easter Island's famous stone statues, the moai, were carved hundreds of years ago. More than 600 of the statues are scattered on the island, some rising 40 feet (12 meters). About 15 moai have been restored to their original positions.

An Easter Islander displays his lobster catch, *above.* Many islanders fish and farm for a living, but tourism and the production of wool for export are the island's main industries.

name. Tourism and the production of wool for export are the main industries.

The Moai

Some scholars believe that the stone statues called *moai* were intended to honor ancestors. The oldest and smallest statues were carved and put in place about A.D. 700, about 1,000 years before the first Europeans arrived. Today more than 600 statues are scattered on the island. Most are from 11 to 20 feet (3.4 to 6 meters) tall, while some rise an awe-inspiring 40 feet (12 meters) and weigh as much as 90 short tons (82 metric tons). The most imposing collection stands where the statues were made, in what was once a quarry on the slopes of an extinct volcano called Rano Raraku in the eastern part of the island.

The islanders used stone hand picks to carve the statues from the lava rock of the extinct volcano. They set the statues on raised temple platforms called *ahu,* and balanced huge red stone cylinders, like hats, on the heads of some of the statues. Even with modern technology, erecting such huge statues and balancing the cylinders on top of them would be a difficult task.

Extinct volcanoes dominate the landscape of Easter Island, *left.* Small lakes within the craters furnish much of the island's fresh water.

Archaeologists hoist fallen moai into their original upright positions. The victors of a bloody war that broke out about 1680 toppled moai from their platforms. The necks of many statues broke when they fell.

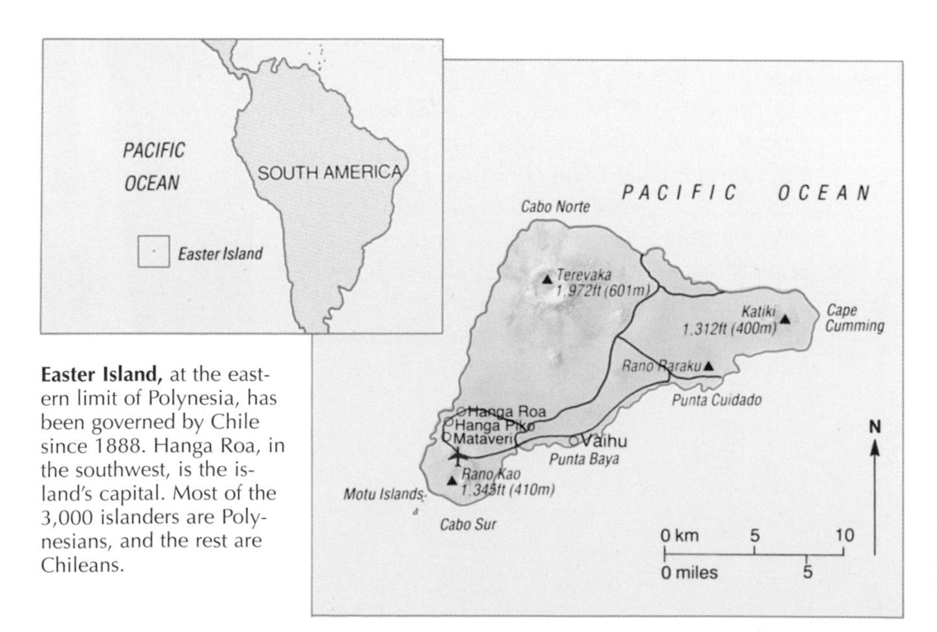

Easter Island, at the eastern limit of Polynesia, has been governed by Chile since 1888. Hanga Roa, in the southwest, is the island's capital. Most of the 3,000 islanders are Polynesians, and the rest are Chileans.

Ecuador

On the weekly market day known as the *feria* in the town of Otavalo near Quito, Ecuador's capital, a woman peers out between the woven panels of cloth that are on display. The bold geometric designs and brilliant colors of the cloth are much like those used long ago by the woman's ancestors—the rich and powerful Inca. During the 1400's, the Inca conquered the Indian tribes who inhabited the region that is now Ecuador.

Inca rulers selected women of beauty and intelligence to be taught the art of weaving. These "chosen women" learned to weave wool and cotton thread into elaborately patterned cloth, which was used by the royal family and in religious ceremonies. The chosen women were also given some education and kept as "gifts" for nobles and other men who had performed a service for the rulers.

Although such customs disappeared when the Inca civilization fell, the skill of turning thread into cloth, of combining color and texture into a unique work of art, was never lost. Passed along from generation to generation, weaving is still practiced today in the Ecuadorean highlands, much as it was centuries ago.

Spanish conquistadors vanquished the Inca in 1534. Most of the Spaniards settled in the Andes Highlands, where they made Quito their capital. They built many beautiful churches and public buildings in the city, and Quito became a great center of religious art. The Spanish settlers also established large farms and estates called *haciendas,* and they forced the conquered Indians to work on them.

During the 1800's, when Ecuador was a Spanish colony, the French emperor Napoleon invaded and conquered Spain. The rulers of the Spanish colonies in South America, including Ecuador, took advantage of Spain's weakness to demand independence.

In 1822, General Antonio José de Sucre defeated the Spaniards, which ended Spanish rule in Ecuador. Ecuador then joined newly

independent Colombia and Venezuela in the confederation of Gran Colombia. In 1830, Ecuador left the confederation and became an independent country.

Rival leaders fought for power in the new nation. Presidents, dictators, and *juntas* (groups of rulers) rose and fell. In their hunger for control, most showed no concern for the rights and needs of the people. One exception was Gabriel García Moreno, who became president in 1861. His government helped develop farming and industry and built roads and railroads. But he was assassinated in 1875, and his plans and policies were not continued.

During a particularly unstable period between 1925 and 1948, Ecuador had 22 presidents and heads of state, and none of them served a complete term. In 1963, armed forces overthrew the civilian government. Civilian and military rulers alternated in control of the government between 1963 and the civilian elections of 1979 when elections began being held regularly.

President León Febres Cordero, elected in 1984, began work to strengthen the economy. However, two disasters aborted his efforts. First, the 1986 drop in worldwide oil prices cut Ecuador's income by about 30 per cent. Then in 1987, a severe earthquake struck northern Ecuador causing extensive damage to an oil pipeline. Ecuador, whose chief export is petroleum, is still trying to recover from the serious financial crisis triggered by these events.

In early 1997, Ecuador's legislature removed President Abdala Bucaram Ortiz from office, claiming he was mentally unfit to serve, appointing an interim president until elections could be held.

Ecuador faced additional economic difficulties in 1999 and early 2000. The nation failed to pay back loans from several international lending organizations, and its currency dropped sharply in value. In January 2000, a coalition of military officers and Indian groups forced Jamil Mahuad from office. Vice President Gustavo Noboa Bejarano then took over as president.

Ecuador Today

Rich in natural beauty, Ecuador is a country of magnificent landscapes as yet unspoiled by mass tourism. The country's highlands, reached by air or railroad, are breathtaking. Delightful beaches line its Pacific coast, while Quito's churches and monasteries offer an abundance of exquisite colonial paintings and sculptures. The offshore Galapagos Islands, about 600 miles (960 kilometers) west of Ecuador, are a living museum of some of the world's most unusual wildlife.

Yet the beauty of this land, with its fiery volcanic peaks, thick tropical forests, and vast mountain plateaus, is a stark contrast to the bleak and often difficult life of most present-day Ecuadoreans. Many Indians, mestizos, and blacks—who together make up 93 per cent of Ecuador's population—live in terrible poverty, earning not much more than one dollar per day when there is work. Often, the poor people in rural areas suffer from malnutrition; *dysentery,* an intestinal disease; and respiratory illnesses.

Many mestizos live in wooden homes with thatched roofs and work as day laborers on large banana or cacao plantations. Most of the blacks live in the northern part of the coastal plain, and many of them fish for a living. A few Indian tribes live in the forests of eastern Ecuador, cultivating small plots of land that they have cleared and moving on as the soil wears out.

Most of Ecuador's Indians live in Andean villages and have little contact with the rest of the people. Some of them work on haciendas, growing crops and herding livestock. Others are shepherds in pastures high in the mountains.

In sharp contrast to these remote areas is the bustling, urban elegance of Quito, Ecuador's capital. From its glass-and-concrete skyline to its baroque churches decorated with gold and silver, its teeming Indian markets, and its Spanish-style palaces, the architecture of Quito reflects its history.

The residents of Quito and other Andean cities include most of Ecuador's whites of European ancestry. In Ecuador, where they make up about 7 per cent of the population, whites are the wealthiest and most powerful group. Some are leaders of business and industry, and some own the land worked by the Indian and mestizo farmers in the rural areas.

These landlords sometimes hire managers to run their large haciendas. Unlike the Andean villagers, who build simple houses of adobe to protect themselves from the mountain cold, many wealthy white people live in

FACT BOX

COUNTRY

Official name: Republica del Ecuador (Republic of Ecuador)
Capital: Quito
Terrain: Coastal plain (costa), inter-Andean central highlands (sierra), and flat to rolling eastern jungle (oriente)
Area: 109,484 sq. mi. (283,561 km^2)
Climate: Tropical along coast, becoming cooler inland at higher elevations; tropical in Amazonian jungle lowlands
Main rivers: Esmeraldas, Daule, Napo, Curaray
Highest elevation: Chimborazo, 20,561 ft. (6,267 m)
Lowest elevation: Pacific Ocean, sea level

GOVERNMENT

Form of government: Republic
Head of state: President
Head of government: President
Administrative areas: 22 provincias (provinces)
Legislature: Chamber of Representatives with 12 members serving four-year terms and 70 members serving two-year terms
Court system: Corte Suprema (Supreme Court)
Armed forces: 46,500 troops

PEOPLE

Estimated 2008 population: 13,832,000
Population density: 126 persons per sq. mi. (49 per km^2)
Population distribution: 63% urban, 37% rural
Life expectancy in years: Male: 68 Female: 74
Doctors per 1,000 people: 1.5
Percentage of age-appropriate population enrolled in the following educational levels: Primary: 117* Secondary: 59 Further: 34

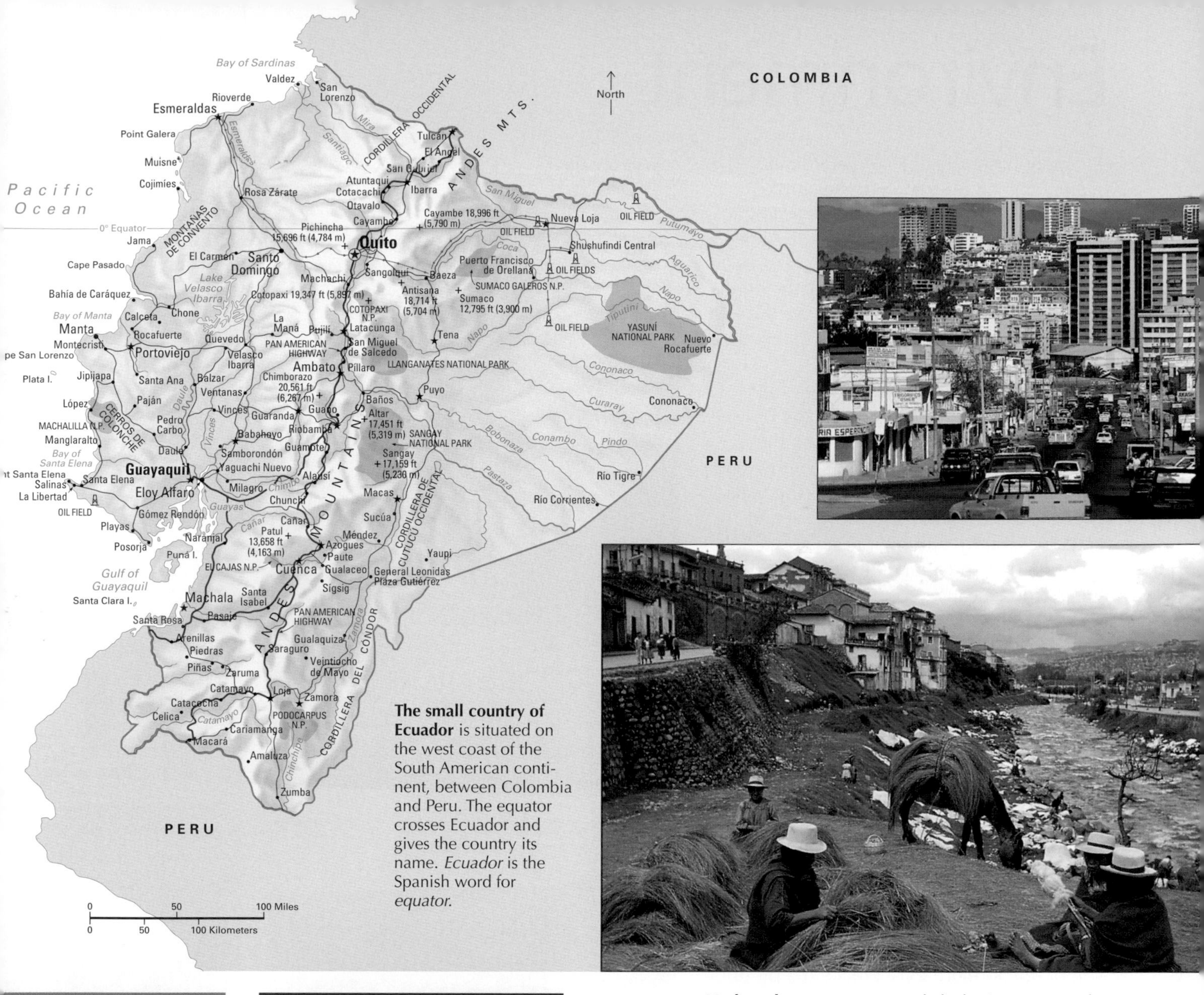

The small country of Ecuador is situated on the west coast of the South American continent, between Colombia and Peru. The equator crosses Ecuador and gives the country its name. *Ecuador* is the Spanish word for *equator.*

Modern skyscrapers, *top,* mark the business center of Quito, the capital of Ecuador. In striking contrast are the colonial-style buildings of Cuenca, *above,* where laborers in Panama hats bale up sisal.

comfortable, stylish city residences and also have weekend country residences.

The uneven distribution of wealth is not unique to this small South American country. Like its neighbors to the north, east, and south, Ecuador struggles with this and other economic problems. However, the government is working to improve the living standards of its people. Several programs have been set up to provide homes, improve medical care, and promote literacy.

Languages spoken:
Spanish (official)
Amerindian languages (especially Quechua)

Religion:
Roman Catholic 90%

*Enrollment ratios compare the number of students enrolled to the population which, by age, should be enrolled. A ratio higher than 100 indicates that students older or younger than the typical age range are also enrolled.

TECHNOLOGY

Radios per 1,000 people: 422

Televisions per 1,000 people: 252

Computers per 1,000 people: 31.1

ECONOMY

Currency: U.S. dollar

Gross domestic product (GDP) in 2004: $49.51 billion U.S.

Real annual growth rate (2003–2004): 5.8%

GDP per capita (2004): $3,700 U.S.

Goods exported: Petroleum, bananas, shrimp, coffee, cocoa, cut flowers, fish

Goods imported: Machinery and equipment, raw materials, fuels; consumer goods

Trading partners: United States, Colombia, Japan, Italy

Environment

Ecuador has a rugged and varied landscape, ranging from the wet, swampy forests of the Eastern Lowland areas to the snow-capped volcanic peaks of the Andes Mountains. Although the equator crosses Ecuador in the northern region of the country, the climate is not always hot because much of the country lies at higher altitudes, where the air is cooler.

Geographers divide the mainland of Ecuador into three land regions: the Coastal Lowland, also known as the Costa; the Andes Highland, also called the Sierra; and the Eastern Lowland, or Oriente. The Galapagos Islands, an offshore province and Ecuador's fourth land region, lie about 600 miles (970 kilometers) off the mainland in the Pacific Ocean.

Land regions

The Coastal Lowland, or Costa, is a large, flat plain that extends along Ecuador's Pacific coast and covers about a fourth of the country. It was formed by mud and sand sediments carried down the mountains by rivers and deposited along the shore. In the north, the Costa is wet and swampy. Dense tropical jungles cover the land, stretching all the way up the Andean slopes to 8,000 feet (2,400 meters) in some places. In the south, near Peru, the Costa is a desert.

The warm, humid conditions and fertile soil of the northern Costa make it a productive farming region. Abundant crops of bananas, cacao, coffee, oranges, palm oil, rice, soybeans, and wheat are grown there for export. Large quantities of balsa wood are harvested from the Costa's forests, making Ecuador the world's largest producer of this lightweight wood.

These products are shipped from Guayaquil, Ecuador's leading commercial center and seaport and its largest city. The area surrounding Guayaquil has two seasons: a hot, rainy period from January to May and a cooler period from June to December when sea breezes from the Peru Current (also called the Humboldt Current) help ease the equatorial heat.

The Andean Highland, or Sierra, makes up another fourth of Ecuador's land area, with two parallel ranges extending from north to south along the entire length of the country. Between them lie a series of high

A group of Andean sheep-herders, *bottom,* wrapped in blankets to shield them against the cold winds of the high pastures, rest for a moment from their work. Nearly all the sheep in Ecuador are grazed at altitudes over 9,000 feet (2,700 meters).

Shepherds lead their flock to pasture on the lower slopes of Chimborazo volcano, *below,* the highest of about 30 Andes mountains that form an "avenue of volcanoes." Chimborazo's tall peak is covered with snow all the time.

An Otavalo Indian woman, *right,* prepares wool before threading her loom. The Otavalos have long been skilled weavers and often travel as far as Colombia, Venezuela, Brazil, and the United States to sell their ponchos, blankets, and textiles.

Ecuador's landscape, as seen on the map, *right,* displays a dramatic contrast in elevation—from sea level along the coast to mountain peaks in the central Andean Highland that reach more than 20,000 feet (6,100 meters).

Mestizo women take shelter from a rainshower in a grove of sugar cane in the northern Costa, *left.* Crops for export are grown mainly in the lowlands. Most products grown in the highlands are used by the local people.

An Indian woman in the Andean Highland does the family wash in the cool waters of a mountain stream, *left.* Living conditions are generally quite simple or primitive in Ecuador's rural areas, and many people use old-fashioned equipment and tools to farm the land. This is because they have very low incomes, commonly less than one dollar per day per person.

plateaus and basins, where Andean farmers grow beans, corn, and potatoes, much as their ancestors did thousands of years ago. Cattle graze in the highland valleys. There are few native trees in the highlands, though the eucalyptus, introduced in the 1860's, is now widely grown in the area. Temperatures on the Sierra plateaus are generally springlike, with colder weather at higher altitudes.

The heavily forested Eastern Lowland, or Oriente, forms part of the Amazon Basin and makes up almost half of Ecuador's land area. Temperatures are high, and rain falls all year long. In the past, the Oriente was virtually undeveloped, but since large petroleum deposits were found there, the area now contains a number of oil fields.

Environmental problems

Ecuador's tropical rain forests have a wide variety of wildlife, including deer, jaguars, monkeys, ocelots, tapirs, and many species of birds. To protect its wildlife, the Ecuadorean government established a broad wildlife-protection program in 1970.

Despite these efforts, extensive clearing of the forests for farmland and oil field development continue to threaten Ecuador's native species. As of 1990, threatened species in Ecuador included two species of falcons and four species of hawks, the brown-headed spider monkey, and the Galapagos giant tortoise. The black caiman, once an inhabitant of Ecuador, is now extinct. A program to reforest cleared areas and maintain existing forests was launched in 1979, but during the early 1980's, only 10,000 acres (4,000 hectares) were reforested annually. Meanwhile, from 1981 to 1985 about 840,000 acres (340,000 hectares) were cleared.

Galapagos Islands

Far out into the Pacific, about 600 miles (970 kilometers) west of the Ecuadorean mainland, lies a chain of islands whose unusual animal inhabitants have fascinated scientists and voyagers for centuries. They are the Galapagos Islands, named for the giant turtles that live there. *Galápagos* is the Spanish word for *turtle.*

The islands are made up of volcanic peaks and cover an area of 3,029 square miles (7,844 square kilometers). Once called the Enchanted Isles, the Galapagos may have been originally settled by Peruvian Indians. During the 1600's and 1700's, pirates and buccaneers used the island as hideouts from which they launched raids on Spanish ships and the coastal towns of South America.

The human population of the Galapagos has increased greatly since the 1980's. Between the censuses of 1982 and 2001, the population grew from about 6,000 to nearly 19,000. Also, more than 60,000 tourists visit each year.

The island is known for its exotic birds and other animals. For example, marine lizards that measure 4 feet (1.2 meters) long slither over the rocks, while 500-pound (230-kilogram) turtles plod along the ground, often with Galapagos hawks perched on their backs.

The giant, plant-eating turtles of the Galapagos graze mainly on the native prickly pear cactus.

The marine iguanas, *right,* can drink seawater, excreting the excess salt through glands in their noses.

Wonders of the natural world

The animals that lived on the islands before people arrived either flew in, swam, or rode in on floating vegetation or on the backs of other animals. The only mammals among these animals were two species of bats, several species of rats, and sea lions and fur seals.

Of the island's many birds, the flightless cormorant is perhaps the most remarkable. It has stunted wings and walks on land with an upright waddle. The cormorant moves very well in water, however, enabling it to compete with penguins for food.

The marine iguanas of Galapagos are the only lizards in the world that live in the sea. They often dive deep into the water to feed on marine algae and shellfish, using their large, clawed feet to dig into the ocean bottom.

Finches

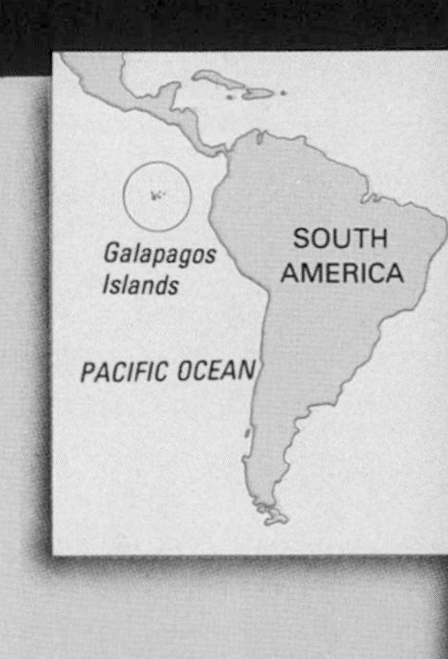

The Galapagos Islands became famous all over the world after Charles Darwin published his book *The Origin of Species* (1859). Darwin was a British naturalist who traveled to the islands on a scientific expedition aboard the H.M.S. *Beagle.*

Darwin studied the finches of the Galapagos Islands. He noted that although the finches were basically similar, a number of different types had evolved. They range from the ground finch, with its thick bill for crunching hard seed, to the tiny warbler finch, which feeds on insects. Darwin realized that this one species had filled all the roles that, in Britain, were filled by a large variety of birds. He concluded that the Galapagos finches had all descended from a single ancestor finch that had flown to the islands from the South American mainland.

Darwin's findings led him to develop his theories on evolution and a process called *natural selection,* or survival of the fittest. *The Origin of Species,* which discusses these theories, sparked heated debate among the biologists and religious leaders of the time.

Pinnacle Rock, on the island of Bartholomé, is a favorite nesting place of the Galapagos penguins. Because of its volcanic origin, the land surface of the islands is made up of basaltic lava. Visitors often compare it to the barren surface of the moon. There are almost no tall trees because the islands have no deep soil where trees can take root. One exception is the prickly pear cactus, which towers up to 30 feet (9 meters) in the arid lowlands of some islands.

Plight of the turtle

These gentle giants of the Galapagos knew no predators until the arrival of the first Europeans. Black rats that darted from sailing ships ate the eggs and hatchlings of these magnificent creatures, reducing their reproduction rate considerably. Sailors craving turtle meat loaded them by the hundreds onto their ships and set sail for long voyages across thousands of miles of ocean. Because the turtles could live for months without food or water, they provided a constant source of fresh meat.

Several hundred thousand turtles may have been taken from the islands in this way. Today the turtle population is nowhere near what it once was, but the rats whose ancestors arrived aboard the sailing ships have reproduced in such numbers that they are now impossible to exterminate. They continue to prey upon turtle eggs and hatchlings—as do the cats, dogs, and pigs brought by settlers from the mainland. In addition, cattle trample turtle nests, and goats compete with them for food.

Four of the 15 subspecies of Galapagos turtles did not survive the sailors' butchery and are now extinct. A single male member of a fifth subspecies, once thought extinct, was discovered in 1971 on Pinta Island.

Widely known as Lonesome George, he now lives in a penned area at the Charles Darwin Research Station, where conservationists provide refuge for turtle eggs and hatchlings. Although George has many turtle friends, efforts to mate him with a female from a similar subspecies have so far been unsuccessful.

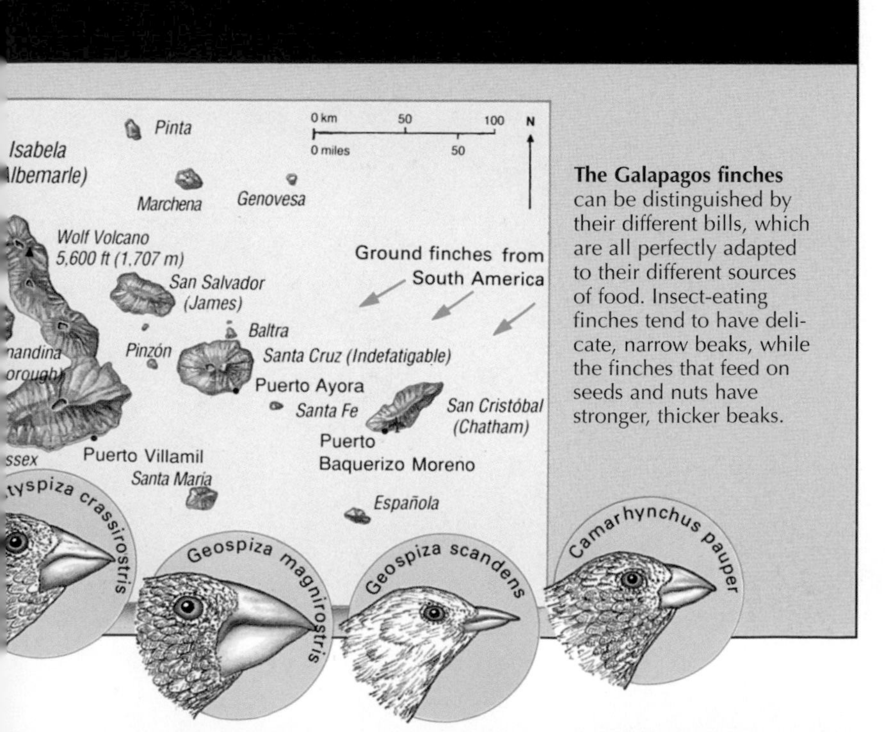

The Galapagos finches can be distinguished by their different bills, which are all perfectly adapted to their different sources of food. Insect-eating finches tend to have delicate, narrow beaks, while the finches that feed on seeds and nuts have stronger, thicker beaks.

Egypt

Perhaps nothing symbolizes Egypt better than the great pyramids at Giza. Built more than 4,500 years ago, the pyramids are the magnificent remains of a great civilization headed by strong rulers called *pharaohs.* Among the outstanding contributions to civilization made by ancient Egyptians are a 365-day calendar, a form of picture writing called *hieroglyphics,* and certain basic forms of arithmetic. They also built great cities in which skilled architects, doctors, engineers, painters, and sculptors worked.

Egypt is no longer a wealthy world power, but much of what was true for ancient Egypt holds true for modern Egypt as well. The ancient Egyptians created the world's first national government, and a central government has controlled the nation almost continuously ever since. The ancient Egyptians' religion influenced their everyday life as well as their political life. They believed their pharaoh was a god, and priests held great power. Today in Egypt, the religion of Islam influences family life, social relationships, business activities, and government affairs.

Just as in ancient Egypt, Egyptians still cluster near the Nile River—on its fan-shaped delta and the two narrow strips of fertile land that lie along the riverbanks. Without the precious waters of the Nile, all of Egypt would be a desert.

West of the Nile Valley lies the Libyan Desert, part of the huge Sahara that stretches across northern Africa. East of the Nile lies the Arabian Desert, also part of the Sahara. Even the triangular Sinai Peninsula to the east, across the Gulf of Suez and the Suez Canal, is arid and desolate.

Egypt's greatest problem today may be overpopulation. More than 75 million people are now crowded into the Nile's narrow valley and delta. The cities of Egypt, in addition to having such typical urban problems as housing shortages and traffic congestion, are overflowing. Many city people live in extreme poverty, while others enjoy all the conveniences of modern life.

Meanwhile, village farmers live much as their ancestors did centuries ago. And the people of Egypt still rely heavily on the crops produced on the fertile Nile land, much as the ancient Egyptians did thousands of years ago. Whether they are city dwellers or rural villagers, whether rich or poor, Egypt's people all share common bonds—the beliefs and traditions of Islam and a rich cultural history.

Egypt Today

On a busy Cairo street, people enter the city's Metro, *right,* an underground railway system opened in 1987. Transportation in Egypt takes a variety of forms, as buses, cars, trucks, and motorcycles share the roads with bicycles, donkeys, and carts.

Egypt is a Middle Eastern country tucked into the northeast corner of Africa. Once a rich and powerful ancient kingdom, it became a modern, independent republic in 1953. Since then, Egypt has played a leading role in the Middle East, one of the world's trouble spots.

Egypt has a strong national government. According to its 1971 Constitution, the country is a democratic and socialist society, and all Egyptian citizens aged 18 or older may vote.

The legislature is called the People's Assembly and has 454 members. At least half the members must be workers or farmers, and all members serve five-year terms.

The president is chosen through multiparty elections. Egypt's president may serve an unlimited number of six-year terms.

The president has great power in all levels of Egypt's government—the People's Assembly does little more than approve the president's policies. The president may even appoint 10 Assembly members, one or more vice presidents, and heads of agencies. The president also appoints the Council of Ministers, including the prime minister, who carries out the policies of the government.

The president appoints a governor to head each of the 27 local governorates. Local districts and villages are also run by appointed leaders, though these officials are assisted by elected council members.

The president and most top government officials belong to the National Democratic Party, Egypt's largest political party. The party supports a mixture of public and private ownership of land and business. Egypt is a developing country in economic terms. One-third of its workers are still employed in

FACT BOX

COUNTRY

Official name: Jumhuriyat Misr al-Arabiyah (Arab Republic of Egypt)
Capital: Cairo
Terrain: Vast desert plateau interrupted by Nile valley and delta
Area: 386,662 sq. mi. (1,001,449 km^2)
Climate: Desert; hot, dry summers with moderate winters
Main river: Nile
Highest elevation: Mount Catherine, 8,651 ft. (2,637 m)
Lowest elevation: Qattara Depression, 436 ft. (133 m) below sea level

GOVERNMENT

Form of government: Republic
Head of state: President
Head of government: Prime minister
Administrative areas: 27 muhafazat (governorates)
Legislature: Majlis al-Sha'b (People's Assembly) with 454 members serving five-year terms and the Majlis al-Shura (Advisory Council) with 264 members
Court system: Supreme Constitutional Court
Armed forces: 450,000 troops

PEOPLE

Estimated 2008 population: 77,243,000
Population density: 200 persons per sq. mi. (77 per km^2)
Population distribution: 57% rural, 43% urban
Life expectancy in years:
Male: 66
Female: 70
Doctors per 1,000 people: 2.1
Percentage of age-appropriate population enrolled in the following educational levels:
Primary: 97
Secondary: 88
Further: N/A

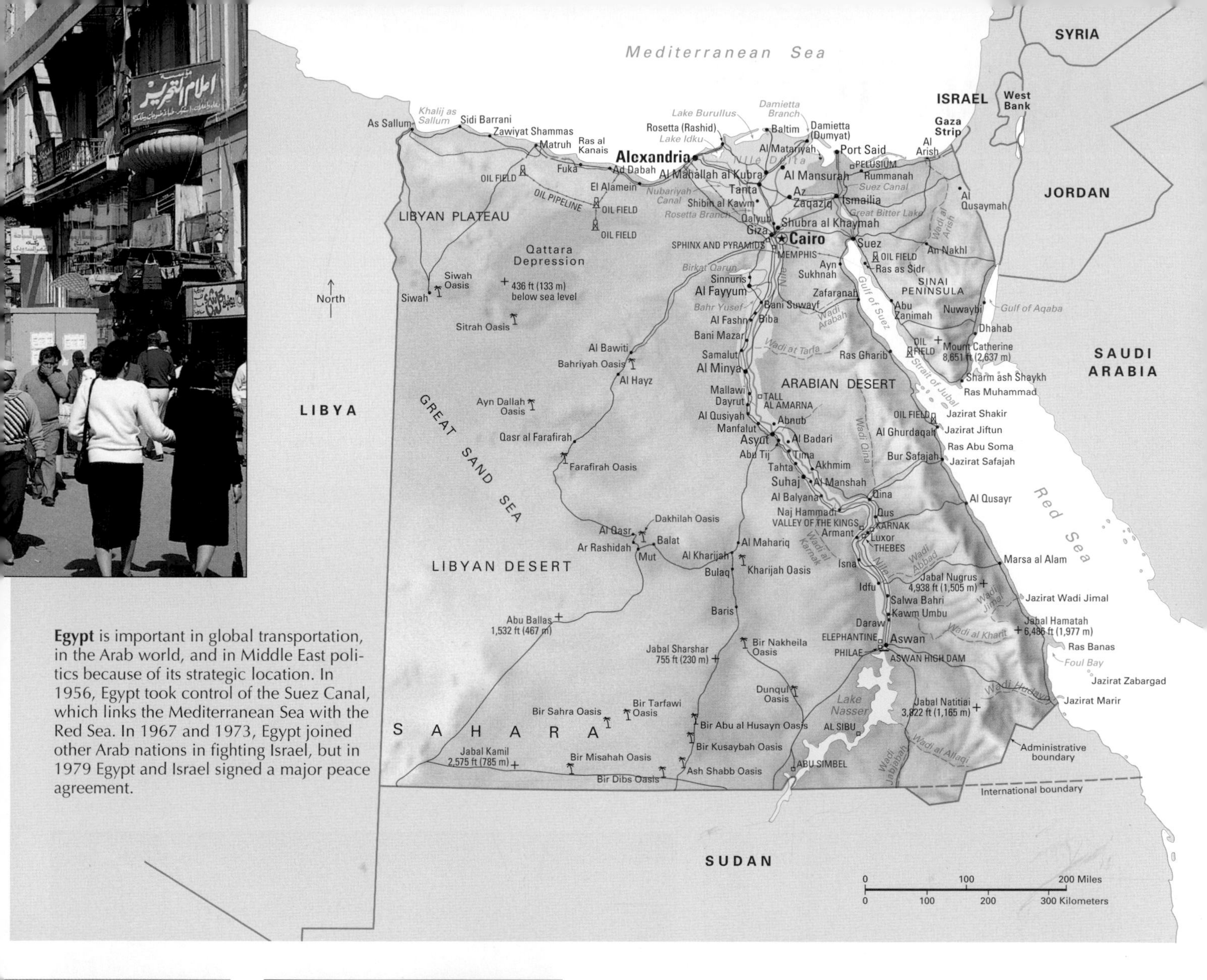

Egypt is important in global transportation, in the Arab world, and in Middle East politics because of its strategic location. In 1956, Egypt took control of the Suez Canal, which links the Mediterranean Sea with the Red Sea. In 1967 and 1973, Egypt joined other Arab nations in fighting Israel, but in 1979 Egypt and Israel signed a major peace agreement.

Languages spoken:
Arabic (official)
English and French widely understood by educated classes

Religions:
Muslim (mostly Sunni) 90%
Coptic Christian

TECHNOLOGY

Radios per 1,000 people: 339

Televisions per 1,000 people: 229

Computers per 1,000 people: 21.9

ECONOMY

Currency: Egyptian pound

Gross domestic product (GDP) in 2004: $316.3 billion U.S.

Real annual growth rate (2003–2004): 4.5%

GDP per capita (2004): $4,200 U.S.

Goods exported: Crude oil and petroleum products, cotton, textiles, metal products, chemicals

Goods imported: Machinery and equipment, foodstuffs, chemicals, wood products, fuels

Trading partners: European Union, United States, Turkey, Japan

agriculture. About one-half of the people work in service industries and one-fifth in industry.

In the 1990's, violence erupted between the government and Islamic fundamentalist extremists. Seeking to topple the government, the extremists attacked Egyptian Christians, foreigners, and government officials. The government responded with mass arrests and executions. National and international agencies accused it of abusing human rights.

Islamic militants attacked tourists in several incidents in 1997. The militants killed the tourists in the hope that tourism, a vital part of Egypt's economy, would be weakened. Most of the violence ended in the late 1990's. Beginning in 1997, Egypt embarked on a 20-year-plan to irrigate desert land and turn it into farmland.

Agriculture

Farming has been the chief economic activity in Egypt for thousands of years. The mighty Nile River has nourished the land and the people of Egypt since ancient times.

Every year in July, the Nile started to flood, swelling with rain water that had fallen in central Africa and flowed downstream. Egyptian farmers trapped the floodwaters in basins on their fields. When the river went down, usually in September, it left a strip of rich, black soil about 6 miles (10 kilometers) wide on each bank. The farmers then plowed and seeded the fertile soil.

In the 1800's, the Egyptians began to replace their seasonal system of basin irrigation with a year-round system of dams, canals, and reservoirs. The changeover was completed in 1968, when the Aswan High Dam began operating. Today, nearly all of Egypt's farmland can be irrigated continuously, and farmers can plant crops the year around.

Cotton is the nation's most valuable crop, and Egypt is one of the world's leading cotton producers. Egyptian farmers grow high-quality, long-staple (long-fibered) cotton, which is strong and durable.

Egyptian rural peasants—called *fellahin*—have farmed small plots of land or tended animals for hundreds of years. However, many fellahin did not own the land they farmed. In the 1950's and 1960's, the Egyptian government tried to help the fellahin by passing land-reform measures that limited the size of farm estates and prevented large, wealthy landowners from controlling so much valuable farmland. But for many fellahin, little has changed. They still rent land or work as laborers in the fields of prosperous landowners.

The government requires farmers to use some of their land for growing the food crops that feed Egypt's increasing population, rather than cash crops such as cotton. Farmers grow corn, oranges, potatoes, rice, sugar cane, tomatoes, and wheat. Clover is grown as animal feed, and dates are grown mainly in desert oases. Farmers and herders raise goats and sheep for meat, milk, and wool. Many farmers also raise chickens for meat and eggs.

Egypt's farmland lies mostly along the Nile River and its delta. Most farms cover about 2 acres (0.8 hectare), and only about 6 million acres (2.4 million hectares) are cultivated—less than 3 per cent of the country's total land area. Land reforms in the 1950's and 1960's changed the pattern of land ownership by redistributing land from the wealthiest owners to middle-sized and small-land owners, but many fellahin are still landless.

Using a primitive water wheel, an Egyptian *fellah* (farmer) draws water to irrigate his field. Irrigation is a necessity in most parts of Egypt, where little rain falls.

A satellite photo of the Nile Delta shows its fan-like shape, which some people think resembles an Egyptian lotus flower. The river forms a narrow life-giving strip in the desert, like the stem of the lotus, then fans out at its delta, like the flower itself.

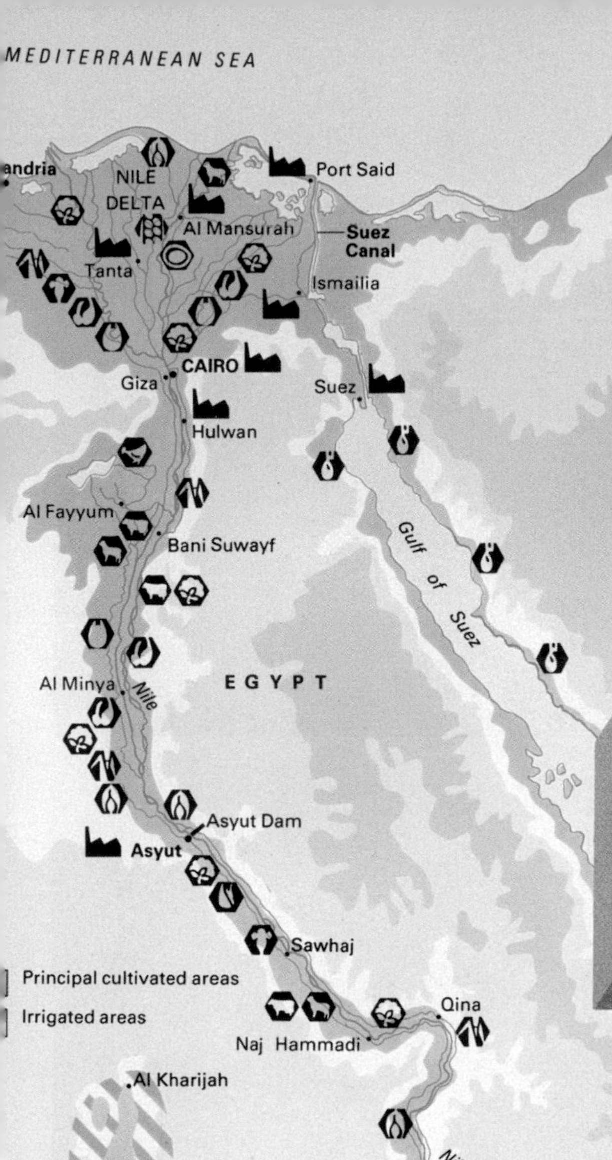

Irrigation

The traditional Egyptian method of irrigation was basin irrigation, *top.* Farmers built canals and small dikes on their land to create basins that would catch the floodwaters of the Nile. When the floodwaters eventually went down, rich black silt remained. As farmers learned to clear more land, *center,* they increased both the number of crops they could raise and the length of the growing season. Devices for lifting water buckets called *shadoofs* or water wheels called *sakiehs* were used to bring water to fields on higher ground. Since the 1800's, modern irrigation—a system of dams, reservoirs, pumps, and canals—has been constructed, *bottom.* But now, modern farmers along the Nile must use chemical fertilizers because the Aswan High Dam holds back the fertile silt.

People

More than 76 million people live in Egypt, and almost all of them are crowded into the narrow Nile Valley and the Suez Canal region. And in these areas, almost half of all Egyptians are further crowded into the cities.

Over hundreds of years, many people from other lands have invaded Egypt and married native Egyptians. As a result, present-day Egyptians can trace their ancestry not only to ancient Egyptians, but also to Arabs, Ethiopians, Persians, and Turks, as well as Greeks, Romans, and other Europeans.

However, most Egyptians consider themselves Arabs. Arab Muslims surged into Egypt and conquered the land in the A.D. 600's. The Egyptian people gradually adopted the Arab language, and most converted from the Coptic Christian religion to Islam.

Arabic is now the official language of Egypt, but people in different regions speak different *dialects* (forms) of Arabic. The dialect spoken in Cairo is the one most widely used today. Many educated Egyptians also speak English or French.

Islam is now the official religion of Egypt. More than 90 per cent of Egypt's people are Muslims who follow the Sunni branch of Islam. Under the law, Coptic Christians and other religious minorities may worship freely, but some Muslim groups have committed acts of violence against Coptic communities.

Egypt has several ethnic minority groups. The Bedouin are Arab nomads who traditionally lived in the desert, though most are now settled farmers. They speak Bedouin dialects. From about 2500 B.C. to the early 1960's, the Nubians lived in southern Egypt and farmed along the Nile. In 1968, when their land was flooded by the Aswan High Dam, the government relocated the Nubians to Kawm Umbu, a town about 30 miles (50 kilometers) north of Aswan. However, like other rural Egyptians, many Nubians have moved to the cities to look for work.

Life in the cities differs greatly from life in the countryside. And there are great extremes in urban life styles because there are such great extremes of wealth and poverty in Egypt's cities.

Some city people live in pleasant residential areas. Many well-to-do Egyptians wear Western-style clothing and buy large quantities of meat and imported fruits and vegetables.

A Cairo coffee house, *top,* provides a pleasant place for city dwellers to relax and chat with friends.

Bedouin women spin wool into yarn, *above,* practicing a craft that was developed about 4000 B.C.

A Nubian woman in Aswan balances a bucket on her head in the traditional method of carrying a burden.

Cairo commuters squeeze on and off their train at a bustling Metro stop. People in Cairo use a variety of forms of transportation, including buses, cars, motorcycles, and trains.

However, many more urban dwellers live in sprawling slums—in crowded, run-down apartments, in makeshift huts, or even on the roofs of buildings. The city poor eat a simple diet of bread and broad beans called *ful* or *fool.* Government-run stores sell meat, cheese, eggs, and other foods at controlled prices, but supplies often run out.

Rural Egyptians generally live much as their ancestors lived hundreds of years ago. The fellahin live in small, mud-brick huts with thatched straw roofs. Most homes have one to three simply furnished rooms and a courtyard that may be shared with the farm animals.

Each member of a fellahin family has certain duties. The husband organizes the planting, weeding, and harvesting, while the wife cooks, carries water, and helps in the field. Children tend animals and help carry water.

Fellahin men traditionally wear pants and a long, shirtlike garment called a *galabiyah,* and the women wear long flowing gowns. The fellahin diet is similar to that of the urban poor.

Islam provides a bond between the people, and rural Egyptians have a strong sense of community. Religion influences many aspects of Egyptian life. Muslims are expected to pray five times a day, give money to the poor, and fast. Some Egyptians dress according to modern Islamic customs. The men wear long, light-colored gowns and skullcaps, and the women wear robes and keep their faces veiled in public.

Traditionally, family roles in Egypt have been rigid, and women have had low status. However, since the 1960's, many more Egyptian women attend school or work outside the home.

Three young women stop for a refreshment offered by a street vendor in Cairo. Most well-to-do city dwellers wear Western-style clothing, while rural villagers and many poorer city dwellers wear traditional clothing.

Al-Azhar University

Egypt's influential position as the traditional center of Arab learning and culture has become even more important since the 1970's, when a wave of strong religious feeling developed among the various segments of many Muslim nations. Much of Egypt's leadership in the Muslim world stems from a historic source—Cairo's Al-Azhar University. Al-Azhar is one of the oldest universities in the world and the most influential religious school in Islam.

Al-Azhar was founded by the Fatimids, a band of Shiah Muslims who invaded Egypt from the west in A.D. 969. The Fatimid *caliphs,* or rulers, claimed they were descended from Fatima, the daughter of the prophet Muhammad, who founded Islam. The Fatimids established a capital city called *Al Qahirah*—perhaps because the planet Mars (*al-Qahir* in Arabic) was rising in the sky when the city was founded. The name *Cairo* came from *Al Qahirah.*

The Fatimids erected beautiful buildings and a great library in Cairo, which soon became one of the most important cities of the Arab world. The Fatimids also built Al-Azhar Mosque, a Muslim place of worship that quickly became the center of Fatimid culture and religion. Eventually, Al-Azhar became a university, a center for the study of Islamic religion and law that attracted students of Islam from many countries.

For hundreds of years, the leading scholars of Al-Azhar had great power and influence in Egypt. Like the priests of ancient Egypt, these religious scholars, called *ulama,* sometimes even controlled Egyptian rulers.

For centuries, almost all of Egypt's educational, legal, social welfare, and health affairs were in the hands of these religious leaders. However, during the 1800's and 1900's, the government brought religious institutions such as Al-Azhar under closer control and limited the role of the ulama in public life.

In 1952, a group of Egyptian military officers seized power and brought about even more dramatic changes. The government took over the appointment of officials to mosques and religious schools. The ulama then came under close government supervision and financial control.

Many changes took place at Al-Azhar University in the 1960's. Along with courses in Islamic religion, Islamic law, and Arabic studies, the school began teaching medicine, engineering, and agriculture. It also began to instruct women.

During the 1970's, however, some Muslim fundamentalists revived a movement which believed the government had strayed from Islamic beliefs. These Muslims want Egypt to return to a true Muslim state—a nation that reflects their interpretation of Islamic teachings. Some Muslim fundamentalists have even resorted to violence to further their cause, and the ulama have at times been pressured to support fundamentalist views.

Nevertheless, Islam is a unifying bond in Egypt. To all Muslims, Islam is a total way of life, not just a religious creed. Muslims believe that the teachings of Islam must guide all of society, not just the individuals within the society.

However, different segments of Egypt's population have different opinions about their religion. The views of a scholarly ulama of Al-Azhar may be worlds apart from those of a poor rural peasant. The ulama follows the *orthodox,* or traditional, Islamic beliefs of the holy book of Islam. Less educated people,

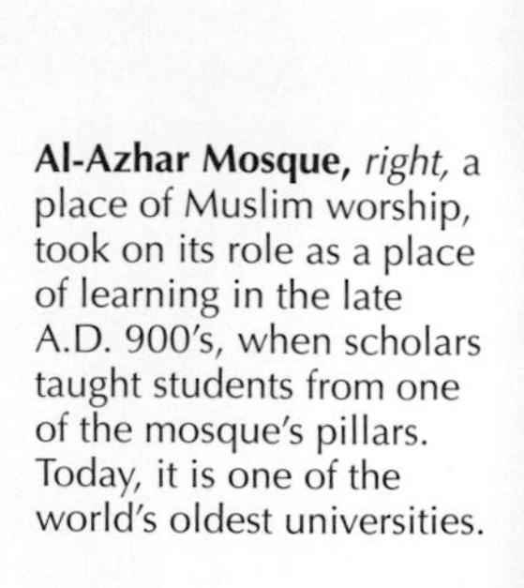

Al-Azhar Mosque, *right,* a place of Muslim worship, took on its role as a place of learning in the late A.D. 900's, when scholars taught students from one of the mosque's pillars. Today, it is one of the world's oldest universities.

The great courtyard of Al-Azhar is surrounded by covered walkways, behind which lie the students' living quarters. The side of the courtyard facing the Arab city of Mecca contains the university's place of prayer because Muslims face Mecca when they pray.

however, may also follow quite nontraditional practices, such as honoring saints or believing in the power of charms and the evil eye. But the fundamentalists insist that all of Egyptian society must follow their view, and many fundamentalists would like to see the Islamic code of laws called *sharia* become the law of all Egypt.

Because the same kind of religious conflict is taking place in other Muslim countries in North Africa and the Middle East today, Egypt—the traditional center of Islamic learning and culture—holds an increasingly important position in today's Muslim world. Now, more than 1,000 years after its inception, Al-Azhar University remains the most influential religious school in Islam.

Students from many countries come to study Islam at Al-Azhar University, where learning and piety go hand in hand. All students are called *mugawireen,* which means *neighbors.*

Cairo

The Egyptian capital of Cairo is the largest city in Africa. More than 6.8 million people live within the city limits.

Cairo is a fascinating mixture of the old and the new. Many of the more modern sections of the city lie on the west bank of the Nile River, on islands in the river, and in Garden City—a narrow strip of land on the river's east bank. Bridges built during the 1900's to connect the riverbanks and islands have made modern urban development possible.

Today, the western part of Cairo includes many government buildings, foreign embassies, hotels, museums, and universities. These newer areas are studded with gardens, parks, and public squares.

In eastern Cairo, by contrast, stand the oldest and most historic sections of the city. In these old quarters, narrow, winding streets meander past centuries-old buildings. On some streets, in colorful outdoor markets called *bazaars,* almost every available space is crowded with goods.

The old sections are also known for their mosques, which include many outstanding examples of Muslim architecture. Some are hundreds of years old, including the mosque of Ahmed Ibn Tulun, built in the A.D. 870's.

At least one minaret can be seen from almost any spot in Cairo's old sections. From these slender towers atop the mosques, Muslim officials called *muezzins* summon the faithful to prayer five times every day.

The residents of Cairo are called Cairenes. Most Cairenes are Arab Muslims, but Christians called Copts also live in the city. The Copts trace their origin back to the Christians who lived in Egypt before the Arabs arrived in the 600's.

Well-to-do Cairenes and poor Cairenes lead vastly different lives. Most middle-class and wealthy people dress in Western clothes and live in Garden City, on the islands, or in the suburbs. These Cairenes are the professional people—doctors, lawyers, teachers, government officials, and managers.

Poor Cairenes are crowded into small apartments or makeshift huts in the old sections. Some of the poorest people have taken refuge in old tombs in an area on the outskirts known as the City of the Dead. Many poor Cairenes do unskilled work in factories or small shops, and others are unemployed. Some dress in traditional Egyptian robes, but most wear Western-style clothing.

The population of Cairo has increased dramatically over the last 100 years. This rapid growth was caused by several factors, including Egypt's high birth rate, the movement of rural people to the city, and the influx of refugees who sought shelter in Cairo after their towns were damaged during Arab-Israeli wars.

The rapid growth of Cairo has intensified the city's problems. As Cairenes cope with housing shortages and traffic jams, the number of poor people continues to grow.

A street scene in one of Cairo's old sections features open-front shops and pedestrians making way for donkey carts.

A young Egyptian girl weaves a rug in Sakkara, near Cairo. The Egyptian people have a rich artistic heritage, dating from the fine paintings and statues created in ancient times.

Cairo sprawls over the banks of the Nile River, covering about 83 square miles (215 square kilometers), but its west and east sides seem worlds apart. Western Cairo is more modern, with many buildings that date from the 1900's. Eastern Cairo is much older and more historic. Its attractions include the Egyptian Museum (1), the Museum of Egyptian Civilization (2), and the Munastirli Palace and Nilometer (3). The citadel of Saladin (4) dates from the late 1100's. Old Cairo (5), Al-Azhar University (6), and the Bazaar Quarter (7) are important centers of city life.

Industry and tourism flourish side by side in Cairo. Modern office towers in the newer sections of the city rise above the Nile, while a large ferryboat carries passengers upstream toward the monuments of ancient Egypt.

In the old sections on the eastern side of Cairo, *above,* many people live in small, crowded apartments on the upper floors of run-down buildings. Numerous minarets rise above the rooftops.

gypt's metropolis

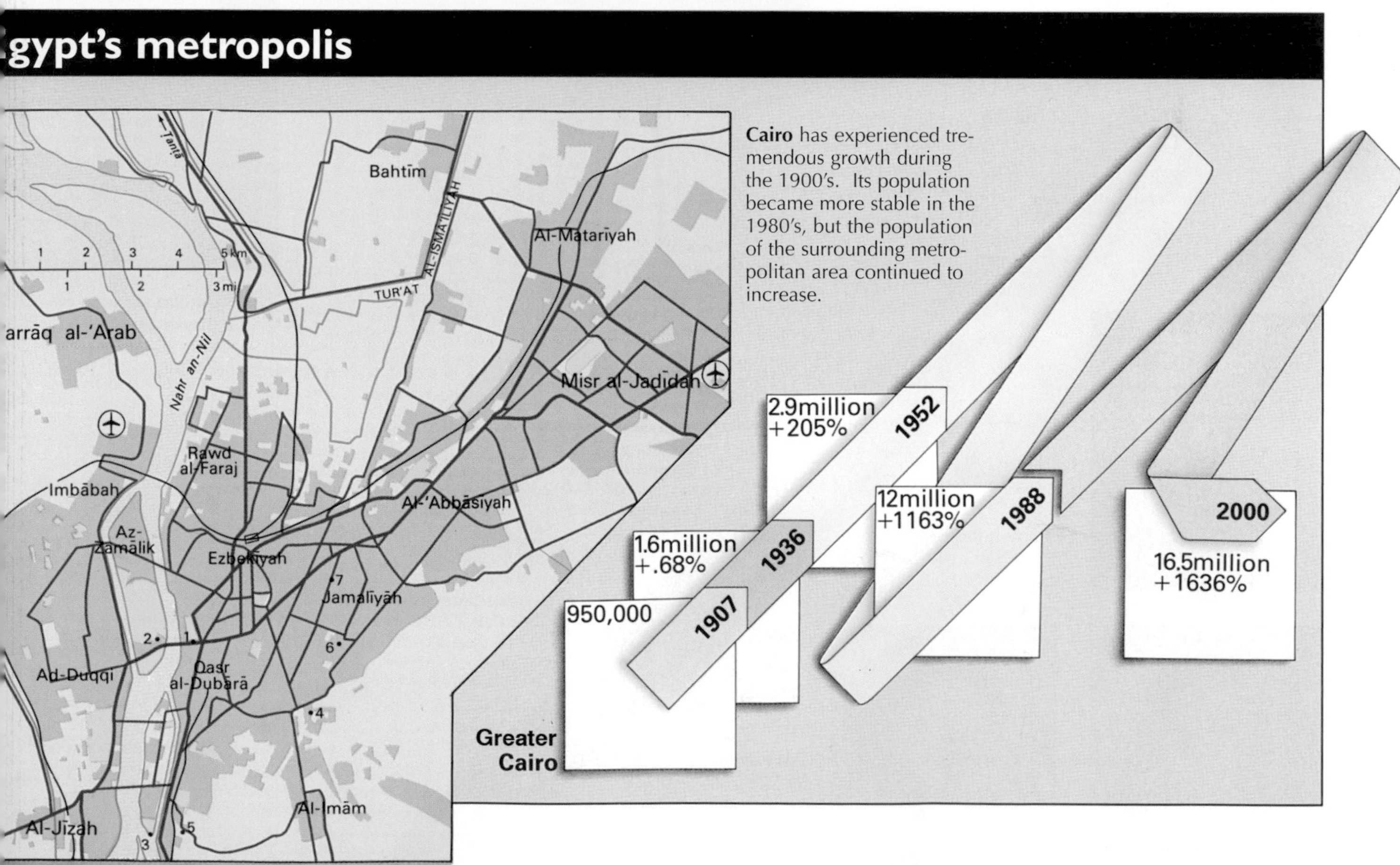

Cairo has experienced tremendous growth during the 1900's. Its population became more stable in the 1980's, but the population of the surrounding metropolitan area continued to increase.

History

According to legend, Egyptian civilization began more than 5,000 years ago when King Menes united two kingdoms on the Nile River and formed the world's first national government.

The age of great pyramid building began about 400 years later, in 2686 B.C. This period of central rule by powerful pharaohs, called the Old Kingdom, lasted for 500 years.

About 1991 B.C., Middle Kingdom pharaohs began to restore the wealth and power that Egypt lost after the decline of the Old Kingdom. But about 1670 B.C., Asian settlers on the Nile Delta, armed with horse-drawn chariots, improved bows, and other "new" weapons, seized control. Their kings ruled Egypt for about 100 years.

Egypt entered the period called the New Kingdom about 1554 B.C. It became the world's strongest power over the next 500 years. Leaders of that time included Queen Hatshepsut, who led armies in battle, and King Thutmose III, who expanded the Egyptian empire into southwest Asia.

Ancient Egypt began to decline rapidly after about 1070 B.C., and in 332 B.C., Alexander the Great of Macedonia added Egypt to his empire. When he died in 323 B.C., Ptolemy, a general in his army, became king.

Queen Cleopatra, a descendant of Ptolemy, lost Egypt to Rome when she married and joined forces with Mark Antony, a co-ruler of Rome. Antony and Cleopatra's navy was defeated by the fleet of Octavian, another Roman co-ruler, and Egypt became a Roman province.

Between A.D. 639 and 642, Arab Muslims began to conquer the Egyptians and convert them to Islam. Thus Egypt became part of the Muslim Empire. It was run by rulers called *caliphs.*

By the mid-1100's, Christian crusaders from Europe were threatening Egypt as part of their effort to recapture the Holy Land from the Muslims. When the Egyptian caliph asked Syrian Muslims for help, Saladin, an officer in the Syrian army, drove the crusaders out and became *sultan,* or prince, of Egypt.

In 1250, military slaves called *Mamelukes* seized control of Egypt. Ruth-

The Ibn Tulun Mosque in Cairo, *above,* built in the A.D. 800's, stands as a reminder of Egypt's Islamic traditions.

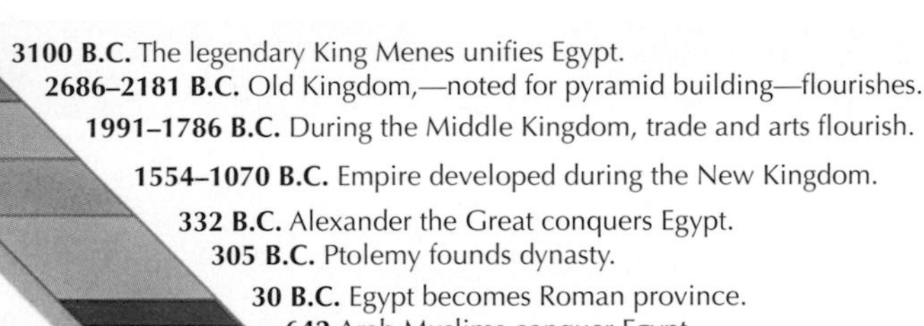

3100 B.C. The legendary King Menes unifies Egypt.
2686–2181 B.C. Old Kingdom,—noted for pyramid building—flourishes.
1991–1786 B.C. During the Middle Kingdom, trade and arts flourish.
1554–1070 B.C. Empire developed during the New Kingdom.
332 B.C. Alexander the Great conquers Egypt.
305 B.C. Ptolemy founds dynasty.
30 B.C. Egypt becomes Roman province.
642 Arab Muslims conquer Egypt.
969–1171 Fatimid dynasty rules Egypt.
1171–1250 Ayyubid dynasty rules.
1250–1517 Mamelukes rule.
1517 Ottoman Turks conquer Egypt.
1798 Napoleon conquers Egypt.
1801 France withdraws from Egypt.
1805 Muhammad Ali seizes control.
1869 Suez Canal opens.
1875 Egypt sells canal shares to Great Britain.
1882 British troops occupy Egypt.
1914 Great Britain declares Egypt a protectorate.
1922 Great Britain grants Egypt independence. Egypt in World War II.
1948–49 Egypt joins in Arab-Israeli war.
1952 King Faruk forced to step down.
1953 Egypt becomes a republic.
1954 Gamal Abdel Nasser comes to power. British agree to withdraw all troops from Egypt by June 1956.
1956 Suez Canal crisis. Israeli forces invade. British and French troops land, but are withdrawn by the end of the year.
1958 Egypt joins Syria in United Arab Republic. Syria withdraws in 1961.
1967 Israel defeats Egypt in Six-Day War.
1968 Aswan High Dam begins operation.
1970 Nasser dies. Anwar el-Sadat becomes president.
1973 Egyptian troops cross Suez Canal and invade Israeli-held Sinai Peninsula in fourth Arab-Israeli war.
1978 Camp David Accords with Israel.
1979 Egypt and Israel sign peace treaty.
1981 Sadat assassinated. Hosni Mubarak becomes president.

Pharaoh Ramses II reigned from 1290 B.C. to 1224 B.C.

Muhammad Ali (1769–1849) worked to modernize Egypt.

Gamal Abdel Nasser (1918–1970) built the Aswan High Dam.

less power struggles followed for more than 200 years. The Ottoman Turks invaded Egypt in 1517, but the Mamelukes were able to keep much of their power until 1798. In that year, France's Napoleon Bonaparte defeated the Mamelukes in the Battle of the Pyramids.

With aid from Great Britain, the Ottomans forced the French out of Egypt in 1801. In 1805, in the disorder that followed the French departure, a Turkish officer named Muhammad Ali made himself ruler. He began an ambitious program to modernize Egypt, but many of his reforms failed.

Under Muhammad Ali's son, Said, French interests began to cut a great canal through the Isthmus of Suez. Said's nephew, Ismail, sold Egypt's interests in the canal to Great Britain in 1875. British interests in Egypt increased, and in 1914 Britain made Egypt a protectorate and kept troops there—even after Egypt was given independence in 1922.

In July 1952, Gamal Abdel Nasser and other army officers led an overthrow of Egypt's King Faruk. Under Nasser, Egypt took control of the Suez Canal and began a period of great economic and social change. The nation fought two wars with Israel and suffered a humiliating defeat in 1967.

Anwar el-Sadat became Egypt's president when Nasser died. Sadat signed a treaty with Israel in 1979. But enraged Muslim fundamentalists killed Sadat in 1981. He was succeeded by Hosni Mubarak, who continued to serve as Egypt's president through the mid-1990's.

The Suez Canal, *far left,* was built by French engineers and completed in 1869. It connected the Mediterranean and Red seas and cut the distance from England to India by 6,000 miles (9,700 kilometers). French and British interests controlled the canal until 1956, when it was seized by Egypt's President Nasser.

The temple of Luxor was the heart of the city of Thebes, capital of ancient Egypt's New Kingdom (1554 B.C. to 1070 B.C.). During this period, Egypt became one of the world's strongest powers and extended its rule into Asia.

El Salvador

The first people to live in what is now El Salvador were Nahua Indians, who came to the area from Mexico as early as 3000 B.C. Later, other Indians settled in the region. The ruins of huge limestone pyramids built by the Maya between A.D. 100 and 1000 still stand in western El Salvador. After the year 1000, the Pipil Indians seized control of the lands west of the Lempa river. The Pipil built cities, raised crops, and produced skilled weaving.

Spanish rule

The first Europeans in the region were Spanish soldiers led by Pedro de Alvarado, who invaded the land in 1524. After a fierce year-long struggle, they conquered the Pipil.

El Salvador was a Spanish colony for the next 300 years, but the land had little gold or silver, so it attracted few colonists. Those who did come to the country mainly farmed the land and raised cattle.

With four other Central American states, El Salvador broke away from Spanish rule in 1821. The five states formed the United Provinces of Central America in 1823. José Matías Delgado, a Salvadoran Catholic priest, led El Salvador's revolt against Spain and drafted the Constitution of the five-state union. However, the union began to break up in 1838, and El Salvador withdrew in 1840. In 1841, the country officially declared its independence.

Struggle for reform

Political violence troubled El Salvador for the rest of the 1800's. Five of the country's presidents were overthrown in revolts, and two others were executed. Dictators from other Central American countries controlled the weak Salvadoran presidents.

After 1900, the political situation became more stable. The country's economy also improved as coffee and other farm products were grown and exported, but great social inequality persisted. Most of the nation's best farmland belonged to a small group of wealthy people, while many rural people had poor farmland—or no land at all.

In 1931, General Maximiliano Hernández Martínez seized control of the government. During his rule, he built public schools, expanded social services, and supported labor reform. However, he ruled as dictator, and a revolution led by soldiers and college students deposed him in 1944.

In 1956, another army officer, Colonel José María Lemus, won the presidency. However, some people disputed the election results. When Salvadorans began to demand election reform, Lemus banned free speech and imprisoned his opponents. A military group overthrew him in 1960. A *junta* (council) then ruled the country until 1961 , when it was ousted by another junta.

In 1962, the Salvadoran legislature adopted a new constitution, and the people elected Colonel Julio Adalberto Rivera as president. Rivera's government raised the income taxes of the wealthy and gave some public and private lands to poor families, but problems continued.

Widespread protests broke out in the late 1970's, with demands for land and jobs for the poor. Some Roman Catholic priests and nuns supported these demands.

In 1979, army officers overthrew the president and replaced him with a junta. Widespread rioting took place when Archbishop Óscar Arnulfo Romero was assassinated. Romero was a champion of El Salvador's poor and had openly criticized the government.

Indian children, *left,* and their families have been plunged more deeply into poverty by the civil war of the 1980's. Many villages have been destroyed in the war, leaving rural families homeless.

Government soldiers search for guerrillas in a poor section of San Salvador, *right.* In November 1989, more than 1,300 people died in San Salvador during large-scale fighting between rebels and government troops.

Coping with natural disasters, such as floods, *below,* strains the Salvadoran government, already pressured by war. In 1986, an earthquake rocked San Salvador, killing more than 1,200. Duarte's critics charged that the government's relief efforts neglected poor neighborhoods devastated in the quake.

With the violence increasing, the junta appointed José Napoleón Duarte as president in 1980. Duarte began reforms, but protests continued. By the early 1980's, a civil war between government troops and rebel guerrilla forces was underway. Under President Ronald Reagan, the United States increased its financial and military aid to the government of El Salvador.

A new Constitution was adopted in 1983. Duarte became president after defeating his opponent, a member of the ARENA party, in the 1984 election. The ARENA party generally has been considered to be behind the death squads responsible for many of the 75,000 deaths during the civil war.

In January 1992, the government and guerrilla leaders signed a peace treaty providing for an end to the 12-year civil war. The war officially ended on December 15, 1992.

El Salvador Today

El Salvador is a tiny, but crowded, Central American country. The smallest nation in Central America in area, El Salvador has more people than any other country in the region except Guatemala and Honduras.

El Salvador ranks as the most densely populated country on the mainland of the Americas, and its population is growing rapidly. Because of this growth, the nation's supply of good farmland is shrinking fast.

The land and economy

Most of the country's fertile farmland lies in the Central Region. This region is a broad plateau of gently rolling land, bordered on the south by the Coastal Range, a rugged mountain range with high, inactive volcanoes. On the range's lower slopes, coffee plantations and cattle ranches sprawl among forests of oak and pine trees. Coffee is El Salvador's leading crop, accounting for almost half the country's earnings from exports.

Past eruptions of the Coastal Range volcanoes helped create the fertile soil of the Central Region plateau, now El Salvador's chief agricultural region. Many Salvadoran farmers own small farms and grow beans, corn, rice, and other crops to feed their families or to trade or sell at local markets. Other farmers work on large commercial plantations called *fincas.*

The Central Region is home to most of the people of El Salvador. About 75 per cent of the Salvadoran people live there, especially in the large cities, such as San Salvador and Santa Ana. Many rural people have moved to the cities looking for work because the supply of farmland has become so limited.

Most of the country's industry is in this area. The federal government has encouraged new industries in an attempt to reduce the country's dependence on agriculture. The leading industries produce chemicals, cigarettes, foods and beverages, leather goods, and textiles. Manufacturing, however, accounts for only a small percentage of the national income.

South of the Coastal Range, the Coastal Lowlands, a narrow strip of fertile land, stretches along the Pacific shore. Grasslands, swamps, and tropical forests once covered much of this plain, but large sections have been developed for farmland. Cotton and sugar cane thrive in these warm, humid lowlands. A number of factories and a fishing industry are located near the town of Acajutla.

The Interior Highlands to the north of the Central Region are made up mainly of a low

FACT BOX

COUNTRY

Official name: Republica de El Salvador (Republic of El Salvador)

Capital: San Salvador

Terrain: Mostly mountains with narrow coastal belt and central plateau

Area: 8,124 sq. mi. (21,041 km^2)

Climate: Tropical; rainy season (May to October); dry season (November to April); tropical on coast; temperate in uplands

Main rivers: Lempa, Río Grande de San Miguel

Highest elevation: Cerro El Pital, 8,957 ft. (2,730 m)

Lowest elevation: Pacific Ocean, sea level

GOVERNMENT

Form of government: Representative democracy

Head of state: President

Head of government: President

Administrative areas: 14 departamentos (departments)

Legislature: Asamblea Legislativa (Legislative Assembly) with 84 members serving three-year terms

Court system: Corte Suprema (Supreme Court)

Armed forces: 15,000 troops

PEOPLE

Estimated 2008 population: 7,218,000

Population density: 888 persons per sq. mi. (343 per km^2)

Population distribution: 60% urban, 40% rural

Life expectancy in years:
Male: 67
Female: 73

Doctors per 1,000 people: 1.3

Percentage of age-appropriate population enrolled in the following educational levels:
Primary: 113*
Secondary: 59
Further: 17

El Salvador's Pacific coast, shown here in the first light of dawn, is only 189 miles (304 kilometers) long. Fishing crews catch shrimp and lobsters in the offshore waters.

El Salvador, *above,* is the smallest Central American country—about the same size as Massachusetts. Guatemala lies to the northwest; Honduras lies east and northeast.

Languages spoken:
- Spanish
- Nahua (among some Amerindians)

Religion:
- Roman Catholic 83%
- Other 17%

*Enrollment ratios compare the number of students enrolled to the population which, by age, should be enrolled. A ratio higher than 100 indicates that students older or younger than the typical age range are also enrolled.

TECHNOLOGY

Radios per 1,000 people: 481

Televisions per 1,000 people: 233

Computers per 1,000 people: 25.2

ECONOMY

Currency: Salvadoran colon and U.S. dollar

Gross domestic product (GDP) in 2004: $32.35 billion U.S.

Real annual growth rate (2003–2004): 1.8%

GDP per capita 2004: $4,900 U.S.

Goods exported: Offshore assembly exports, coffee, sugar, shrimp, textiles, chemicals, electricity

Goods imported: Raw materials, consumer goods, capital goods, fuels, foodstuffs, petroleum, electricity

Trading partners: United States, Guatemala,

mountain range called the Sierra Madre. Hardened lava, rocks, and volcanic ash cover much of this area. The highlands are El Salvador's most thinly populated region, with only a few small farms and ranches.

Civil war

A civil war between the government and the leftist Farabundo Martí National Liberation Front (FMLN) claimed about 75,000 lives between 1979 and 1992. Many atrocities were committed during the war, especially by the military. In January 1992, the government and FMLN signed a peace agreement.

In 1994, Armando Calderón Sol of the right-wing ARENA party was elected president. Members of the new administration received death threats, and Calderón unveiled a security plan that relied on the military and ARENA-dominated national police. FMLN opposed the plan because it gave only a small role to a new, civilian police force formed to replace the national police.

People

More than 6.7 million people live in El Salvador, and that number is growing rapidly. About 92 per cent of all Salvadorans are *mestizos,* or people of mixed European and Indian descent. Nearly 5 per cent are of European ancestry, and another 3 per cent are Indians.

Two cultures

The people of El Salvador form two distinct cultural groups—Ladinos and Indians. Ladinos follow Spanish-American ways of life and speak Spanish. All mestizos and whites are Ladinos, but Indians can also become Ladinos if they adopt the Spanish-American culture, as have many Indians.

Many other Indians still follow the ways of their ancestors. El Salvador's Pancho Indians, who are descended from the Pipil Indians, live in villages built by the Pipil and speak a form of Nahua, the Pipil language. The Pipil were the largest Indian group in El Salvador when the Spaniards first arrived. Most Pancho Indians now live in the southwest.

Social class and family life

Almost half of all Salvadorans live on farms or in rural towns. Some farmers live in adobe houses with dirt floors and thatched roofs. Many poor people in rural areas live in *wattle huts.* The walls of such huts are made of tree branches woven together and covered with mud. In sharp contrast, some wealthy owners of coffee plantations live in luxury on their scenic estates.

Since the 1940's, hundreds of thousands of rural Salvadorans have moved to the cities to look for jobs, but there is not enough work for them all. Entire families often live in one-room apartments in crowded, run-down buildings. Middle-class city residents live in row houses or in comfortable apartments. The rich live in the suburbs of the cities in luxurious, modern houses with landscaped gardens.

In El Salvador, however, few people are middle class or wealthy, and the widespread poverty has disrupted family life for many people. Some men move from place to place seeking work, and many of El Salvador's children live in orphanages or wander the streets or countryside with no one to care for them.

The law requires all children between the ages of 7 and 12 to attend school. But El Salvador does not have enough schools or teachers, so many children do not go to school at all. Of those who do, only about half graduate. Students who complete nine years of elementary school can go on to secondary school for three years and then attend a university.

Leisure

The people of El Salvador love to spend their leisure time outdoors. Soccer is the national sport, and a game is often in progress in neighborhood fields. Families like to spend their weekends at resorts near lakes or on Pacific beaches. Los Chorros, a popular national park near Nueva San Salvador, has four swimming pools surrounded by tropical gardens and waterfalls.

About 80 per cent of all Salvadorans, including most Indians, are Roman Catholic, and religious festivals are popular. The most colorful festival of the year celebrates the Feast of the Holy Savior of the World. From July 24 to August 6, the people celebrate this festival with carnival rides, fireworks, folk dancing, and colorful processions.

Food

The diet of most Salvadorans consists mainly of beans, breads, corn, and rice. Dairy products and meat are occasional luxuries when the people can afford to buy them.

Salvadorans have bread and coffee for breakfast and eat their main meal at midday. In the late afternoon, many people snack on *pupusas,* which are corn meal cakes stuffed with chopped meat, beans, and spices.

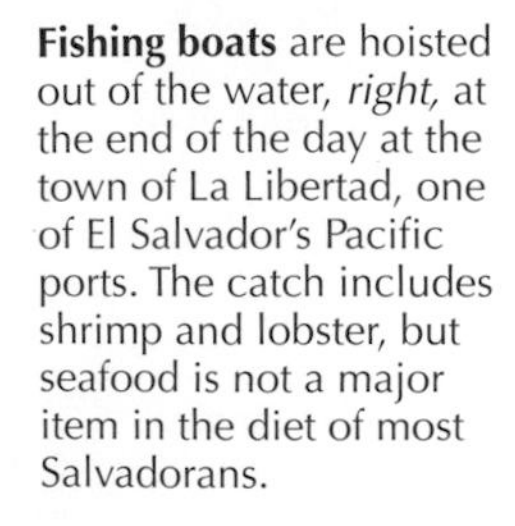

Fishing boats are hoisted out of the water, *right,* at the end of the day at the town of La Libertad, one of El Salvador's Pacific ports. The catch includes shrimp and lobster, but seafood is not a major item in the diet of most Salvadorans.

Faces in the crowd at an election rally in San Salvador show the Salvadoran people's mixed Indian and Spanish ancestry.

An Easter procession, *below,* reflects the importance of religious festivals in El Salvador.

A guerrilla fighter, *above,* guards a rebel base during the Salvadoran civil war of the 1980's and 1990's. In a country where most of the farmland traditionally has been controlled by a few wealthy landowners, the rebels sought greater equality in land ownership.

Equatorial Guinea

The small west African country of Equatorial Guinea consists of a mainland region and five islands in the Atlantic Ocean. The mainland territory, called Río Muni, is home to most of the nation's people. The largest island, Bioko, lies about 100 miles (160 kilometers) northwest of Río Muni, while the other four islands—Corisco, Elobey Chico, Elobey Grande, and Annobon—lie southwest of Río Muni.

People

Equatorial Guinea has a little more than half a million people, and about 55 per cent live in rural areas. Most rural people are farmers, raising such food crops as bananas, cassava, and sweet potatoes. Some farms produce cacao for export, especially on Bioko. Coffee also thrives in Bioko's rich volcanic soil.

Some rural people fish for a living, especially on the islands, and others work in lumber camps. Dense tropical rain forests cover much of the land, and forestry is important to the country's economy. The okoumé tree, for example, is cut down to make plywood.

About 45 per cent of the people of Equatorial Guinea live in urban areas such as the largest city, Bata, and the capital, Malabo, and many work in small industries or in trade. The country has some food-processing plants, but little manufacturing. Petroleum production is the chief economic activity in Equatorial Guinea and accounts for more than 90 per cent of the country's exports. Large oil and gas deposits were discovered near Bioko in the mid-1900's.

More than half of the people of Equatorial Guinea live in rural areas.

The people of Equatorial Guinea belong mainly to various black African ethnic groups. In Río Muni, most people are members of the Fang group. Their language—also called Fang—is the most widely spoken in the country, though Spanish is the official language. Most of the people on the island of Bioko belong to the Fernandino or Bubi ethnic groups.

Equatorial Guinea suffers from poor educational and medical services. Many children do not attend school because of a shortage of teachers, and diseases such as malaria and measles spread quickly due to the lack of medical care.

History

Pygmies who inhabited the area before the 1200's were probably the first people to live in what is now mainland Río Muni. Then various other groups began to move into the region, including the Fang, the Benga, and the Bubi. Bubi from the mainland region also settled on Bioko during the 1200's.

FACT BOX

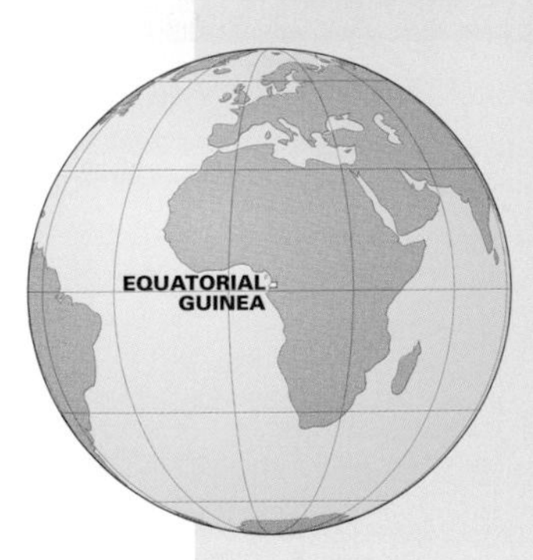

COUNTRY

Official name: Republica de Guinea Ecuatorial (Republic of Equatorial Guinea)
Capital: Malabo
Terrain: Coastal plains rise to interior hills; islands are volcanic
Area: 10,831 sq. mi. (28,051 km^2)
Climate: Tropical; always hot, humid
Main river: Mbini
Highest elevation: Santa Isabel Mountain, 9,869 ft. (3,008 m)
Lowest elevation: Sea level along the coast

GOVERNMENT

Form of government: Republic
Head of state: President
Head of government: Prime minister
Administrative areas: 7 provincias (provinces)
Legislature: Camara de Representantes del Pueblo (House of People's Representatives) with 80 members serving five-year terms
Court system: Supreme Tribunal
Armed forces: 1,320 troops

PEOPLE

Estimated 2008 population: 538,000
Population density: 50 persons per sq. mi. (19 per km^2)
Population distribution: 61% rural, 39% urban
Life expectancy in years: Male: 47 Female: 50
Doctors per 1,000 people: N/A
Percentage of age-appropriate population enrolled in the following educational levels: Primary: N/A Secondary: N/A Further: N/A

The first Europeans in the region were Portuguese, who landed on the island of Annobon in 1471. Portugal later claimed that island, along with Bioko and part of the mainland coast. Spain gained control of these territories in the mid-1800's, and the region became a Spanish colony in 1959.

Equatorial Guinea became an independent nation on Oct. 12, 1968, and later that year, Francisco Macias Nguema took control as president and dictator. During his rule, Macias Nguema had many people killed or imprisoned. In 1979, army officers overthrew Macias Nguema and established a military government. Macias Nguema was executed, and Lieutenant Colonel Teodoro Obiang Nguema Mbasogo became president.

Government by a group of military officers continued in the 1980's. In 1991, a new Constitution was approved. In multiparty elections held in 1993, 1996, and 2002, Mbasogo was reelected president. However, the elections were widely viewed as fraudulent.

A tree-lined riverbank is silhouetted by light from the setting sun. Dense tropical rain forests cover much of the country.

Languages spoken:
- Spanish (official)
- French (official)
- Pidgin English
- Fang
- Bubi
- Ibo

Religions:
- Roman Catholic 75%
- Indigenous beliefs

TECHNOLOGY

Radios per 1,000 people: N/A

Televisions per 1,000 people: N/A

Computers per 1,000 people: N/A

ECONOMY

Currency: Communaute Financiere Africaine franc

Gross national income (GNI) in 2002: $1.27 billion U.S.

Real annual growth rate (2001–2002): 20%

GNI per capita (2002): $2,700 U.S.

Goods exported: Petroleum, timber, cocoa

Goods imported: Petroleum, manufactured goods and equipment

Trading partners: United States, Spain, France, China

Equatorial Guinea consists of Río Muni on the west African mainland and five offshore islands, including Bioko, the largest.

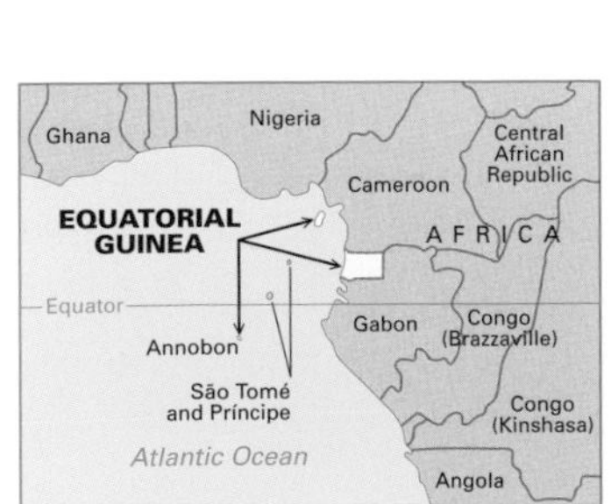

Eritrea

Eritrea is a small country situated on the east coast of Africa. It is bordered to the north and west by Sudan, to the east by the Red Sea, and to the south by Ethiopia. Formerly a province of Ethiopia, Eritrea gained its independence in 1993 after a 30-year struggle for independence.

Eritrea's coastal plain is one of the hottest and driest spots in Africa. Although less than 5 per cent of the land is under cultivation, about 80 per cent of the people work in agriculture. The nation's chief products include barley and lentils.

Eritrea was colonized by Italy in the late 1800's. The Italians used the region as a base from which they established their brief rule over Ethiopia. They built railways and other commercial enterprises in Eritrea. This period of colonization helped the Eritreans develop many skills—and an important trading edge over Ethiopia.

The United Kingdom drove Italy from Eritrea in 1941, during World War II (1939–1945). In 1952, British administration of the territory ended, and it was joined with Ethiopia as part of a federation. But the region governed itself.

Eritrea became an Ethiopian province in 1962, and civil war broke out between Eritrean rebels, who wanted independence for the province, and Ethiopian government troops. Over the next 30 years, the rebels developed into the highly organized Eritrean People's Liberation Front, led by Issaias Afwerki. The separatists maintained that Eritrea had been unfairly handed back to Ethiopia in 1952. Ethiopia resisted the move for Eritrean independence because it feared becoming completely landlocked and losing access to the Red Sea.

After 30 years of violence and bloodshed, Eritrean fighters defeated the Ethiopian army—the largest in sub-Saharan Africa. When the struggle ended in May 1991—shortly after the Ethiopian army's last strongholds in the cities of Asmara and Keren fell—it had become modern Africa's longest and deadliest war. It is estimated that about 160,000 Ethiopian and Eritrean fighters were killed, as well as 40,000 civilians in Eritrea.

When the war ended, Issaias did not immediately declare Eritrea's independence. Instead, he chose to hold a popular referendum for independence within two years, a process that would be more well accepted in the international community.

In the meantime, the Eritreans began the long and difficult task of rebuilding a country devastated by three decades of war. Schools, roads, and medical facilities were rebuilt.

FACT BOX

COUNTRY

Official name: Hagere Ertra (State of Eritrea)

Capital: Asmara

Terrain: Dominated by extension of Ethiopian north-south trending highlands, descending on the east to a coastal desert plain, on the northwest to hilly terrain and on the southwest to flat-to-rolling plains

Area: 45,406 sq. mi. (117,600 km^2)

Climate: Hot, dry desert strip along Red Sea coast; cooler and wetter in the central highlands; semiarid in western hills and lowlands; rainfall heaviest June to September except in coastal desert

Highest elevation: Mount Soira, 9,885 ft. (3,013 m)

Lowest elevation: Near Kulul within the Denakil Depression, 360 ft. (110 m) below sea level

GOVERNMENT

Form of government: Transitional government

Head of state: President

Head of government: President

Administrative areas: 8 awraja (provinces)

Legislature: National Assembly with 150 members

Court system: Supreme Court; 10 provincial courts; 29 district courts

Armed forces: 201,750 troops

PEOPLE

Estimated 2008 population: 4,886,000

Population density: 108 persons per sq. mi. (42 per km^2)

Population distribution: 81% rural, 19% urban

Life expectancy in years:
Male: 52
Female: 55

Doctors per 1,000 people: Less than 0.05

Percentage of age-appropriate population enrolled in the following educational levels:
Primary: 63
Secondary: 28
Further: 2

Eritrea, which officially became a nation in 1993, lies along the western shore of the Red Sea in northeast Africa. Its capital is Asmara.

Elections at the village, local, and provincial level were held, and work on a constitution began. Eritrean leaders also announced that their country would have a market-driven economy.

After an April 1993 referendum, in which the Eritreans overwhelmingly voted for independence, the nation declared its independence on May 24, 1993. A four-year transitional period, during which the country moved toward a constitutional political system, was also declared. In 1997, Eritrea adopted a new constitution.

A section of the border between Eritrea and Ethiopia wasn't clearly defined when Eritrea achieved independence. In May 1998, the two nations began warring over this disputed area. The two sides signed a cease-fire agreement in 2000, and later that year they signed a formal peace treaty. A commission identified the border between the countries in 2002. Droughts struck Eritrea in the early 2000's. As a result, more than 2 million people faced starvation.

A forbidding landscape of rocky desert studded with volcanic cones, marks the Denakil Depression, the lowest spot in Eritrea. Daytime temperatures may rise above 120°F. (49°C) in the country's lowland deserts.

Languages spoken:
- Afar
- Amharic
- Arabic
- Tigre and Kunama
- Tigrinya
- Other Cushitic languages

Religions:
- Muslim
- Coptic Christian
- Roman Catholic
- Protestant

TECHNOLOGY

Radios per 1,000 people: 464

Televisions per 1,000 people: 53

Computers per 1,000 people: 2.9

ECONOMY

Currency: Nakfa

Gross domestic product (GDP) in 2002: $1.27 billion U.S.

Real annual growth rate (2001–2002): 20%

GDP per capita (2002): $2,700 U.S.

Goods exported: Livestock, sorghum, textiles, food, small manufactures

Goods imported: Processed goods, machinery, petroleum products

Trading partners: Ethiopia, Sudan, Italy, Saudi Arabia, United Arab Emirates

Jubilant Eritreans, *below,* celebrate the passage of an April 1993 referendum that established Eritrea as an independent nation. *Bottom,* Eritreans seek shelter from the hot sun in Asmara, the country's capital and industrial center.

Estonia

Estonia is a European nation that gained its independence from the Soviet Union in 1991. Estonia's landscape consists chiefly of a low plain covered by farmland, forests, and swamps. The country has 481 miles (774 kilometers) of coastline along the Baltic Sea, the Gulf of Finland, and the Gulf of Riga. Estonia also includes several Baltic islands, the largest of which is Saaremaa Island.

When the Soviet Union took over Estonia in 1940, the Estonians—a people related to the Finns—made up about 90 per cent of the population. However, many Russians emigrated to the republic to fill jobs created by rapid industrialization following World War II. As a result, native Estonians now account for only about 60 per cent of the population.

History

The ancestors of present-day native Estonians settled in the area several thousand years ago. By the 1500's, German nobles owned much of Estonia. Sweden took over northern Estonia in 1561, and Poland conquered the southern part of the country.

Sweden controlled all of Estonia between 1625 and 1721, when Russia took over the area. In the mid-1800's, a movement for Estonian independence gained strength, and in 1918, Estonia proclaimed its independence and established a democratic government.

Estonia's landscape is studded with woodlands, plains, and lakes, but intensive manufacturing and mining development have led to environmental problems.

From independence to annexation

In 1939, Soviet dictator Joseph Stalin and Adolf Hitler, the leader of Nazi Germany, secretly agreed to divide up between themselves a number of Eastern European countries. In August 1940, the Soviet Union forcibly annexed Estonia and turned the nation's private factories and farms into government-controlled enterprises.

The Estonians bitterly opposed the takeover. However, tens of thousands of people who joined the Estonian resistance movement were deported to Siberia. Resistance to Soviet control grew during the late 1980's, and Estonia began to press the Soviet Union for restoration of the republic's independence. In January 1991, the Soviet government responded to Estonia's suspension of compulsory military service by sending in troops to enforce the draft.

FACT BOX

COUNTRY

Official name: Eesti Vabariik (Republic of Estonia)
Capital: Tallinn
Terrain: Marshy, lowlands
Area: 17,413 sq. mi. (45,100 km^2)
Climate: Maritime, wet, moderate winters, cool summers
Main rivers: Gauja, Jagala, Lasam, Narva, Plyussa
Highest elevation: Suur Munamagi, 1,043 ft. (318 m)
Lowest elevation: Baltic Sea, sea level

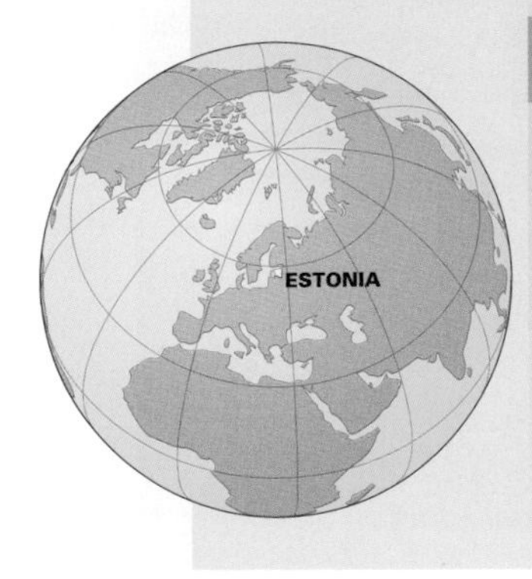

GOVERNMENT

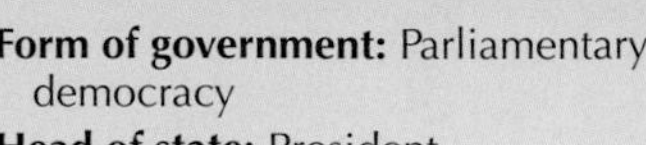

Form of government: Parliamentary democracy
Head of state: President
Head of government: Prime minister
Administrative areas: 15 maakonnad (counties)
Legislature: Riigikogu (Parliament) with 101 members serving four-year terms
Court system: National Court
Armed forces: 4,980 troops

PEOPLE

Estimated 2008 population: 1,334,000
Population density: 77 persons per sq. mi. (30 per km^2)
Population distribution: 69% urban, 31% rural
Life expectancy in years:
Male: 65
Female: 77
Doctors per 1,000 people: 3.2
Percentage of age-appropriate population enrolled in the following educational levels:
Primary: 101*
Secondary: 96
Further: 64
Languages spoken: Estonian (official), Russian, Ukrainian, English, Finnish

The spires and turrets rising above Tallinn are typical of the impressive architecture for which the city is renowned.

Freedom reclaimed

After becoming independent in 1991, Estonia moved forward with economic reform and reduced government control of most economic activities. By the mid-1990's, most businesses had become privately owned. Estonia sought to strengthen its ties with western Europe and to reduce Russian influence over its affairs. In 2004, Estonia joined both the European Union (EU) and the North Atlantic Treaty Organization (NATO).

Religions:
Evangelical Lutheran 80%
other Eastern Orthodox

*Enrollment ratios compare the number of students enrolled to the population which, by age, should be enrolled. A ratio higher than 100 indicates that students older or younger than the typical age range are also enrolled.

TECHNOLOGY

Radios per 1,000 people: 1,136
Televisions per 1,000 people: 507
Computers per 1,000 people: 440.4

ECONOMY

Currency: Estonian kroon
Gross domestic product (GDP) in 2004: $19.23 billion U.S.
Real annual growth rate (2003–2004): 6%
GDP per capita (2004): $14,300 U.S.
Goods exported: Machinery and appliances, wood products, textiles, food products, metals, chemical products
Goods imported: Machinery and appliances, foodstuffs, chemical products, metal products, textiles
Trading partners: Finland, Sweden, Russia

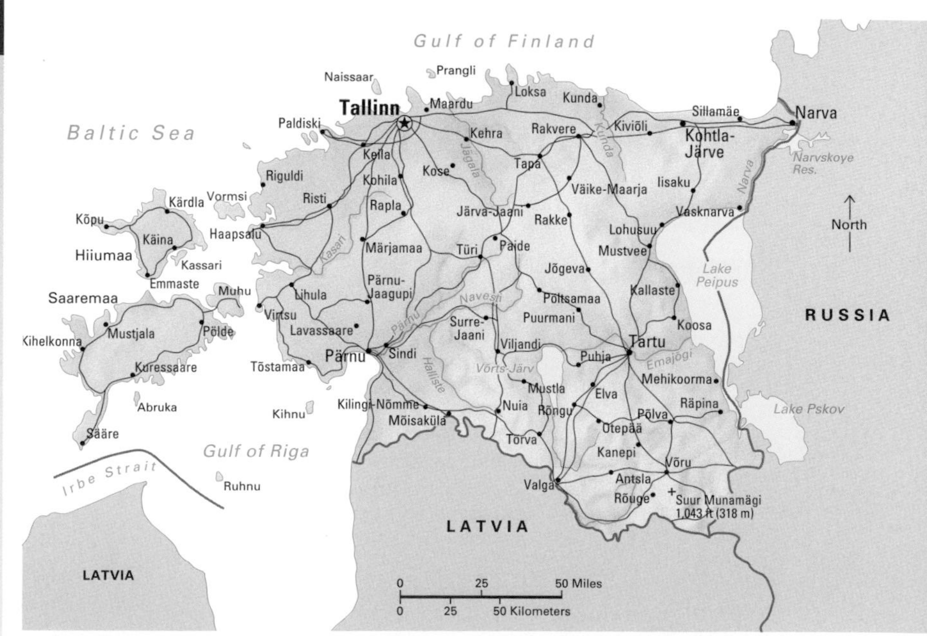

Estonia, *map above,* lies on the Baltic Sea in northern Europe. It is one of the three Baltic States, along with Lithuania and Latvia.

Ethiopia

Ethiopia is one of the oldest nations in Africa. By the A.D. 200's, a powerful kingdom became established in Ethiopia. Its capital, Aksum, grew wealthy by trading gold, ivory, and spices with Arabia, Egypt, Greece, India, Persia, and Rome.

The power of Aksum fell sharply in the 600's, when Muslims took control of Arabia, the Red Sea, and the coast of northern Africa. About 500 years later, in 1137, the Zagwé dynasty rose to power but was overthrown in 1270. After the 1500's, the Ethiopian Empire broke up into a number of small kingdoms.

The empire was rebuilt in the late 1800's, when Menelik II became emperor. In 1916, Menelik's daughter, Zauditu, became empress, ruling with the help of Ras Tafari, the son of Menelik's cousin. When Zauditu died in 1930, Tafari became emperor, taking the title Haile Selassie I. In 1974, army officers overthrew Haile Selassie.

Under Lieutenant Colonel Mengistu Haile-Mariam, the government adopted socialism and began land reform. The military leaders also began killing many of their opponents. In the late 1970's, rebellion broke out in the northern region of Tigre, and Ethiopia went to war over Ogaden with its neighbor Somalia, which also claimed the region. The war with Somalia ended in 1988. Meanwhile, Eritrean rebels set up their own government in the Eritrean province and in 1993 gained independence from Ethiopia. From 1998 to 2000, Eritrea and Ethiopia battled over the location of their shared border. Although a commission issued a border ruling in 2002, Ethiopia objected to the ruling, leaving the matter unsettled.

In 1991, a group of rebel armies called the Ethiopian People's Revolutionary Democratic Front (EPRDF) overthrew Mengistu's military government. The rebels formed a transitional

An Ethiopian child is sheltered in the arms of an elderly man. Many Ethiopians suffer from extreme poverty, and severe droughts have also plagued the country, causing famine and death.

Rolling green hills in northwestern Ethiopia stretch to Lake Tana, the source of the Blue Nile. About 85 per cent of Ethiopia's people live in rural areas, either in villages or isolated homesteads, much as their ancestors did.

FACT BOX

COUNTRY

Official name: Ityop'iya Federalawi Demokrasiyawi Ripeblik (Federal Democratic Republic of Ethiopia)
Capital: Addis Ababa
Terrain: High plateau with central mountain range divided by Great Rift Valley
Area: 426,373 sq. mi. (1,104,300 km^2)
Climate: Tropical monsoon with wide topographic-induced variation
Main rivers: Awash, Baro, Blue Nile (Abay), Genale, Omo, Wabe Shebele
Highest elevation: Ras Dashen Terara, 15,158 ft. (4,620 m)
Lowest elevation: Denakil, 381 ft. (116 m) below sea level

GOVERNMENT

Form of government: Federal republic
Head of state: President
Head of government: Prime minister
Administrative areas: 9 ethnically-based stedader akababiwach (administrative regions), 2 chartered cities
Legislature: Parliament consisting of the Council of the Federation or upper chamber with 118 members serving five-year terms and the Council of People's Representatives or lower chamber with 548 members serving five-year terms
Court system: Federal Supreme Court
Armed forces: 100,000 troops

PEOPLE

Estimated 2008 population: 78,326,000
Population density: 184 persons per sq. mi. (71 per km^2)
Population distribution: 84% rural, 16% urban
Life expectancy in years:
Male: 45
Female: 47
Doctors per 1,000 people: Less than 0.05
Percentage of age-appropriate population enrolled in the following educational levels:
Primary: 66
Secondary: 19
Further: 2

government made up of an 80-member Council of Representatives chosen by various political organizations. The council elected a president, who, along with the council, ran the government until elections took place in June 1994. Voters elected a Constituent Assembly to write a new constitution. Although the vote was peaceful, opposition parties boycotted it and the ruling Ethiopian People's Revolutionary Democratic Front (EPRDF) won a majority. Free elections took place in May 1995.

Ethiopia lies in northeastern Africa, extending far into the interior of the continent in the south and west. It ranks third in population among African nations.

Languages spoken:
Amharic
Tigrinya
Oromifa
Guaraginga
Somali
Arabic
Other local languages
English (major foreign language taught in schools)

Religions: Muslim 45%-50%
Ethiopian Orthodox 35%-40%
Animist 12%

TECHNOLOGY

Radios per 1,000 people: 189

Televisions per 1,000 people: 6

Computers per 1,000 people: 2.2

ECONOMY

Currency: Birr

Gross domestic product (GDP) in 2004: $54.89 billion U.S.

Real annual growth rate (2003–2004): 11.6%

GDP per capita (2004): $800 U.S.

Goods exported: Coffee, gold, leather products, oilseeds

Goods imported: Food and live animals, petroleum and petroleum products, chemicals, machinery, motor vehicles

Trading partners: Germany, Japan, Italy, United States

Although much of Ethiopia is fertile with usually adequate rainfall, farmers make use of only a small part of this good farmland. Most farmers work the land with wooden plows pulled by oxen.

Nevertheless, agriculture is the nation's chief economic activity, employing about 85 per cent of Ethiopian workers. Farmers' main food crops for their own use include corn, sorghum, teff (a grain), and wheat, and most farmers also raise cattle, chickens, goats, and sheep. Cash crops include oilseeds, sugar cane, and coffee.

Ethiopia suffered famine in 1994. At least 5,000 people died due to drought, cattle disease, and poor distribution of food. Another drought in the early 2000's threatened millions of people with starvation.

Ethiopia has a developing economy. Only about 5 per cent of the workers have jobs in manufacturing, and about 10 per cent are employed in service industries.

Land and People

Ethiopia is a land of rugged mountains with a high, fertile plateau that stretches across about two-thirds of the country, covering much of the western and central regions. The Ethiopian Plateau, which lies between 6,000 and 10,000 feet (1,800 and 3,000 meters) above sea level, is crossed by deep river gorges and mountain ranges that rise more than 14,000 feet (4,300 meters).

Most Ethiopians live on the plateau, which has the best farmland in the country and usually receives more than 40 inches (102 centimeters) of rain a year. However, severe droughts occur from time to time, sometimes causing famine.

The plateau is split by the Great Rift Valley, which runs north and south through eastern Africa. A series of lakes extends through the valley in southern Ethiopia, but the country's largest lake, Tana, lies in the northwest.

The Blue Nile—called *Abay* in Ethiopia—flows out of Lake Tana for 20 miles (32 kilometers) before spilling over the spectacular Tisissat Falls. Eventually, the Blue Nile joins the White Nile in Sudan, west of Ethiopia, where the two branches form the mighty Nile River.

The Ethiopian Plateau slopes downward in all directions toward lowland regions. The lowlands are thinly populated because of the hot, dry climate and because the soil is poor for farming.

Grasslands cover much of the plateau, as well as many lowland areas to the south and east. The grasslands are home to such wild animals as antelope, lions, elephants, and rhinoceroses. Tropical rain forests thrive in the southwest.

Ancestry

Ethiopians are descended mainly from three groups of people: brown-skinned people who resembled Europeans, black Africans, and Arabs. Most Ethiopians today have brown skin and features that range from European to black African.

Although Amharic is the country's official language, more than 70 languages are spo-

An ancient church cut from solid rock is a reminder of Ethiopia's rich history. Much of the country's art—especially early paintings and writing—is related to the Ethiopian Orthodox religion.

Famine

The devastating droughts that hit Ethiopia in the 1970's and 1980's led to widespread famine, and thousands of Ethiopians died of starvation, malnutrition, or disease. Many crowded into refugee camps where food from other countries was distributed. The United Nations estimated that in 1988 about 7 million Ethiopians were dependent on such foreign aid, but their military government placed more importance on fighting rebels in northeastern Ethiopia than in aiding famine victims. The government ordered international relief agencies to leave Eritrea and Tigre, even though those areas were hardest hit by famine. Other countries charged that the Ethiopian government was using starvation as a way to win the war against the rebels.

Religious leaders of the Ethiopian Orthodox Church wear magnificent robes to mark their rank. About 40 per cent of all Ethiopians follow that faith.

An Afar woman prepares *injera,* a slightly sour pancake-shaped bread eaten by many Ethiopians. They often use injera to scoop up portions of *wat,* a spicy stew.

Red Cross camps, like the one at the left, were crowded with hungry refugees from the Ethiopian highlands during the mid-1980's, when famine gripped the country. Ethiopia continues to depend on emergency aid to feed its people. In 1989, crop failures in Eritrea and Tigre left nearly 4 million people with insufficient food.

ken in Ethiopia. In addition to their own language, many Ethiopians speak English or Arabic. Ge'ez, an ancient Ethiopian language, is still used in Ethiopian Orthodox religious ceremonies.

Most Ethiopians can be classified into two large groups based on the kind of native language they speak. The Semites—those who speak Semitic languages—live mainly in northern and central Ethiopia and include such ethnic groups as the Amhara and the Tigre. The Cushites—those who speak Cushitic languages—live chiefly in southern and eastern Ethiopia and include the Afar (or Danakil) and Somali.

In addition, people of mostly black African ancestry speak languages of the Nilo-Saharan family. They live along the western border and make up about 5 per cent of the population.

Modern life

Nearly 85 per cent of the Ethiopian people live in the countryside. Poverty is widespread in these rural areas, and each year, large numbers of Ethiopians move to urban areas hoping to find work. Schools, medical care, and such modern conveniences as electricity are also more available in the cities.

Ethiopia's cities have many modern buildings, and several skyscrapers rise above Addis Ababa, the capital. However, most Ethiopians—whether urban or rural—live in traditional round houses. These homes have walls made of wooden frames plastered with mud and cone-shaped roofs of thatched straw or metal sheeting. Where stone is plentiful, many Ethiopians live in rectangular stone houses.

Falkland Islands

A dependency of Great Britain, the Falkland Islands lie about 320 miles (515 kilometers) east of the southern coast of Argentina. They form the southernmost part of the British Empire outside the British Antarctic Territory.

History and government

English explorer John Davis sighted the Falklands in 1592. British captain John Strong landed on the islands in 1690 and named them for Viscount Falkland, the British treasurer of the navy.

Because they lie on the important sea route around the tip of South America, the Falklands became an important stop for vessels rounding Cape Horn. Sailing vessels could drop anchor in the superb natural harbor of Stanley, take on supplies, and repair damage caused by the southern gales. These advantages led France, Spain, and Argentina to also lay claim to the islands, but Britain established control there in 1833.

The Falklands are now an important British base. Nevertheless, Argentina has continued to claim the islands, which it calls Islas Malvinas.

In April 1982, Argentine troops invaded and occupied the islands. Britain responded by sending troops, ships, and planes to the Falklands. Air, sea, and land battles then broke out between Argentina and Britain. The Argentine forces surrendered in June 1982, the economy considerably damaged.

A governor, aided by an executive and a legislative council, rules the British dependency. The government provides schools, which children must attend. Traveling teachers instruct children in isolated settlements.

Environment and people

The British dependency includes two large islands, East and West Falkland, and about 200 smaller ones. East Falkland covers 2,580 square miles (6,682 square kilometers), and West Falkland covers 2,038 square miles (5,278 square kilometers). Together, the islands have a coastline of 610 miles (982 kilometers). A vast area of islands and ocean became dependencies of the Falkland Islands Colony in 1908. The principal islands included South Georgia, South Orkney, South Shetland, and South Sandwich. The South Orkney and South Shetland island groups became part of the British Antarctic Territory in 1962. The territory includes the area south of 60° south latitude, between 20° and 80° west longitude.

The landscape of the Falkland Islands resembles that of Scotland or Ireland, with many small bays and inlets along the coast, and low, rolling, grass-covered hills in the interior. Small streams wind through the shallow valleys, where many wildflowers grow. The temperature seldom falls below 18° F. (?8° C) in winter. In summer, temperatures rarely rise above 77° F. (25° C). Annual rainfall averages about 25 inches (65 centimeters), and snow falls occasionally in winter. Strong winds limit the growth of trees on the islands.

Small outlying communities dot the landscape of the Falklands, below. While sheep farming is still a major activity, the sale of fishing licenses to foreign fishing fleets provides the main source of income.

Stanley, the capital and main settlement of the Falklands, sits on a fine, natural harbor. Strong winds limit the growth of trees on the islands, which consist mainly of rolling pastures.

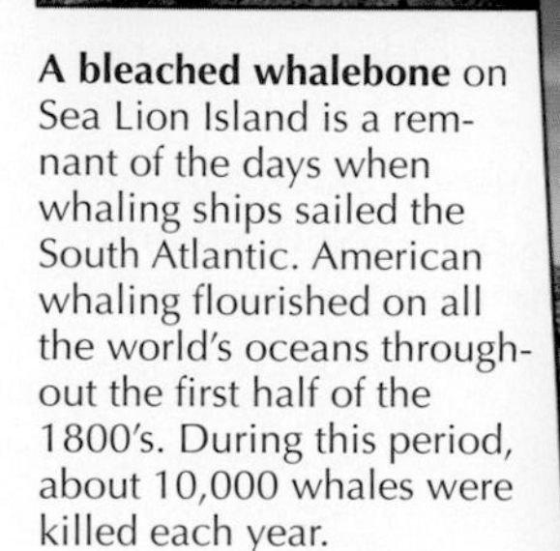

A bleached whalebone on Sea Lion Island is a remnant of the days when whaling ships sailed the South Atlantic. American whaling flourished on all the world's oceans throughout the first half of the 1800's. During this period, about 10,000 whales were killed each year.

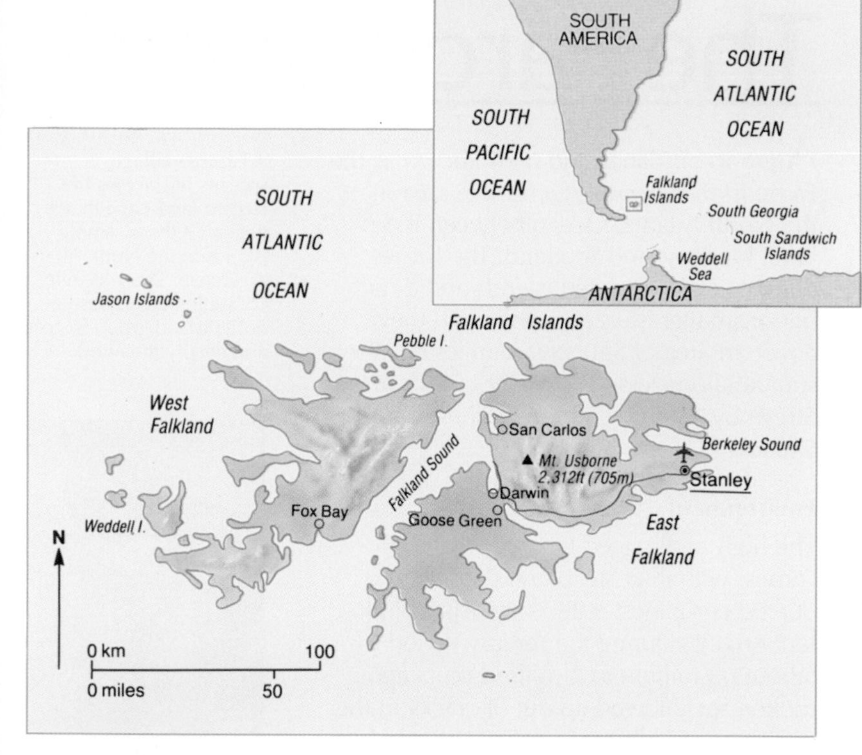

The Falkland Islands lie in the South Atlantic, off the southern tip of South America. Although Argentina claims them, the British established rule there in 1833, and most of the inhabitants are of British origin.

New Island, left, off West Falkland, is home to large colonies of penguins, including rockhoppers. The wildlife of the islands includes seals and many varieties of sea birds. Small islands set aside as nature preserves help protect the animals.

Spinning wool, an ancient British craft, is practiced by Falkland Islanders a world away from Great Britain.

The Falklands are a major breeding ground for about 60 species of birds, including many varieties of penguins. The coastline also supports large populations of marine mammals, including certain types of porpoises, dolphins, seals, and sea lions.

Most of the approximately 2,000 inhabitants are of British origin. About 50 per cent of the people live in Stanley, the capital and chief town, which lies on East Falkland Island.

Raising sheep was once the main economic activity of the islanders. While many of the islanders still raise sheep and export wool, the Falkland Islands' main source of income comes from the sale of fishing licenses to foreign fishing fleets. The sale of postage stamps and coins, primarily to collectors, also contributes to the economy.

The Faroe Islands

A group of islands and reefs known as the Faroe Islands occupies a remote area in the North Atlantic Ocean between Iceland, Norway, and Scotland. The Faroes consist of 18 inhabited islands and a number of smaller islets and reefs, and they cover an area of 540 square miles (1,399 square kilometers). The major islands are Streymoy, Eysturoy, Vágar, Sudhuroy, and Sandoy.

Environment

The high and rugged landscape of the Faroes, with their steep and deeply indented coastlines, reflects the islands' violent origins. During the Tertiary Period (about 63 million to 2 million years ago), molten rock flowed up out of cracks in the earth's crust and spread out over the seabed. The molten rock solidified in layers of *basalt* (black volcanic rock) and red ash.

Later, Ice Age glaciers shaped the rock by gouging out steep-sided valleys and smoothing the basalt. The Faroes took their present shape about 10,000 years ago, when the Ice Age ended, causing the glaciers to disappear and the sea to flood into the valleys. Today, the Faroes' angular landscape consists of steep cliffs (known locally as "hammers"), numerous tiny rivers and waterfalls, chains of valleys dotted with lakes, and broad, green meadows.

Strong, gusting winds, raging storms, and crashing breakers add to the Faroes' rugged, desolate environment, while treacherous currents along the shores of the islands make navigation difficult. But, despite the Faroes' position near the Arctic Circle, the flow of the warm Gulf Stream guarantees surprisingly mild weather, with frequent fog and rain and relatively ice-free harbors. During the brief but beautiful summer, the midnight sun barely sinks below the horizon, creating glistening points of light on the deep waters of the fiords.

History

The inhabitants of the Faroes are hardy people of Norse origin. They fish and raise sheep for a living, and they also sell the

A Faroese village spreads out across the rugged landscape that is typical of these remote islands in the North Atlantic Ocean. The 140-mile (225-kilometer) coastline of the island group is steep and deeply indented.

A young boy earns pocket money by cleaning fish. The Faroes' busy fishing industry has created a high standard of living for its people.

Faroese farm workers gather hay on a slope above a fishing village. The hay is used for animal feed during the winter. Although the islanders raise sheep for their wool and meat, livestock production no longer has the economic importance it once did.

Dock workers unload supplies from a boat in the harbor of Mikladur, *right,* on the island of Kalsoy. Strong winds and powerful currents create rough waters in the narrow channels that separate the Faroe Islands, making navigation difficult.

The Faroe Islands, *map below,* are situated in the North Atlantic Ocean between Iceland, Scotland, and Norway. Like Greenland, the island group once belonged to Norway, but it is now a self-governing territory of Denmark. The national symbol of the Faroes is a ram, a reminder that raising sheep was once the major economic activity of the islands.

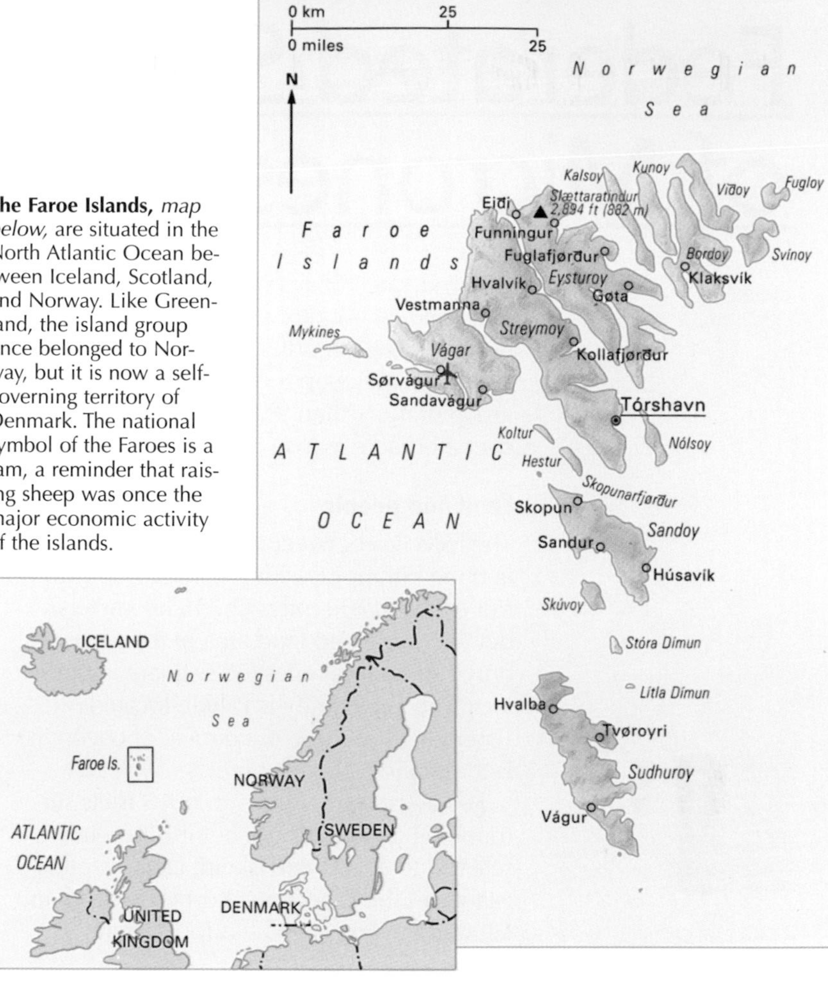

eggs and feathers of the many sea birds that nest on the cliffs. Descended from the Vikings, the present-day islanders are peace-loving people who are proud of their Norse heritage.

Among the first settlers to arrive in the Faroes were Irish monks who settled in this remote corner of the world about 700. About 100 years later, Norse Vikings colonized the islands. Many of the settlers were refugees escaping the hard rule of Harold I (Fairhair), the brutal leader who unified Norway around 900. Others were drawn to the islands by a sense of adventure.

Norway ruled the Faroes from the 800's until 1380, when the islands came under the control of Denmark. Danish then replaced the Faroese tongue of the islanders as the official language, but the people held on to their national pride and cultural identity during the years of Danish rule. Faroese—which is related to Icelandic and Norwegian, and preserves many features of the Old Norse tongue—was restored as the territory's official language in 1948.

In 1948, the Faroes became a self-governing part of the kingdom of Denmark, with their own *Lagting* (parliament), flag, and currency. The Lagting sends representatives to the Danish parliament in Copenhagen.

Way of life

Today, about 48,000 Faroese live in the islands—most of them in Tórshavn, which is the capital of the Faroes and also the economic and cultural center. Fishing and fish processing are the most important economic activities, with fish and fish products making up about 80 per cent of the island's exports. Through their Lagting, the Faroese have established their own policy on fishing, choosing not to join the European Union (EU), which sets fish quotas.

Federated States of Micronesia

The Federated States of Micronesia (FSM), an independent country since 1986, consists of 607 islands just north of the equator. The FSM and Palau make up the Caroline Islands, a group of more than 930 islands scattered over a broad expanse of the Pacific Ocean.

Land and people

The FSM lies between the Marshall Islands and the Philippines. The island group extends more than 1,700 miles (2,736 kilometers), but the combined land area of the islands is only 271 square miles (702 square kilometers). The capital city is Palikir, located on Pohnpei. *Copra* (the dried meat of coconuts) is the islands' chief export.

Kosrae Island is made up of 15 islets surrounding a main island, but it is generally referred to as a single island. Like the other islands in the FSM, most Kosraeans work in agriculture. The island has four natural harbors.

Pohnpei, the largest island of the FSM, is made of volcanic rock and surrounded by coral reefs. The shores are lined with mangrove swamps, but firm, fertile ground lies inland. Mountains rise more than 2,300 feet (700 meters). Pohnpei is famous for its fine yams. Other crops include bananas, breadfruit, coconuts, limes, and *taro* (the starchy, rootlike stem of the taro plant, used as food).

The Chuuk Islands, a large island group, lie west of Pohnpei. About 48 of these islands are surrounded by a barrier coral reef that forms a lagoon 40 miles (64 kilometers) wide. About 38,000 people live on the islands. Weno, the largest city in the FSM, lies on one of the Chuuk Islands.

The Yap islands, which lie west of the Chuuks, include 4 large islands and 10 smaller islands separated by narrow channels and ringed by coral reefs. They are composed of ancient crystalline rocks and have a rugged surface. Most of the Yap Island people make a living by farming. Fishing is also important.

History

The Yap Islands were among the first island groups settled in Micronesia. Archaeologists believe that people from Asia moved there thousands of years ago. Settlers arrived on

FACT BOX

COUNTRY

Official name: Federated States of Micronesia
Capital: Palikir
Terrain: Islands vary geologically from high mountainous islands to low, coral atolls; volcanic outcroppings on Pohnpei, Kosrae, and Truk
Area: 271 sq. mi. (702 km²)
Climate: Tropical; heavy year-round rainfall, especially in the eastern islands; located on southern edge of the typhoon belt with occasionally severe damage
Highest elevation: Totolom, 2,595 ft. (791 m)
Lowest elevation: Pacific Ocean, sea level

GOVERNMENT

Form of government: Constitutional government
Head of state: President
Head of government: President
Administrative areas: 4 states
Legislature: Congress with 14 members
Court system: Supreme Court
Armed forces: U.S. is responsible for the Federated States of Micronesia's defense

PEOPLE

Estimated 2008 population: 112,000
Population density: 413 persons per sq. mi. (160 per km²)
Population distribution: 78% rural, 22% urban
Life expectancy in years: Male: 67 Female: 69
Languages spoken: English (official and common language), Trukese, Pohnpeian, Yapese, Kosraean
Religions: Roman Catholic 50%, Protestant 47%

ECONOMY

Currency: United States dollar
Gross domestic product (GDP) in 2002: $222 million U.S.
Real annual growth rate (2001–2002): 0.1%
GDP per capita (2002): $2,066 U.S.
Goods exported: Copra, fish, garments, bananas, black pepper
Goods imported: Food, manufactured goods, machinery and equipment, beverages
Trading partners: Japan, United States, Guam, Australia

No data are available for doctors, education, and technology.

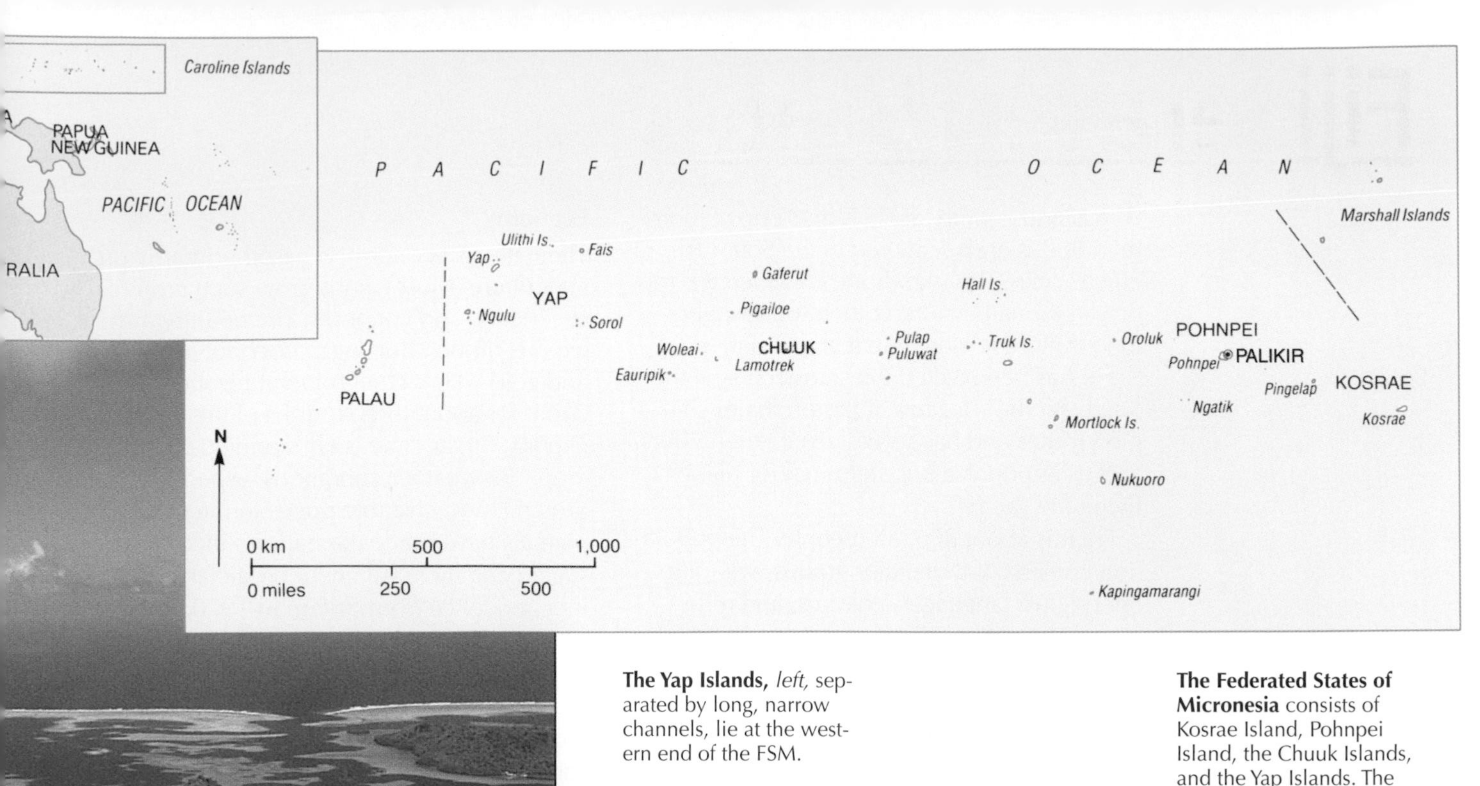

The Federated States of Micronesia consists of Kosrae Island, Pohnpei Island, the Chuuk Islands, and the Yap Islands. The island group extends over 1,700 miles (2,736 kilometers) of the western Pacific.

The Yap Islands, *left,* separated by long, narrow channels, lie at the western end of the FSM.

Kosrae, Pohnpei, and the Chuuk Islands later. The ruins of Nan Madol on Pohnpei are the remnants of an advanced civilization dating from some time after the original Micronesians arrived from Asia. Mysterious structures made of huge blocks are all that remain of palaces, administrative buildings, houses, and places of worship.

In the 1500's, Spanish explorers were the first Europeans to reach the islands. Spain formally claimed them in 1885 and sold them to Germany in 1899. In 1905, the Yap Islands became internationally important as a cable station between the United States, the Netherlands Indies (now Indonesia), and Japan.

Japan captured the islands during World War I (1914–1918). After the war, the League of Nations gave them to Japan as mandates. During World War II (1939–1945), U.S. forces captured some of the islands, and in 1947 the United Nations made the United States trustee of the islands as part of the Trust Territory of the Pacific Islands.

In 1980, Kosrae, Pohnpei, the Chuuks, and the Yaps formed the Federated States of Micronesia in agreement with the United States. In 1986, these islands became an independent country. The United States provides economic aid and defends the islands in emergencies. In November 1990, Typhoon Owen left about 4,500 islanders homeless and destroyed about 90 per cent of crops.

Chuuk Islanders weigh copra in preparation for market. Copra, which contains oil used to make such products as margarine and soap, is the chief export of the FSM.

Young Pohnpei Islanders participate in a formal ceremonial dance, *above.* Pohnpei is the largest island of the FSM.

Fiji

The island country of Fiji is made up of more than 800 islands scattered in the South Pacific Ocean. The islands are considered part of Melanesia because of their location, but their culture is more like that of Polynesia.

Fiji has been called the "crossroads of the South Pacific" because it lies on major shipping routes and has several excellent harbors, and its airport is a busy terminal for planes flying the Pacific.

Fiji has about 866,000 people. The population consists of two major groups with different origins, languages, cultures, and religions. The native Fijians are Melanesians whose ancestors arrived in Fiji thousands of years ago, probably from Indonesia. The other group is made up of descendants of laborers brought from India between 1879 and 1916 to work on sugar plantations. A small number of Fijians are of Chinese, European, Micronesian, or Polynesian ancestry.

English is the official language of Fiji and is used in the schools. But Fijian and Hindi are also widely spoken. More than 85 per cent of those from 6 to 13 years old attend school, although it is not required by law. Most Fijian and Indian youngsters attend separate schools. The University of the South Pacific in Suva, the country's only university, serves students from many of the Pacific Islands.

Economy

The nation's economy is based primarily on agriculture. Most Fijians grow such crops as sugar cane and coconuts. The country also exports timber. But sugar, coconut products, and gold—Fiji's chief mineral resource—account for about three-fourths of the country's exports. Sugar cane is Fiji's principal cash crop, but weather conditions, especially drought, and unstable prices in international markets have made the nation's income from sugar cane increasingly unpredictable. Since independence from Britain in 1970, the government has encouraged tourism and the development of manufacturing and forestry. It has also promoted the production of new crops to reduce Fiji's dependence on sugar cane and coconuts. In 1978 an Australian aid program was launched to turn hilly, undeveloped Fijian land into grazing land. In addition, Australia and Fiji expanded a jointly owned Fijian gold mine, thus increasing the nation's gold output. In 1987, construction of new lumber mills expanded the timber industry. After 1982, tourism overtook the sugar industry as the major source of foreign earnings.

Government

Fiji was a British colony between 1874 and 1970. After the nation became independent,

FACT BOX

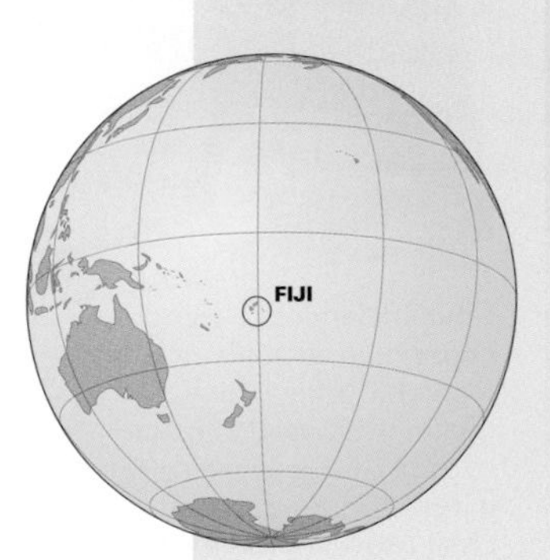

COUNTRY

Official name: Republic of the Fiji Islands
Capital: Suva
Terrain: Mostly mountains of volcanic origin
Area: 7,056 sq. mi. (18,274 km^2)
Climate: Tropical marine; only slight seasonal temperature variation
Main river: Rewa
Highest elevation: Tomanivi, 4,341 ft. (1,323 m)
Lowest elevation: Pacific Ocean, sea level

GOVERNMENT

Form of government: Republic
Head of state: President
Head of government: Prime minister
Administrative areas: 4 divisions, 1 dependency
Legislature: Parliament consisting of the Senate with 32 members and the House of Representatives with 71 members serving five-year terms
Court system: Supreme Court
Armed forces: 3,500 troops

PEOPLE

Estimated 2008 population: 876,000
Population density: 124 persons per sq. mi. (48 per km^2)
Population distribution: 51% rural, 49% urban
Life expectancy in years: Male: 65 Female: 69
Doctors per 1,000 people: N/A
Percentage of age-appropriate population enrolled in the following educational levels: Primary: N/A Secondary: N/A Further: N/A

Workers harvest sugar cane by hand under the bright Fijian sun. Sugar cane is one of Fiji's chief agricultural products. Bananas and coconuts also grow well in Fiji's tropical climate.

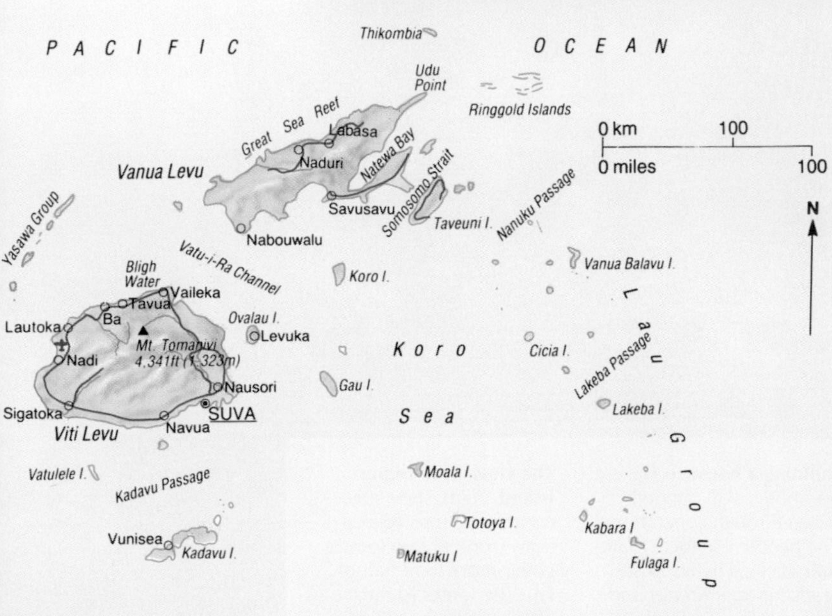

Fijians held more power in government than Indians. In 1987 a Fijian prime minister appointed a multiracial Cabinet. Military officers seized power in protest. Later in 1987, Fiji was returned to civilian rule.

In 1990, Fiji approved a new constitution, which established a two-house legislature. Elections for legislators took place in 1992.

In 1997, the Constitution was amended to grant political power to all races. In 1999, Fiji elected its first Indian prime minister.

In May 2000, a group of rebels, claiming to represent the native Fijians, stormed Parliament and held the prime minister and most members of the Cabinet hostage. Military leaders took control of the country, setting up an interim government and revoking the 1997 constitution. In November 2000, Fiji's High Court ruled that the 1997 constitution was still in effect. A new interim government was established in March 2001 with an ethnic Fijian as prime minister. The prime minister appointed a cabinet that excluded representatives of parties dominated by ethnic Indians. In 2002, Fijian courts ruled that such parties must be included.

Fiji is often called the "crossroads of the South Pacific" due to its excellent location, harbors, and airport.

Languages spoken:
English (official)
Fijian
Hindustani

Religions:
Christian 52%
(Methodist 37%, Roman Catholic 9%)
Hindu 38%
Muslim 8%

TECHNOLOGY

Radios per 1,000 people: N/A

Televisions per 1,000 people: N/A

Computers per 1,000 people: N/A

ECONOMY

Currency: Fijian dollar

Gross domestic product (GDP) in 2004: $5,173 billion U.S.

Real annual growth rate (2003–2004): 13.6%

GDP per capita (2004): $5,900 U.S.

Goods exported: Sugar, clothing, gold, processed fish, lumber

Goods imported: Machinery and transport equipment, petroleum products, food, chemicals

Trading partners: Australia, New Zealand, United Kingdom, Japan

A group of Fijian children, *below,* reflects the multiracial nature of the islands.

Environment

Fiji has a total land area of 7,056 square miles (18,274 square kilometers). The Fijian island of Viti Levu (Big Fiji) accounts for about half the land area of the country. Suva, Fiji's capital and largest city, lies on Viti Levu's southern coast. A second island, Vanua Levu (Big Land) occupies about a third of Fiji's land area. Only about a hundred of the more than 800 islands of Fiji are suitable for human habitation. Many of the other islands are little more than piles of sand on coral reefs.

Cool winds make Fiji's tropical climate relatively comfortable. Temperatures range from about 60° to 90° F. (16° to 32° C). Heavy rains and tropical storms occur frequently between November and April.

Volcanic islands

Most of the islands of Fiji consist of lava built up from the ocean floor by eruptions of oceanic volcanoes. The volcanoes are no longer active, but the heat trapped under the islands now causes hot springs to flow, for example, on Vanua Levu.

The larger volcanic islands have fertile soil that has developed from the weathered volcanic rock. The larger islands also have high volcanic peaks, rolling hills, rivers, and grasslands. The mountainous interiors of these islands form a barrier that keeps most moisture of the southeasterly trade winds on the eastern side, leaving the west drier and typically covered by grasslands. On the southeastern slopes of the islands, where rainfall is heaviest, dense rain forests thrive. These tropical rain forests cover more than half the total area of Fiji.

Coral reefs—limestone formations composed of tiny sea animals and plants and their remains—surround nearly all the islands. Fiji also has coral islands, which consist chiefly of coral reef material piled up by ocean waves and winds. Coral islands tend to be flat with sandy beaches and lie only a few feet above sea level. The larger coral islands often have rich vegetation, including vines, grasses, broad-leafed trees, and coconut palms.

Fiji's fertile soils support the growth and export of sugar, coconuts, and bananas.

Fiji also has small deposits of gold and manganese.

Building a house in Fiji is a traditional skill handed down through generations. The people use local materials such as bamboo and reeds for their homes and mangrove wood for their cooking fires.

The view from Mana Island, *right,* shows the country's dense vegetation. Tropical rain forests cover more than half of Fiji. The larger islands also have high volcanic peaks.

Wildlife

Mangrove swamps cover the coastal flats and help to build up the Fiji islands. Mangrove trees usually grow in places by quiet ocean water. Their thousands of stiltlike roots catch silt, which piles up to form new land. The roots also form a breeding place for many fish and other creatures. The clear seas around the coral reefs support a wide variety of marine life.

Over the generations, Fijians in certain areas have developed the custom of calling up sea creatures from the deep. For example, on Kadavu and Koro, people chant a traditional song to lure turtles into shallow water. On Vatulele Island, people summon red prawns with a certain call. And in Lake-

ba, an island in the Lau group, people still practice the art of calling sharks.

In contrast to the richness of marine life, few land animals inhabit the islands. Originally, volcanic and coral islands have no land animals or plants, but these islands become inhabited by birds that fly across the sea and by other animals that swim to the islands. Some animals and insects are carried to the islands on logs or other debris.

In Fiji, the only native mammals are six species of bats and a small Polynesian rat. Pigs and dogs came to the islands with the first settlers, but all other mammals have arrived since the Europeans began to settle there in the 1800's.

Cleared for use as farmland, many inland parts of the Fiji Islands now produce sugar cane, rice, and other crops. Coconuts are an important cash crop, and pine forests have also been planted to provide valuable timber for export.

An industrial area near Fiji's capital of Suva produces cement, beer, and cigarettes. Suva is Fiji's chief port and commercial center. Ships stop at the city to load copra and tropical fruit. Factories in Suva make coconut oil and soap.

History and People

In 1643, Abel Tasman, a Dutch navigator, became the first European to see Fiji. Captain James Cook, a British explorer, visited Vatoa, one of the southern islands, in 1774. During the 1800's, traders, Methodist missionaries, and escaped Australian convicts came to visit or settle there.

The Fijians were cannibals. Various tribes fought one another, and Fijians had a reputation as the most savage and warlike of all the Melanesian peoples. In 1871, however, a chief named Cakobau gained power and brought peace to Fiji. To protect his country from outside interference, Cakobau asked the United Kingdom to make Fiji a crown colony. The United Kingdom did so in 1874.

Indians and Fijians

Soon after Fiji became a crown colony, the colonial government decided to import about 60,000 laborers from India to work on Fiji's plantations. The traditional religious and caste systems of the Indians were ignored, and they often had to work under very harsh conditions. They were looked down upon by both the Fijians and the Europeans, and they had to struggle to achieve political and social recognition.

Fiji became an independent nation at its own request in 1970. Throughout the years, the differences between the native Fijians and the Indians have led to racial violence and political upheaval. Although Indians control most of Fiji's economy, native Fijians traditionally have held more power in the government. In April 1987, however, an Indian-backed coalition led by Timoci Bavadra won a majority in Parliament. Bavadra became prime minister and appointed Indians to a majority of the cabinet posts.

Many Fijians resented this increase of Indian political power, and military officers overthrew the government. The new military leader, Colonel Sitiveni Rabuka, abolished the Constitution, named himself head of state and government, and declared the right of Fijians to govern the nation. In December 1987, Rabuka appointed a president and returned Fiji to civilian rule.

In July 1990, Fiji approved a new Constitution, establishing a two-house legislature. In 1997, the Constitution was amended to grant political power to all races. And in 1999,

Fijian fishermen display their catch. Some still use a variety of magical techniques, including "calling," to attract fish.

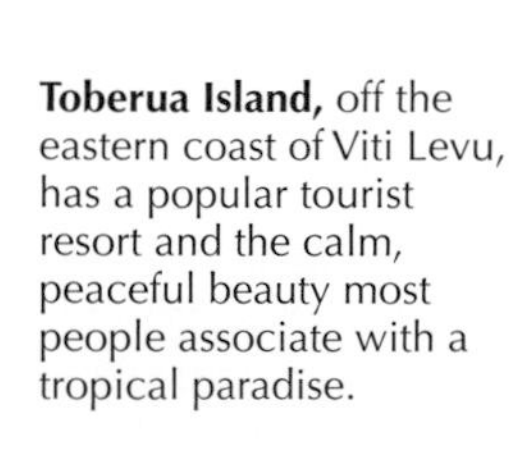

Toberua Island, off the eastern coast of Viti Levu, has a popular tourist resort and the calm, peaceful beauty most people associate with a tropical paradise.

Fiji's Indian children are descended from plantation workers brought to Fiji between 1879 and 1916. Today, Indians control much of Fiji's economy.

A spectacular tropical sunset, *far left,* shows off Fiji's beauty. Fiji lies south of the equator. Although it has a hot, tropical climate, cool winds make it relatively comfortable.

Mahendra Chaudhry became Fiji's first prime minister of Indian descent.

In 2000, a group of rebels stormed Parliament, seized Chaudhry and most members of the Cabinet, and held them hostage. The rebels claimed to represent the interests of native Fijians. Chaudhry was eventually removed from office, and military leaders took control of the country. The military rulers set up an interim government and revoked the 1997 constitution.

Modern life

Many Fijians of Indian descent still work in the cane fields, but others have become prosperous shopkeepers or business people. The Indian women wear *saris,* the traditional dress of Indians. Most Indians are Hindus or Muslims.

Slightly more than half of the native Fijians live in rural areas, mainly in small farming or fishing villages. Chiefs still play an important part in the affairs of many villages—a village chief is expected to advise and lead the people, show hospitality to visitors, and uphold the good name of the community. In Fiji, the office of chief passes from father to son.

On most islands, the villagers celebrate such occasions as births and marriages by traditional feasting, dancing, and singing. Important festivals include the ceremonial drinking of *kava,* a drink made from pepper plants. The men often wear a cloth skirt called a *lava-lava* or *sulu,* and some of the women wear long, loose-fitting cotton dresses called *muumuus.* Fijian clothing is often made of *tapa,* a cloth made by soaking and pounding the inner bark of paper mulberry trees.

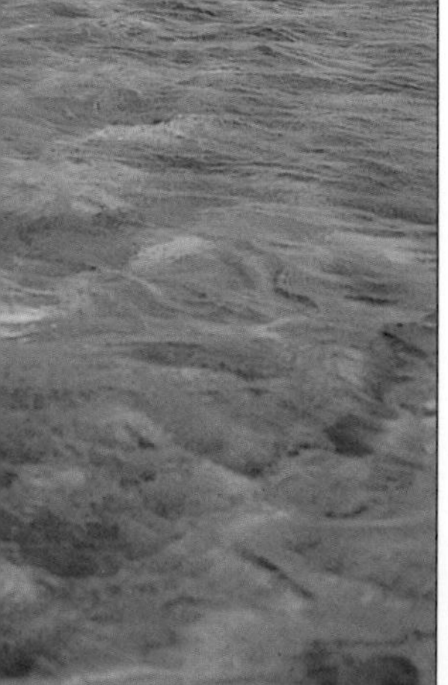

Fire walking, *above left,* is a tradition among men of the Sawau tribe from the island of Beqa. By tradition, the men must avoid women and coconuts for two weeks before the ceremony. They believe that if they do not, they may burn their feet.

Catamarans, raftlike boats with two hulls, allow visitors to enjoy Fiji's beautiful coastal waters. The modern version has been adapted from a traditional Polynesian outrigger design developed thousands of years ago.

Finland

Finland, a country famous for its scenic beauty, lies in northern Europe north of the Baltic Sea and east of Sweden. About one-third of the country lies within the Arctic Circle in an area known as the Land of the Midnight Sun. In this region, the sun shines 24 hours a day for long periods each summer. During the endless winter nights, the *aurora borealis* (Northern lights) fill the sky with colorful curtains of light.

Finland is a prosperous nation, largely because of the huge forests that cover almost two-thirds of the land. As a result of the nation's thriving economy, the Finnish people enjoy a high standard of living.

The Finns differ from the Scandinavians and the Slavs — the other peoples of northern and eastern Europe — in language and culture. Although Finland was dominated for centuries by its powerful neighbors, Sweden to the west and Russia to the east, the Finns never lost their national identity. Their rich folk culture is reflected today in the country's crafts, literature, music, and painting.

The Finns' contribution to world culture includes the *Kalevala,* a huge collection of song-poems and chants of Finnish peasants. The *Kalevala,* which became the Finnish national epic, inspired the works of other Finnish masters, such as the sym-phonic poems of Jean Sibelius and the paintings of artist Akseli Gallen-Kallela.

The earliest Finns

The Sami, who now make up a small minority of the Finnish population, were Finland's earliest known inhabitants. The ancestors of the present-day Finns began to move into the country from the south shores of the Gulf of Finland thousands of years ago.

In the A.D. 1000's, both Sweden and Russia began a struggle for control of Finland. By the 1200's, Sweden was the dominant power, and Swedish was the nation's official language. Lutheranism became the official religion about 1540.

However, Russia feared Swedish expansion and sought to protect its western border by regaining Finland. From the 1500's to the 1700's, Russia fought a series of

wars with Sweden. In 1808, Russia invaded Finland, and it finally conquered the country in 1809. Finland was made an independent grand duchy, with Russia's czar as grand duke. The duchy had local self-rule based on government systems developed while Finland was under Swedish control.

The rise of nationalism

In 1899, Czar Nicholas II took away most of Finland's powers of self-rule and began an attempt to force the Finns to accept Russian government and culture. But the Finns opposed the czar's actions, and resistance to Russian control reached a peak in 1905 with a six-day nationwide strike. The strike forced the czar to restore much of Finland's self-government, but he continued his efforts to "Russianize" the people.

In 1917, when a revolution in Russia overthrew the czar, Finland declared its independence. However, a civil war broke out in Finland between the socialist Red Guard, aided by the new Soviet regime, and the antisocialist White Guard, assisted by German troops. The war ended in a White victory in May 1918, and in 1919, Finland adopted a republican Constitution.

The new country's relations with Sweden and Russia remained unsettled throughout the early 1900's. Finland and Sweden quarreled over possession of the Aland Islands until 1921, when the League of Nations awarded the islands to Finland. Disputes with Russia centered on the eastern Finnish region of Karelia. In November 1939, Soviet troops invaded Finland, and the Finns were forced to admit defeat after the 15-week "Winter War." A peace treaty signed in March 1940 gave southern Karelia—about 10 per cent of Finland's total land area—to the Soviet Union.

During World War II (1939-1945), Finland allowed Nazi Germany to move troops through its northern regions to attack the Soviet Union, and the Soviets then bombed Finland. As the Germans retreated, they burned towns, villages, and forests behind them.

Finland Today

After World War II, which caused a devastating loss of life and property in Finland, the country's leaders adopted a policy of neutrality in international politics. Finland also developed close economic and cultural ties with the Soviet Union and the countries of Scandinavia. On Jan. 1, 1995, Finland became a member of the European Union (EU).

Finland is a democratic republic, with a president and a parliament called the *Eduskunta.* Parliamentary elections are based on a system of *proportional representation,* which means that the number of seats gained by a political party depends upon a share of the total number of votes cast.

Under this system, it is often difficult for any single political party to establish an overall majority. Thus, Finland is generally governed by a *coalition,* in which several parties join together to form a government.

Like Sweden and Norway, Finland has an extensive welfare system, which provides the people with many services. The system has developed gradually in response to social needs.

Since the 1920's, maternity and child welfare centers have given free health care to pregnant women, mothers, and children. In 1939, Finland established an old-age and disability insurance program, and since 1948, families have received an allowance when they have a new baby, as well as a yearly allowance for each child under 16.

Finnish children are required to attend *basic school* for nine years beginning at the age of 7. Almost all students attend public school, where they receive one free meal a day, as well as free books and medical and dental care. After basic school, students may choose an *upper secondary school,* which emphasizes academic subjects, or a *vocational school,* which offers training in skilled manual work. Vocational institutes and universities offer higher-education programs.

More than 90 per cent of Finland's people are Finnish by descent, and most of the rest are Swedish. The majority of the people are tall, with fair skin, blue or gray eyes, and blond or light brown hair. Finnish and Swedish are both official languages. About 93 per cent of the people speak Finnish, and about 6 per cent speak Swedish. Most of the Swedish-speaking people live on the south and west coasts and on the offshore Aland Islands.

The majority of Finland's people live in the south, and about two-thirds live in cities and towns. In urban areas, most Finns live in

FACT BOX

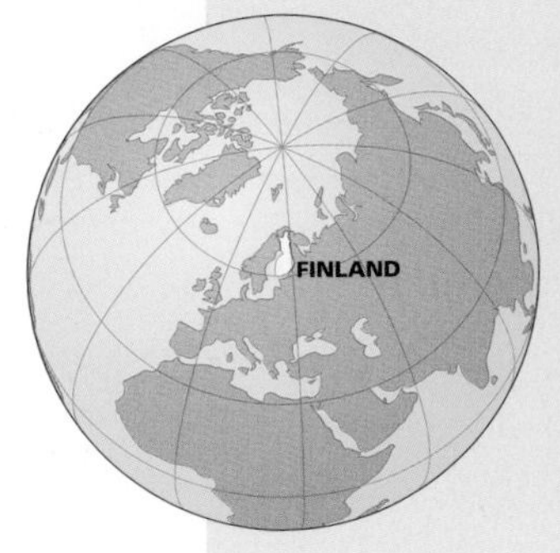

COUNTRY

Official name: Suomen Tasavalta (Republic of Finland)

Capital: Helsinki

Terrain: Mostly low, flat to rolling plains interspersed with lakes and low hills

Area: 130,559 sq. mi. (338,145 km^2)

Climate: Cold temperate; potentially subarctic, but comparatively mild because of moderating influence of the North Atlantic Current, Baltic Sea, and more than 60,000 lakes

Main rivers: Kemijoki, Ounasjoki, Muonio

Highest elevation: Mount Haltia, 4,344 ft. (1,324 m)

Lowest elevation: Baltic Sea, sea level

GOVERNMENT

Form of government: Democratic Republic

Head of state: President

Head of government: Prime minister

Administrative areas: 12 laanit (provinces)

Legislature: Eduskunta (Parliament) with 200 members serving four-year terms

Court system: Korkein Oikeus (Supreme Court)

Armed forces: 34,400 troops

PEOPLE

Estimated 2008 population: 5,285,000

Population density: 40 persons per sq. mi. (16 per km^2)

Population distribution: 61% urban, 39% rural

Life expectancy in years:
Male: 75
Female: 82

Doctors per 1,000 people: 3.1

Percentage of age-appropriate population enrolled in the following educational levels:
Primary: 102*
Secondary: 126*
Further: 86

A statue of Finnish soldier and statesman Carl Gustaf Mannerheim stands in front of Helsinki's parliament building. Mannerheim served as Finland's president between 1944 and 1946.

Finland, *below,* is noted for the scenic beauty of its sparkling waters, its forests of pine, spruce, and birch, and its many islands. It is often called the *land of a thousand lakes.*

apartments, while people in rural areas live in single-family homes on farms or in villages.

In 2000, Finland elected Tarja Halonen as its first woman president. In 2006, she was reelected for a second 6-year term.

Languages spoken: Finnish 95% (official)
Swedish 5% (official)
Small Sami- and Russian-speaking minorities

Religions:
Evangelical Lutheran 89%
Greek Orthodox 1%
None 9%

*Enrollment ratios compare the number of students enrolled to the population which, by age, should be enrolled. A ratio higher than 100 indicates that students older or younger than the typical age range are also enrolled.

TECHNOLOGY

Radios per 1,000 people: 1,624

Televisions per 1,000 people: 679

Computers per 1,000 people: 441.7

ECONOMY

Currency: Euro

Gross domestic product (GDP) in 2004: $151.2 billion U.S.

Real annual growth rate (2003–2004): 3%

GDP per capita (2004): $29,000 U.S.

Goods exported: Machinery and equipment, chemicals, metals; timber, paper, and pulp

Goods imported: Foodstuffs, petroleum and petroleum products, chemicals, transport equipment, iron and steel, machinery, textile yarn and fabrics, fodder grains

Trading partners: European Union, United States, Russia

Environment

Finland stretches about 640 miles (1,030 kilometers) from the Arctic Circle in the north to the Baltic Sea in the south. Much of the country consists of a low-lying plateau broken by small hills and valleys and low ridges and hollows.

The land rises gradually from south-southwest to north-northeast, averaging only about 400 to 600 feet (120 to 180 meters) above sea level. It is covered mainly by dense forests of pine, spruce, and birch trees, and about 60,000 lakes scattered throughout the countryside.

Finland's present landscape was formed more than a million years ago, when continental glaciers covered the land, gouging the surface as they advanced. When the glaciers receded about 10,000 years ago, they left deposits of rocks that formed *end moraines* or *terminal moraines*—irregular ridges with hills and hollows.

Finland enjoys a surprisingly mild climate for its northern latitude, largely because of the warm ocean current known as the Gulf Stream. The country's many lakes, as well as the gulfs of Bothnia and Finland also help give the country a warmer climate. However, winters are long and often harsh. Snow covers southern Finland from November to April, while northern Finland is snowbound from October to April.

Land regions

Finland's four main land regions are the Coastal Lowlands, the Coastal Islands, the Lake District, and the Upland District. The Coastal Lowlands, a region of many small lakes, lie along the Gulf of Bothnia and the Gulf of Finland. The lowlands have less forested land and a milder climate than the Lake and Upland districts, and they rank as Finland's most densely populated area.

The Coastal Islands consist of thousands of islands in the Gulf of Bothnia and the Gulf of Finland. The majority are small, uninhabited islands with thin, stony soil that cannot support much plant life. However, many kinds of plants grow on a few of the larger islands. Some islands are inhabited by people who fish for a living, but most are used as summer recreation areas.

A lake near Kuopio, in the heart of Finland's Lake District, reflects the natural beauty of central Finland's landscape, with its interplay of lakes, islands, inlets, and forestland.

A Sami from northern Finland takes her child out for some fresh air in a baby stroller equipped with sled runners instead of wheels. Most of Finland is icebound in winter, and ice-breaking ships work continuously to keep the ports open.

The Aland Islands are the most important of Finland's Coastal Islands. About 80 of the approximately 6,500 Aland Islands are inhabited, mostly by Swedish-speaking people. The main island in this group, also called Aland, is Finland's largest island.

Low hills rise above the forests of Finland's Upland District, *far top right.* Plant life becomes sparse as one travels north into the treeless, frozen tundra.

A trapper's hut, with a motorized sled parked outside, stands alone in the wilderness of the Upland District, *far bottom right.* Few people live in this region, where winter temperatures sometimes drop as low as –22° F. (–30°C).

The Lake District extends across central Finland north and east of the Coastal Lowlands. Island-dotted lakes, connected by narrow channels or short rivers, cover about half the total area of the district. Finland's largest lake, Saimaa, covers about 680 square miles (1,760 square kilometers) in the southeastern part of the region.

The Upland District

The Upland District, Finland's northernmost region, covers about 40 per cent of the country and ranks as the nation's most thinly populated region. The district has a harsher climate and less fertile soil than other regions. Hilly areas are separated by swamps and marshlands, and in the northernmost parts, the land is covered by frozen, treeless tundra.

The district's marshlands are the habitat of Finland's largest animal—the moose. This magnificent creature, a member of the deer family, has been reduced in numbers by hunting and deforestation. However, today moose are carefully protected, and their numbers are increasing.

Finland's only mountains are located in the extreme northwest of the Upland District, where the country's tallest peak, Mount Haltia, rises 4,344 feet (1,324 meters) along the Norwegian border. The Kemijoki, the longest river in Finland, begins in the Upland District near the Russian border and winds 340 miles (547 kilometers) southwestward to the Gulf of Bothnia.

The Finnish sauna

A tradition in Finland for more than 1,000 years, the dry-heat bath known as the *sauna* is used for cleansing and relaxation. In a sauna, bathers sit or lie on wooden benches, perspiring freely from the warmth created by stones heated on top of a furnace. The room is made even hotter when water is thrown on the stones, causing steam (löly) to rise. Bathers also beat their bodies gently with birch twigs (vihta) to stimulate circulation. Some people end a sauna bath with a roll in the snow, but most settle for a cold shower. Then they lie down to rest until their body returns to normal temperature.

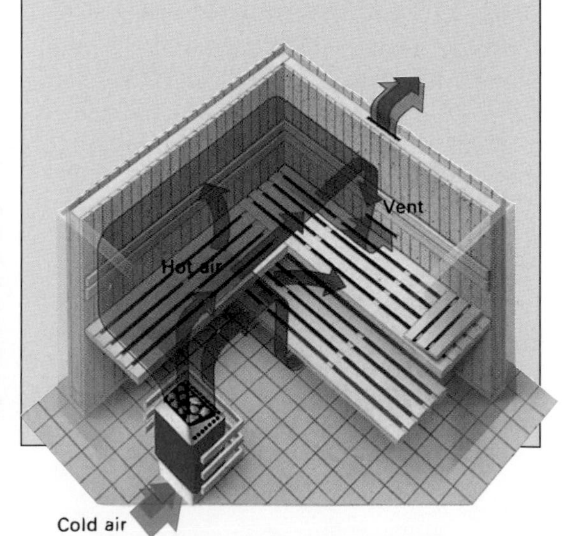

Helsinki

Helsinki, Finland's capital and largest city, is a thriving seaport on the Gulf of Finland, on the country's southern coast. Helsinki covers a peninsula and several islands. Founded in 1550 by King Gustavus I of Sweden, the city is one of Finland's chief ports as well as the nation's commercial and cultural center. About one-fifth of the Finnish people live in Helsinki and its suburbs. The city is home to Finland's main university as well as to theaters, art galleries, and museums.

During its early days, Helsinki suffered a series of disasters, including fires, plagues, war, and famine. Between the 1500's and the 1700's, Helsinki was almost destroyed twice in the fighting between Russia and Sweden over control of Finland. Later, the Great Fire of 1808 burned more than two-thirds of the city to the ground.

In 1809, the city passed from Swedish to Russian control, and in 1812, Czar Nicholas II made Helsinki the capital of Finland. The royal order also included a large-scale program of city planning and construction, designed to reflect the city's new political importance.

The German-born architect Carl Ludwig Engel was directed to create a city that combined innovative architecture with a spacious layout of broad avenues and numerous parks. Present-day Helsinki's reputation as a showpiece of modern architecture is largely due to the efforts begun by Engel and his partner, Johan Albrekt Ehrenström.

Architectural wonders

Engel's inspiration was the neoclassical architecture of St. Petersburg in Russia. His impressive works can be seen in Helsinki's imposing Senate Square, surrounded by the Government Palace, the University of Helsinki, and the Lutheran Cathedral with its gleaming white columns and shining domes.

The tradition of dazzling architectural design begun by Engel continued in Aleksander Gornostayev's Uspensky Cathedral, completed in 1868. This cathedral is a vast Byzantine structure of red brick and golden cupolas, with a lavishly decorated interior. Standing in dramatic contrast to the work of Engel and Gornostayev is the pink granite and art deco styling of Eliel Saarinen's railway station, which in 1914 seemed shockingly innovative.

Almost every street in Helsinki features works by distinguished Finnish architects, including Alvar Aalto, Herman Gesellius, and Saarinen. Notable among the most recent examples of Finnish architecture is Aalto's white marble and limestone Finlandia Hall, which opened in 1971.

A city of the sea

Helsinki is surrounded by the sea on three sides, and fishing boats set sail from its harbors every morning. Their catch of Baltic herrings (silakka) is sold directly from the boats at South Harbor quay.

Near the fish stalls stands the bronze statue of the sea maiden Havis Amanda, the center of traditional May Day Eve festivities. On this night, students scramble into the protective moat surrounding the statue and crown Havis Amanda with their white caps.

Helsinki's main street typifies the city's spacious atmosphere, with the sea never far away. Helsinki is a city of scenic bays and broad, treelined streets, with the contrast of old and new architecture adding to its unique charm.

The magnificent Lutheran Cathedral, begun in 1830, *far right top,* dominates Senate Square, Helsinki's administrative center. In the middle of the square stands Walter Runeberg's statue of Alexander II of Russia. A market, *far bottom right,* supplies fresh fruits, vegetables, and flowers.

Finlandia Hall—a marble masterpiece on the shores of Töölönlahti Bay—is part of Alvar Aalto's bold new city plan. The spires of the National Museum rise in the background.

Across from the statue are the flower, fruit, and vegetable stalls of Market Square, Helsinki's morning market. In the summer months, the market reopens in the late afternoon, and street vendors sell local handicrafts to tourists. Also in the summer months, the fruit stalls are laden with wild berries ripened by the Midnight Sun—cranberries, whortleberries, rowanberries, and the delicately flavored Arctic cloudberry known as suomuurain.

Helsinki is justly famous for its beautiful buildings designed by Finnish architects, including Engel's Lutheran Cathedral, Gornostayev's Uspensky Cathedral, and Saarinen's Central Station. Among Helsinki's most unusual structures is the Temppeliaukio Church, designed by the brothers Timo and Tuomo Suomalainen. The church is built into solid granite, about 40 feet (12 meters) above the street. Vertical strip windows around the base of the dome ceiling admit shafts of natural light.

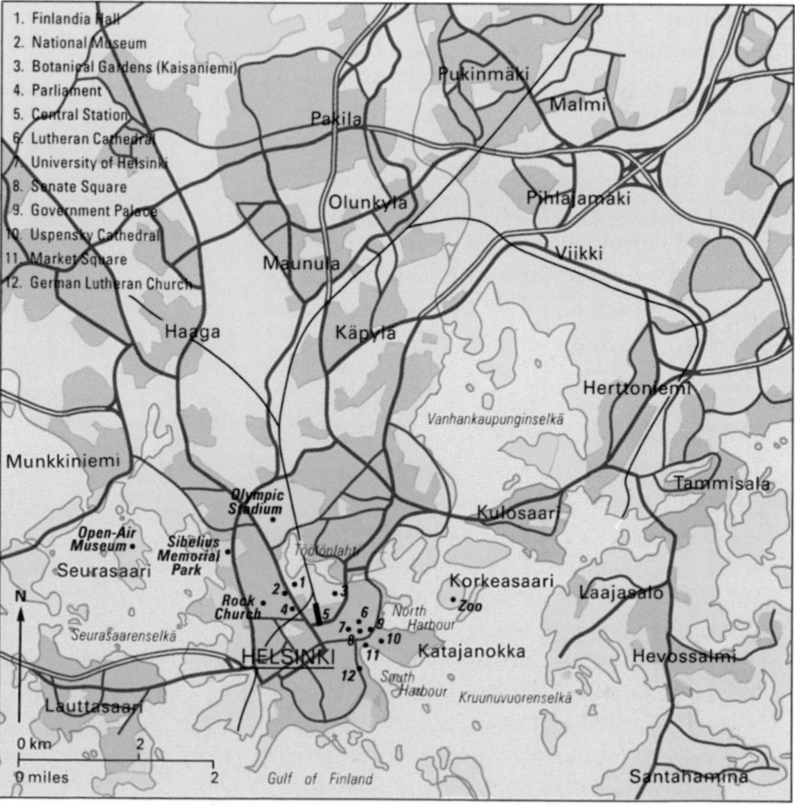

Economy

Before the outbreak of World War II (1939–1945), the economy of Finland depended largely on agriculture, forestry, and fishing. The war left the nation's economy in a desperate condition. About 10 per cent of Finland's productive capacity was lost to the Soviet Union, and the country was forced to take in about 400,000 war refugees.

Between 1944 and 1952, Finland also had to pay the Soviet Union $300 million in war damages. The cost of these massive reparations, in addition to rising inflation and rapid increase in the Finnish population, severely burdened the economy. However, the demands of reparation payments also helped fuel industrial growth, and the country's metal-working industry developed rapidly.

By 1947, Finland's gross national product (GNP) had reached its prewar level, and since then the economy has enjoyed consistent growth. The Finnish people have built a productive and diversified economy despite Finland's relatively poor soil, harsh climate, and lack of coal, oil, and most other mineral resources.

Finland's prosperity depends heavily on international trade. The nation's most valuable exports are forestry and metal-working products, particularly paper and paper products, ships and other transport equipment, and industrial machinery. Finland belongs to the European Union (EU), a group of European nations striving to form a single economic market. Members of the EU have ended all trade tariffs among themselves and have a common tariff on goods imported into the EU.

Forestry and agriculture

Finland is the most heavily forested of all European countries, and its widespread forests remain its greatest natural resource. As a result, forestry plays a leading role in Finland's economy, providing about 35 per cent of Finland's exports.

The Finnish government owns about a third of the nation's forests, chiefly in the north. The southern forests, owned mostly by private farmers, are the most productive because of the longer growing season and the cheap, efficient transport provided by the nation's network of lakes and water-

Huge rafts of timber are floated down rivers and assembled on Finland's Lake Saimaa, *right,* the largest of Finland's extensive lake systems. Steamships travel through the lakes, stopping at towns along the way.

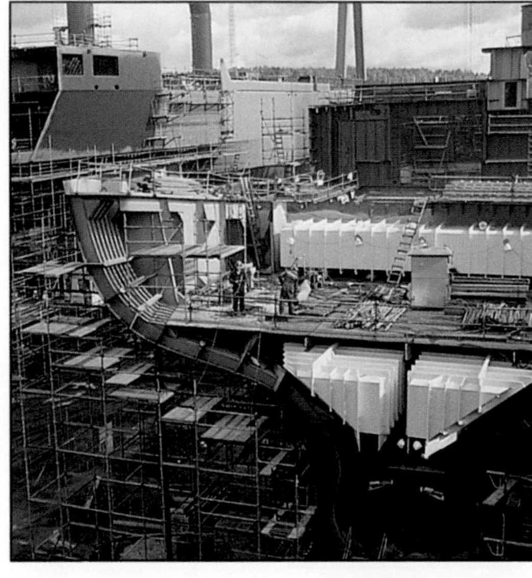

Turku, *above,* on the southwest coast, is known for its shipyards. The city is important to Finland's economy because its harbor is ice-free and more easily kept open in winter than the nation's other seaports.

Papermaking *far bottom right,* ranks as one of Finland's most important economic activities.

The process of transforming trees into forest products begins with felling, after which a skidding tractor delivers the trunks to the processor. Floating logs down rivers provides the cheapest transport from upland forests to sawmills. Finland's many forest products include rounded timbers, *lumber* (sawed timbers), *plywood* (thin layers of wood glued together under pressure), *veneers* (thin layers with a fine finish) and *particle boards* (wood shavings mixed with adhesive and compressed into panels).

Hayfields, *left,* provide food for Finland's cattle. Livestock production accounts for about 70 per cent of the nation's farming income.

ways. Careful conservation and replanting have largely made up for the loss of southeastern Finland's Karelian forests to the Soviet Union after World War II.

Today, Finland is the world's leading producer of plywood, and the manufacture of paper, paperboard, cellulose, and wood paneling are major industries. Finnish birch furniture is also prized throughout the world for its quality and craftsmanship.

Agriculture in Finland is limited by the relatively small proportion of suitable farmland. Most Finnish farms lie in the south and west, and their average size is about 29 acres (12 hectares). Nevertheless, Finland's farmers produce all the milk, eggs, meat, and bread grains needed by the people.

Manufacturing and energy

The metal-working industry has become as important to Finland's economy as its forest-based industries. The nation's metal-working industry produces electric motors and generators, farm equipment and machinery, and machinery used in the paper and lumber industries. Other manufactured products include chemicals, metals, processed foods, telephones, and textiles and clothing. Finnish shipyards, which build vessels suitable for cold northern waters, are noted for their powerful ice-breakers. The Finns also produce high-quality recreational yachts that are sold throughout the world.

Domestically produced power for Finland's industries comes from hydroelectric plants on the major rivers. During the 1970's and 1980's, Finland completed the construction of four nuclear power plants that now supply more than a third of the nation's energy needs.

France

High atop Montmartre, the tallest hill in Paris, stands the Basilique du Sacré Coeur (Basilica of the Sacred Heart). With its huge bell tower and onion-shaped dome, this gleaming white church is one of the city's most famous landmarks.

As an international capital of art, fashion, and learning, Paris ranks among the world's great cities. But although many people have come to believe that Paris is France and France is Paris, the beautiful City of Light, as it is known the world over, is only one of France's treasures.

From the rocky, mist-covered cliffs of Normandy to the sunny beaches of Nice, France is a country of great diversity and charm. With its colorful apple orchards, dairy farms, and vineyards, the French countryside has a charm all its own. Historic *châteaux* (castles) still dominate the Loire Valley, and picturesque fishing villages line the Atlantic coast in the northwest.

With its fertile soil and mild climate, France is a leader in agriculture—growing wheat, vegetables, and many other crops. The country is also an important manufacturing center, with thriving automobile, chemical, and steel industries. In addition, France plays an important role in world politics and economic affairs.

France has a long and colorful history. Julius Caesar and his Roman soldiers conquered the region before the time of Christ. Then, after Rome fell, the Franks and other Germanic tribes invaded the region. France was named for the Franks. By the A.D. 800's, the mighty Charlemagne, king of the Franks, had built the area into a huge kingdom.

In 1792, during the French Revolution, France became one of the first nations to overthrow its king and set up a republic. A few years later, Napoleon Bonaparte seized power. He conquered much of Europe before he finally was defeated. During World Wars I and II, France was a bloody battleground for Allied armies and the invading German forces.

Throughout their history, the French people have been among the leaders of European culture. French styles in painting, music, drama, and other art forms have often inspired countries throughout the Western world.

The French people are also noted for their *joie de vivre*—their celebrated enjoyment of life. They especially value good food and wine—fine French cooking has been raised to an art form. Their relaxed way of life also includes the conversation and the company of friends at a sidewalk cafe.

France Today

On July 14, 1989, France celebrated its bicentennial, the 200th anniversary of the French Revolution—a time of reform and violent change. The revolution introduced democratic ideals to France and ended the days of *absolute monarchy,* in which the king had almost unlimited power.

The French Revolution was a historic turning point for the people of France. It paved the way for the development of a unified state, a strong central government, and a free society dominated by the middle class and the landowners.

Today, more than 200 years after the start of the revolution, the goals of the revolutionaries still ring out in France's national motto—*Liberté, Egalité, Fraternité* (Liberty, Equality, Fraternity)—and endure in its way of life. French society is no longer divided into the estates that gave the monarchy, clergy, and nobles all the political power. And the French people no longer suffer from widespread famine, as they did in the dark days before the revolution.

France now prospers as a *parliamentary democracy* with a strong, centralized government whose president is elected by the people. The president manages the nation's foreign affairs. The prime minister, who is appointed by the president, directs the day-to-day operations of the government.

France is now a world leader in both industrial production and agricultural output. As a result, its people enjoy a high standard of living that includes such modern conveniences as automobiles, refrigerators, telephones, and washing machines. Social security laws give French workers some protection from the economic effects of unemployment, illness, and old age. And France is a member of the European Union, a group of western European nations working to form a single market for their economic resources.

France also has a well-developed educational system. French children between the ages of 6 and 16 must attend school. After five years of elementary school, students enter a four-year school known as a *collège.* After collège, students enter a *lycée,* a vocational or general high school.

General high schools prepare students for study at a university. Special schools called *Grandes Écoles* (Great Schools) equip students for high-ranking careers in the civil and military services, commerce, education, industry, and other fields.

Like all modern nations, France has its share of economic and social problems. Inadequate housing affects large numbers of immigrant workers and their families from Africa and southern Europe who are crowded in city slums. Unemployment, especially

FACT BOX

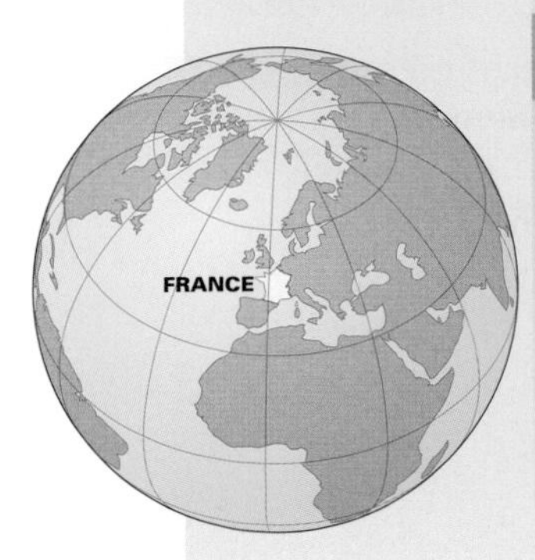

COUNTRY

Official name: Republique Francaise (French Republic)
Capital: Paris
Terrain: Mostly flat plains or gently rolling hills in north and west; remainder is mountainous, especially Pyrenees in south, Alps in east
Area: 212,935 sq. mi. (551,500 km^2)
Climate: Generally cool winters and mild summers, but mild winters and hot summers along the Mediterranean
Main river(s): Loire, Garonne, Rhône, Seine
Highest elevation: Mont Blanc, 15,771ft. (4,807 m)
Lowest elevation: Rhone River delta, 7 ft. (2 m) below sea level

GOVERNMENT

Form of government: Parliamentary democracy
Head of state: President
Head of government: Prime minister
Administrative areas: 22 regions
Legislature: Parlement (Parliament) consisting of the Senat (Senate) with 319 members serving nine-year terms and the Assemblee Nationale (National Assembly) with 577 members serving five-year terms
Court system: Cour de Cassation (Supreme Court of Appeals), Conseil Constitutionnel (Constitutional Council), Conseil d'Etat (Council of State)
Armed forces: 400,000 troops

PEOPLE

Estimated 2008 population: 61,225,000
Population density: 288 persons per sq. mi. (111 per km^2)
Population distribution: 77% urban, 23% rural
Life expectancy in years:
Male: 76
Female: 83
Doctors per 1,000 people: 3.3
Percentage of age-appropriate population enrolled in the following educational levels:
Primary: 105*
Secondary: 108*
Further: 54
Languages spoken:
French
Rapidly declining regional dialects and

France is a major political and economic power in today's world. Its foreign policies affect millions of people in other countries. France also has a long and colorful history. It was one of the first European nations to overthrow its king and set up a republic.

languages (Provencal, Breton, Alsatian, Corsican, Catalan, Basque, Flemish)

Religions:
Roman Catholic 80%
Muslim 7%
Protestant 2%
Jewish 1%

*Enrollment ratios compare the number of students enrolled to the population which, by age, should be enrolled. A ratio higher than 100 indicates that students older or younger than the typical age range are also enrolled.

TECHNOLOGY

Radios per 1,000 people: 950

Televisions per 1,000 people: 632

Computers per 1,000 people: 347.1

ECONOMY

Currency: Euro

Gross domestic product (GDP) in 2004: $1.737 trillion U.S.

Real annual growth rate (2003–2004): 2.1%

GDP per capita (2004): $28,700 U.S.

Goods exported:
Machinery and transportation equipment, chemicals, iron and steel products, agricultural products, textiles and clothing

Goods imported:
Crude oil, machinery and equipment, chemicals, agricultural products

Trading partners: European Union, United States

among young people, is also a problem, while many older people on fixed incomes have financial troubles.

An underwater railway connecting France and the United Kingdom opened in May 1994. The 31-mile (50-kilometer) Channel Tunnel, nicknamed "the chunnel," runs beneath the English Channel and cost between $15 billion and $16 billion. In January 1995, northern France was hit by severe floods that left about 40,000 houses under water. Damage was estimated at $564 million.

Environment

From the snowy peaks of the French Alps to the rolling plains and sunny beaches of the Aquitanian Lowlands, France is noted for its splendid scenery. Each of the country's diverse geographical areas has its own natural beauty.

The Brittany-Normandy Hills of northwest France have low, rounded hills and rolling plains with some rocky land. This region is actually a peninsula that juts out into the Atlantic Ocean. Steep cliffs on the northern coastline along the English Channel plunge almost vertically into the sea.

The southern coast of the peninsula, which faces the Bay of Biscay, has both sandy beaches and rocky cliffs. Many charming fishing villages line both coasts. Farther inland, small dairy farms as well as apple orchards and grasslands dot the landscape, their fields protected from the sea breezes by thick hedges called *bocage.*

Northeast of the Brittany-Normandy Hills lie the fertile and productive Northern France Plains—a landscape of flat and gently rolling land, broken up by forest-covered hills and plateaus. This region includes the Paris Basin, or *Île-de-France,* a large, circular area surrounded by lush, forested hills and drained by the Seine and other major rivers.

The gentle, rolling landscape of the plains gives way to the forests of the Northeastern Plateaus. Here, France shares with Belgium the Ardennes Mountains and their large deposits of iron ore. This wooded region becomes a little more rugged to the southeast in the Vosges mountains.

The Rhine River, Europe's most important inland waterway, flows along France's eastern border through the rich farmland of the Rhine Valley. Trees and vines cover the higher slopes.

Farther south, between the Rhine and Rhône rivers, lie the Jura Mountains. Much of the valuable forestland that once blanketed the higher slopes has been cleared for pasture, and vineyards cover many of the lower slopes. The Juras form part of the boundary between Switzerland and France.

The French Alps begin south of the Juras, forming the border between France and Italy. Their majestic peaks provide some of

NORTHERN FRANCE PLAINS
Seine
NORTHEASTERN PLATEAUS
Moselle
RHINE VALLEY
BRITTANY–NORMANDY HILLS
Loire
CENTRAL HIGHLANDS (MASSIF CENTRAL)
JURA MOUNTAINS AND FRENCH ALPS
AQUITANIAN LOWLANDS
Garonne
Rhône
MEDITERRANEAN LOWLANDS AND RHÔNE–SAÔNE VALLEY
PYRENEES MOUNTAINS
CORSICA

Elevation:
below 1,640 feet | 500 meters
1,640-3,280 feet | 500-1,000 m
above 3,280 feet | 1,000 meters

Red-roofed cottages, *right,* nestle comfortably around the chapel of Sons Brancion in western France. The rustic charm of these old buildings amid the rolling fields of the lowlands typifies the unspoiled beauty of rural France. About 26 per cent of the French people live in rural areas.

The varied landscapes of France, *left,* include soaring mountains, deep river valleys, vast plateaus, and rolling plains. The Atlantic coast features steep, rugged cliffs, while the French Riviera on the Mediterranean shore provides miles of sun-drenched beaches. France's beautiful scenery attracts visitors from all over the world.

France's most spectacular scenery.

Mont Blanc, which rises 15,771 feet (4,807 meters), is the highest mountain in the Alps. Thick woods and swift streams cover its lower slopes, but there is always a thick blanket of snow above 8,000 feet (2,400 meters). Mont Blanc also has huge valley glaciers, including the famous Mer de Glace (Sea of Ice).

West of the French Alps, the fertile land of the Mediterranean Lowlands and Rhône-Saône Valley produces fruits, vegetables, and wine grapes. The Central Highlands *(Massif Central)* of southeastern France features rolling hills and plateaus.

Along the country's southeastern border are the pine forests, rolling plains, and sand dunes of the Aquitanian Lowlands. The Pyrenees Mountains extend along France's border with Spain.

About 100 miles (160 kilometers) off the southeast coast of mainland France, the high, craggy coastline of the French island of Corsica rises out of the Mediterranean Sea. Sheep graze on the island's mountainous interior, and crops are grown in its narrow, fertile valleys.

French "cowboys" herd the famous wild horses of the Camargue, a vast marsh area at the delta of the Rhône River. The Camargue was once an unspoiled wilderness, but much of the region's marshland has been drained.

The top half of Mont Blanc is almost always covered by a thick blanket of snow. Many fashionable ski resorts can be found on the lower Alpine slopes, which offer some of the world's finest skiing.

People

While the people of France take great pride in being "French," the broad ethnic and cultural variety in France's population gives the country a unique and charming character. The influence of many traditions and cultures can be seen in regional differences around the country.

The provinces of France are home to many ethnic groups whose ancestors came to France from many different lands. Newcomers arrived during various periods of the country's history.

Many languages and cultures

The people of Normandy trace their heritage to a group of Vikings from Scandinavia. These *Norsemen* raided and then settled along the French coasts and river valleys in the A.D. 800's and early 900's.

After colonizing this area, they invaded England under the leadership of William, Duke of Normandy, who felt he was the rightful heir to England's throne. Norman warriors crossed the English Channel and conquered the English forces at the Battle of Hastings in 1066. In time, the Norse influence spread throughout the British Isles.

The Norse heritage of the people of Normandy is still evident. They tend to be taller than most French citizens, with blue eyes and light-colored hair.

Along France's border with Belgium, many people in the province of Picardy speak the Flemish dialect of Dutch. Southeast of Picardy, the region of Alsace-Lorraine has a strong German heritage. This region has long been a prize in wars between France and Germany, and the people of Alsace-Lorraine have been part German and part French for hundreds of years. They speak a dialect that is closely related to Swiss-German.

High on the slopes of the Pyrenees Mountains live the Basques. Little is known about the ancestors of the Basques, except that they settled in the Pyrenees long before many other early settlers arrived in France.

Today, their descendants speak a language called Euskera or Euskara. Many Basques still wear the traditional red or black beret. The Basques are also famous for their lively, colorful folk dances.

Farther north along the Atlantic coast, some people in Brittany still speak Breton, a language related to Welsh. The area was settled by Celts, who came from what is now Great Britain in the A.D. 400's to 600's and named their new home Brittany (Little Britain). For many years, the Bretons tried to stay independent from France. They isolated themselves from the rest of France and developed their own culture.

Although life in Brittany today is more like that of the rest of France, many Bretons carry on the old customs and wear traditional costumes on special occasions. Breton women, for example, wear satin or velvet aprons and fine lace headdresses.

On the island of Corsica, most of the people speak a dialect similar to Italian. However, as in all these regions, French is taught in the schools.

Cheered on by fans, a cyclist completes a leg of the Tour de France. This bicycle race, one of the top sporting events in France, lasts nearly a month and covers 2,500 miles (4,000 kilometers).

A long religious procession, winding through the streets of St.-Tropez, shows the strength of Roman Catholic tradition in today's France.

A group of Bordeaux residents, *below,* enjoy a quiet game of cards under a shady tree. The Bordeaux region is noted for its vineyards, and the city is a leading center of wine shipping.

Immigrants to France

Today, many foreign residents add to France's ethnic mix. The largest foreign groups are people from Algeria, Indochina, Italy, Morocco, Portugal, Spain, Tunisia, and Turkey. Hundreds of thousands of refugees from former French colonies in Africa and Indochina have arrived in recent years. Immigrants now make up about 7 per cent of the country's population. The status of these immigrants is a controversial issue in the country. They may be the first to be laid off during periods of slow economic activity.

Together as one nation

Although a number of dialects and languages are spoken throughout France, the official language is French. Now that French is taught in all the nation's schools, fewer people in each new generation speak these local dialects or languages. The French government has now started to encourage its people to keep alive the age-old regional traditions.

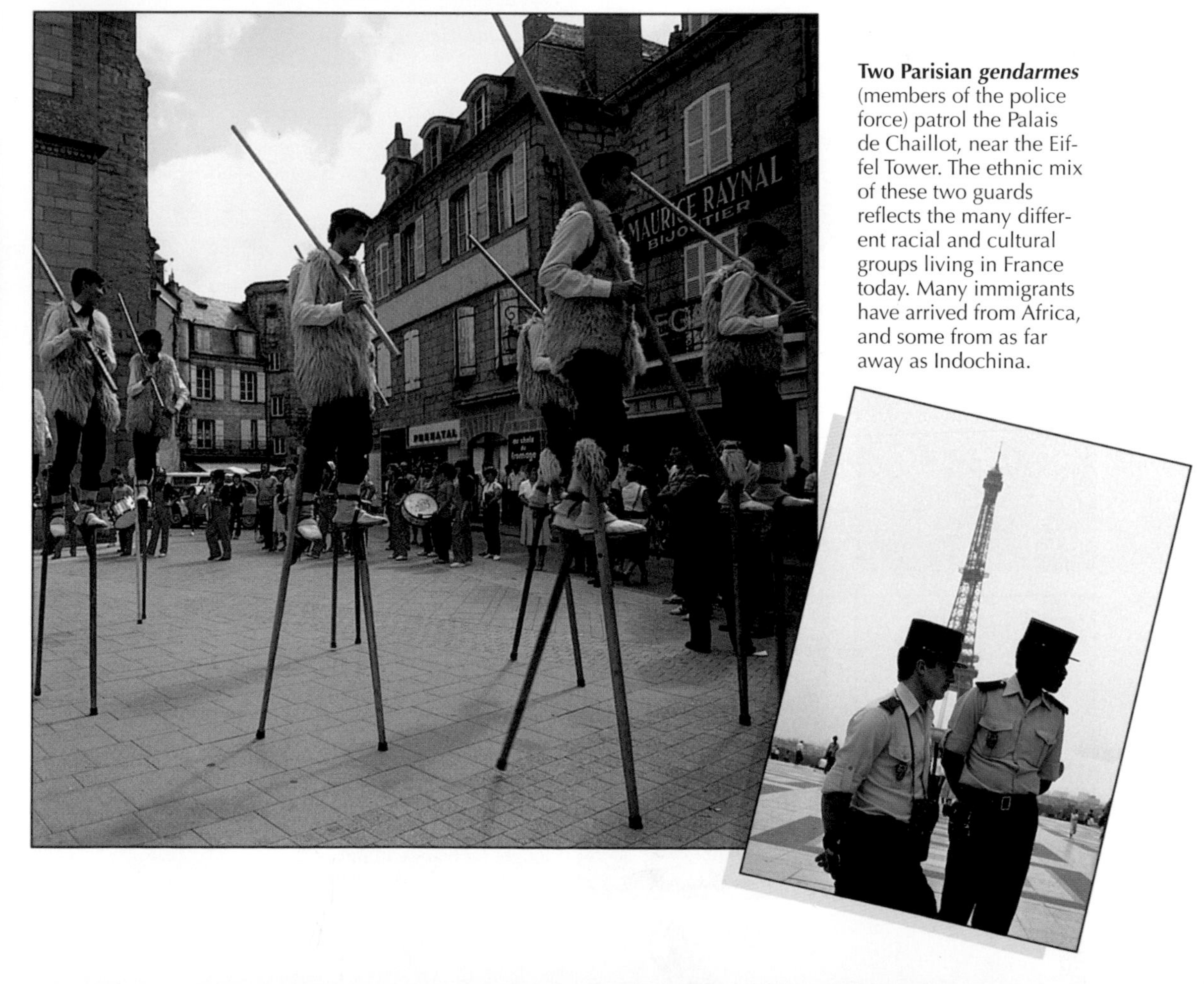

Folk dancers on stilts perform at a summer festival in Brive, a town in the province of Limousin. Such colorful festivals are held in many southern cities during July and August to entertain French families on vacation.

Two Parisian *gendarmes* (members of the police force) patrol the Palais de Chaillot, near the Eiffel Tower. The ethnic mix of these two guards reflects the many different racial and cultural groups living in France today. Many immigrants have arrived from Africa, and some from as far away as Indochina.

Early History

Human beings have inhabited what is now France for more than 100,000 years. Little is known of these early inhabitants, but they left a remarkable reminder of their presence in the wallpaintings of the Lascaux Cave in southwestern France. These realistic paintings of hunters, bulls, horses, and reindeer date from about 15,000 B.C.

By about 500 B.C., tribes of Celts had settled in the region that is now France. The Romans, who began to invade the region about 200 B.C., called it *Gallia* (Gaul). Julius Caesar conquered the entire region between 58 and 51 B.C.

The people, who were called Gauls, soon adopted Roman ways of life. For example, the Gauls used the Latin language, which would later have a basic influence on the French language.

Many structures built during the Roman period still stand today. These include the large Roman bath at Cluny in Paris, as well as the Pont du Gard, an aqueduct near Nîmes, and amphitheaters at Nîmes and Arles.

The Merovingians

In the A.D. 400's, the border defenses of the West Roman Empire began to weaken. Germanic tribes from the east, including Burgundians, Franks, and Visigoths, crossed the Rhine River and entered Gaul. After a series of battles with Rome and with the other Germanic tribes, Clovis, the king of the Salian Franks, founded the Merovingian *dynasty.* (A dynasty is a series of rulers from the same family.)

During the Merovingian dynasty, the rulers began to establish *manors* throughout France. These huge estates were governed by *landlords* or *lords,* who offered military protection to the peasants, called *serfs.*

In the 700's, a system called *feudalism* began to develop. Under this system, a lord gave land to his noble subjects in return for military and other services. This land was called a *fief.* Some fiefs were quite large, such as the province of Normandy.

By the mid-600's, however, the Merovingian kings had become weak rulers. Pepin of Herstal, the chief royal adviser, gradually gained most of the royal power. His son,

Royal power in France began to grow in the late 900's during the Capetian dynasty. Hugh Capet, the first of this long line of kings, firmly controlled only a small territory around Paris. Later Capetian kings enlarged the royal holdings, increased the power of the rulers, and established a strong, centralized government.

Royal lands in 987
Under English control 11
Royal lands in 1180
Dependent lands (1180)
Under English control (12
Additions to royal lands (c.1270)
Additions to royal lands (c.1314)
Additions to royal lands (c.1328)

c. 15,000 B.C. Cro-Magnon people live in what is now southwestern France.
c. 600 B.C. Greeks found the city of Marseille and name it *Massalia.*
c. 500 B.C. Celts settle in France.
121 B.C. Rome gains control of southern France.
58-51 B.C. Julius Caesar conquers Gaul.
A.D.400's Germanic tribes (Franks, Burgundians, and Visigoths) enter Gaul.
486 Clovis, king of the Franks, defeats Romans and founds Merovingian dynasty.
507 Clovis makes Paris the capital of the Frankish kingdom.
751 Pepin the Short becomes king of the Franks and founds Carolingian dynasty.
800 Charlemagne becomes emperor of the Romans.
800's–900's Norse people conquer Normandy in northwestern France.
814 Charlemagne dies, and his kingdom is divided. Charles the Bald receives most of what is now France.
987 Hugh Capet is crowned king of France.
1066 William the Conqueror, a Norman duke, invades England.
1100–1300 French nobles fight in the Crusades.
1100's–1400's Cathedrals at Paris, Chartres, and Reims are built.
1152 Eleanor of Aquitaine marries Henry II of England, which begins English control of the duchy of Aquitaine in western France.
1180–1223 Philip II (Philip Augustus) rules France.
1302 Philip IV (Philip the Fair) calls together the first Estates-General, the ancestor of the French Parliament.
1309–1377 Pope Clement V moves the pope's court from Rome to Avignon, where it remains until 1377.
1337–1453 Hundred Years' War is fought between France and England.
1346 Battle of Crécy is won by the English in the greatest victory of the Hundred Years' War.
1429 Joan of Arc rescues Orléans from the English. Charles II is crowned king at Reims.
1453 English are driven out of France.

Charlemagne, *left,* ruled the Franks from 768 until his death in 814.

Huge Capet, *far left,* was king of France from 987 to 996.

Joan of Arc (1412?-1431) fought for France in the Hundred Years' War.

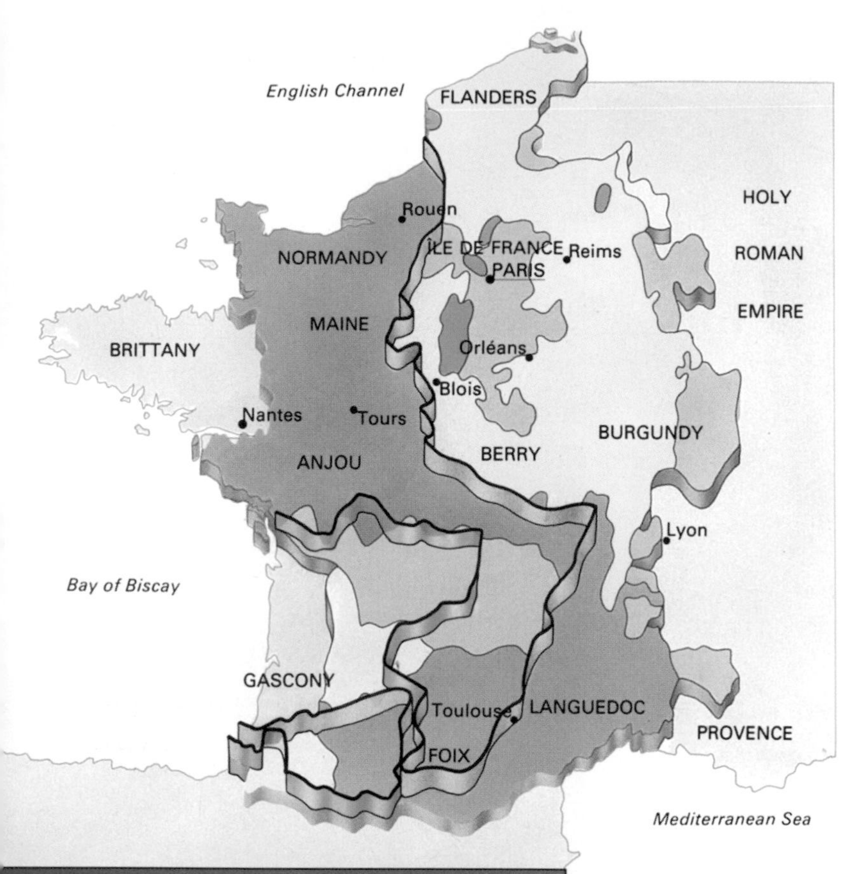

Charles Martel, increased the family's power and became king of the Franks in all but title.

Charles Martel's son, Pepin the Short, overthrew the last Merovingian ruler and became king of the Franks in 751. He founded the Carolingian dynasty, and his son, Charlemagne, became king in 768.

Charlemagne

Charlemagne, or Charles the Great, was the most famous ruler of the Middle Ages. Through a series of military campaigns, he expanded his territories far beyond what is now France. Charlemagne united these territories to create a great empire, and revived political and cultural life in Europe.

After the death of Charlemagne in 814, most of what is now France went to his grandson Charles the Bald. By the late 900's, the feudal nobles gained more political power. In 987, they ended the Carolingian dynasty and chose Hugh Capet as their king. He started the Capetian dynasty.

Capetian kings

The first great Capetian king was Philip II, called Philip Augustus, who came to the throne in 1180. He developed Paris as a center of culture, government, and learning. Two of the city's greatest structures—the Cathedral of Notre Dame and the Louvre—were largely built during his reign.

The last Capetian king, Charles IV, died in 1328 without leaving a male heir to the throne. The ensuing dispute over the crown led to the Hundred Years' War (1337–1453) between England and France. At first, England won most of the battles. Later, the French king Charles II regained power after the young peasant girl Joan of Arc led his troops to victory at Orléans. France eventually won the war.

The cathedral at Rouen is a magnificent example of Gothic architecture. Many cathedrals were built in France during the Middle Ages when architects believed that large, impressive buildings inspired greater faith.

The Pont du Gard, is an aqueduct built by the Romans to carry water to the city of Nîmes in southern France. The ancient Romans laid out a network of roads and built many public buildings, including some that still stand today.

Loire Châteaux

The Renaissance, a great cultural movement that began in Italy, reached France in the late 1400's. Francis I, who was king of France between 1515 and 1547, became very interested in Renaissance art and literature. He brought Leonardo da Vinci and many other Italian artists and scholars to France. Francis I also tried to surround himself with the work of the Italian Renaissance masters.

The most magnificent *châteaux* of the Loire Valley were built during this period. The châteaux were used as elegant hunting lodges and country residences by the French royalty and nobility.

Today, more than 1,000 of these châteaux, France's lasting contribution to Renaissance architecture, dot the landscape along the Loire River. The architectural heritage of this peaceful, wooded region of the Loire Valley makes it one of France's most famous regions.

During the 1500's, Francis I's Château d'Amboise was a magnet for scholars, poets, musicians, and artists. In 1517, Leonardo da Vinci came to Amboise at the invitation of the king. Scale models of his most noted scientific inventions are in the collection at Le Clos-Luce, the house where Leonardo spent the last two years of his life, just southeast of the Château d'Amboise.

Visitors often begin their tour of the Loire châteaux at the mouth of the Loire, in the Breton seaport of Nantes. A bridge over the river offers the best view of Château d'Ussé, perhaps the most romantic and fanciful of these Renaissance castles. It is said that Ussé, inspired Charles Perrault, who once lived there, to write the fairy tale *sleeping Beauty.*

To the east of Ussé, Château d'Azay-le-Rideau, among the most beautiful and harmonious of the Loire châteaux, stands on an island in the Indre, a branch of the Loire. Its massive corner towers soar to the sky. Colorful presentations in period costumes take place at this château on summer evenings.

West of Tours is the Château de Villandry, famous for its terraced gardens. These Renaissance gardens have been restored according to the original plans dating from the 1500's.

Visitors can admire three tiers of gardens, including an herb garden, as well as a lake and about 7,500 square feet (700 square meters) of ponds and basins, all restored to their former beauty. In the Garden of Love on the lowest tier, beds of flowers represent the many forms of love in beautiful designs.

East of Villandry lies the Château de Chenonceau. Smaller than most other castles of time, Chenonceau has a light and graceful quality. Chenonceau was Henry II's gift to his mistress, Diane de Poitiers, but after Henry died, his widow, Catherine de Médicis, forced the woman to give up the château. Driven from Chenonceau, Diane de Poitiers took up residence at the Château de Chaumont. Today, Chaumont contains a fine collection of tapestries and Renaissance furniture.

No visit to this land of fairy-tale castles would be complete without a journey to Chambord, the largest of the Loire châteaux. Chambord has 440 rooms, and

The largest of the Loire châteaux, Chambord was built for Francis I beginning in 1519. Much of its furniture was destroyed during the French Revolution. Chambord became the property of the French government in 1932.

Azay-le-Rideau, *bottom right,* is known for the charm of its location. On summer evenings, a *son et lumière* (sound and light) spectacle tells the story of Francis I's treasurer, who supervised the building of the château. He was later found to be stealing royal funds.

its grounds were said to have once been as large as all of Paris. Here, in 1539, Francis I greeted the Holy Roman Emperor Charles V.

The builders of Chambord even altered the course of the Loire to enhance these 13,600 acres (5,500 hectares) of forests, gardens, lakes, and ponds. This wooded area, surrounded by a wall 20 mile (32 kilometers) long, is now a reserve for deer and wild boar.

During the Renaissance, the Loire châteaux were the center of social and intellectual life for the French nobility. However, an occasional dark deed occurred amid the splendor. In 1588, King Henry III's rival, Duke Henry of Guise, was murdered in the Château de Blois. It is said that the king himself, hidden behind a curtain, witnessed the killing. Today, Blois is most noted for the magnificent spiral staircase that many people believe was designed by Leonardo da Vinci.

The Loire Valley, *maps below,* contains some of France's greatest Renaissance architecture. The magnificent châteaux are set like jewels along the Loire River, as it flows peacefully through the countryside. The construction of the Château d'Amboise, which Charles VIII began in 1492, marked the beginning of this period. Amboise reached its height in 1517 when Francis I, a patron of Renaissance artists, brought Leonardo da Vinci to work there.

Villandry, *left,* is best known for its gardens. In the 1500's, rare species of vegetables, mostly imported from other countries, were grown in vegetable gardens near the château, where they could be guarded.

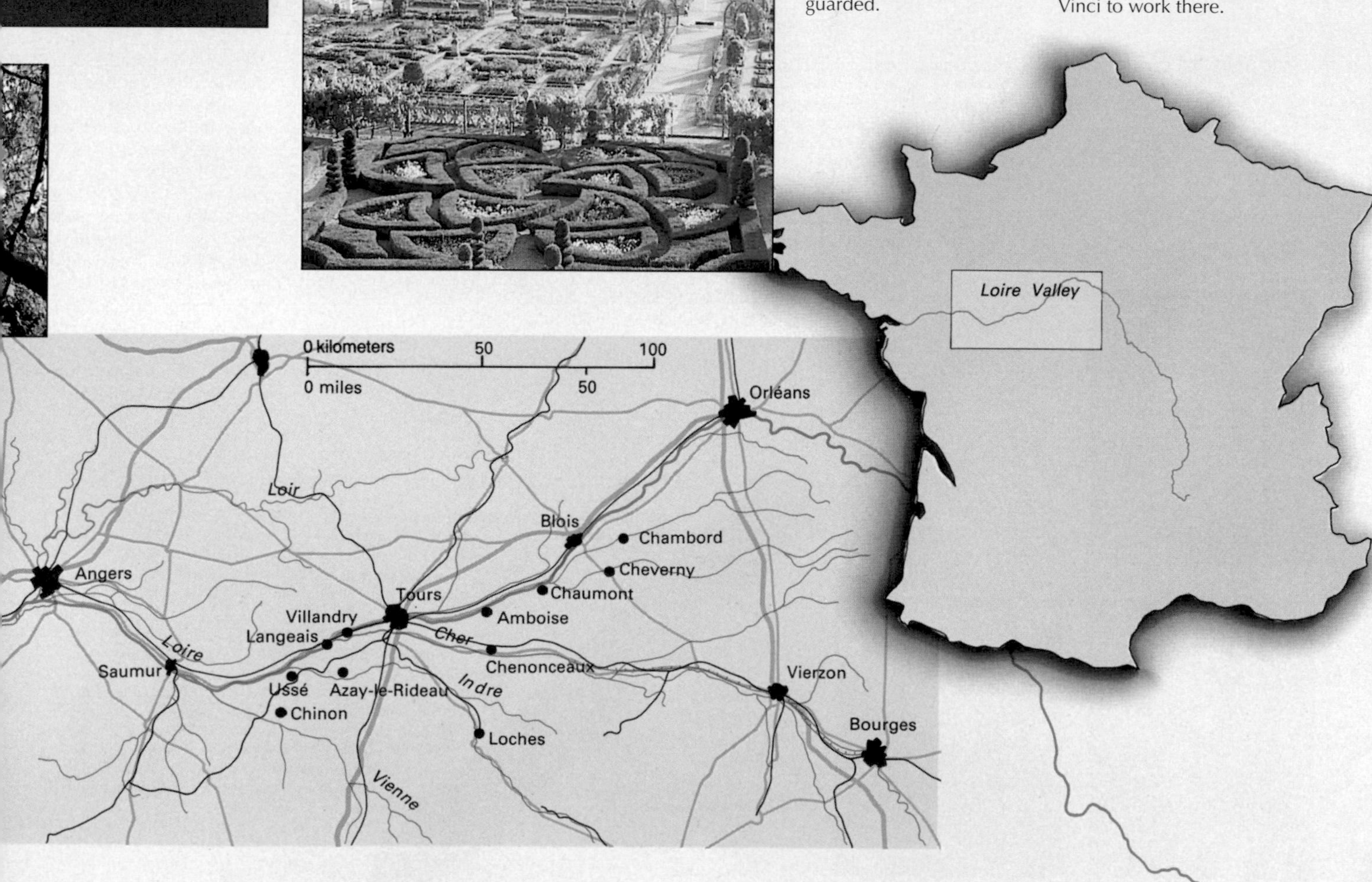

Modern History

During the late 1500's, French Roman Catholics and Huguenots (French Protestants) fought a series of civil wars. In 1598, King Henry IV signed the Edict of Nantes, which granted limited freedom of religion to the Huguenots.

Between the 1500's and 1700's, the power of the French kings and their *ministers* (high government officials) grew steadily. The ministers were largely responsible for France becoming a strong nation during this period.

Louis XIV came to the throne of France in 1643, when he was only 4 years old. He was king for 72 years—the longest reign in European history. Under his rule, France ranked above all other nations in art, literature, war, and statesmanship.

Louis XIV believed in the complete political authority of the king. Although not well educated, Louis XIV was clever in choosing wise counselors, including Jean Baptiste Colbert, his minister of finance. Colbert promoted a strong economy, but the cost of Louis XIV's wars of expansion, as well as the construction of his Palace of Versailles, seriously drained the country's finances.

By the 1700's, a corrupt, bureaucratic government had developed, and an unfair system allowed lawyers and nobles to avoid paying taxes. Another series of expensive wars forced the government to impose even heavier taxes on the overburdened middle class. Out of this financial crisis, the French Revolution was born.

The French Revolution

To gain support for more taxes, King Louis XVI called a meeting of the Estates-General on May 5, 1789. The Estates-General was made up of representatives of the three classes of French society—the clergy, the nobility, and the commoners. The following month, members of the third estate—the commoners—declared themselves a National Assembly. They drafted a constitution that made France a limited monarchy, with a one-house legislature.

The French Revolution was a period of great violence for the French people. It

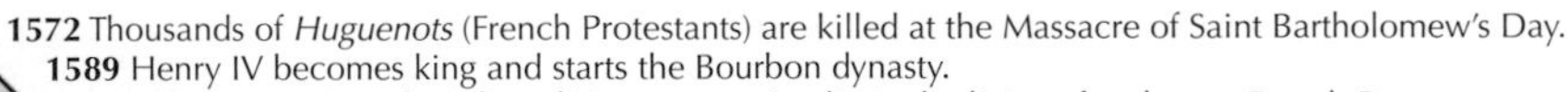

1572 Thousands of *Huguenots* (French Protestants) are killed at the Massacre of Saint Bartholomew's Day.
1589 Henry IV becomes king and starts the Bourbon dynasty.
1598 Henry IV signs the Edict of Nantes, granting limited religious freedom to French Protestants.
1643–1715 Louis XIV rules France as absolute monarch and builds the Palace of Versailles.
1789 A crowd of Parisians storms the Bastille fortress, setting off the French Revolution and ending the absolute rule of kings.
1792 The First Republic is established.
1799 Napoleon seizes control of France.
1804 Napoleon founds the First Empire.
1814 Napoleon is exiled; Louis XVIII comes to power.
1815 Napoleon returns to power but is defeated at Waterloo. Louis XVIII regains the throne.
1830 Charles X is overthrown in the July revolution.
1848 Revolutionists establish the Second Republic.
1852 Louis Napoleon Bonaparte, nephew of Napoleon Bonaparte, declares himself emperor and founds the Second Empire.
1870–1871 Prussia defeats France in the Franco-Prussian War. The Third Republic begins.
1870's–1914 France establishes a vast colonial empire in Africa and Asia.
1914–1918 France fights on the Allied side in World War I.
1939–1940 France fights on the Allied side in World War II until defeated by Germany.
1940–1942 Germany occupies northern France.
1942–1944 The Germans occupy all of France.
1945 Germany is defeated by Allied forces.
1946 France adopts a new constitution, establishing the Fourth Republic.
1946–1954 France gives up French Indochina after a revolution in the colony.
1954 Revolution breaks out in the French territory of Algeria.
1957 France joins the European Economic Community, also called the European Common Market.
1958 A new constitution is adopted, marking the beginning of the Fifth Republic. Charles de Gaulle is elected president.
1962 France grants independence to Algeria.
1969 De Gaulle resigns as president.
1981 Socialist victories in presidential and parliamentary elections result in France's first leftist government since 1958.
1994 A railway tunnel under the English Channel between France and England opened.

The colorful parades celebrating the 200th anniversary of the French Revolution, *center,* are a marked contrast to the violence of Louis XVI's execution at the Place de la Concorde in 1793, *top.* The revolutionary government executed many people with the guillotine. They said that terror was necessary to preserve liberty, About 150 years later, smiling faces greet the Allied troops, *bottom,* as the French are freed from a modern-day terror—the Nazi occupation.

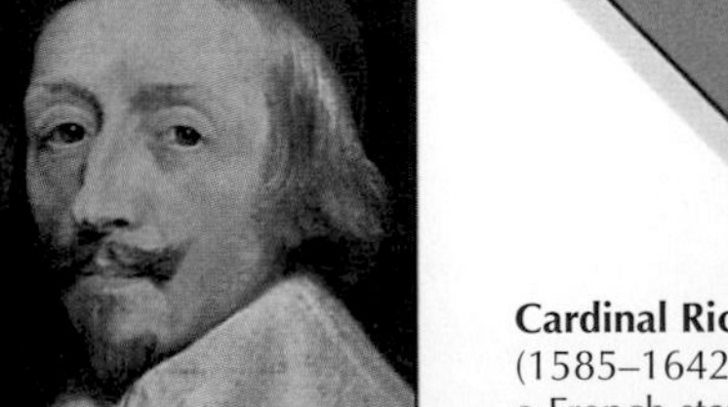

Cardinal Richelieu (1585–1642), *left,* was a French statesman under Louis XIII.

Napoleon I (1769–1821), *far left,* was emperor of France.

Charles de Gaulle (1890–1970) was president of France's Fifth Republic.

began on July 14, 1789, when huge crowds of Parisians stormed the Bastille fortress. It continued for many years as the revolutionary governments executed thousands of people they considered "enemies of the republic," including the French king, Louis XVI.

During the revolution, a young officer named Napoleon Bonaparte became a general in the army. In 1799, he overthrew the revolutionary government and seized control of the nation. In 1804, he crowned himself emperor of France. Napoleon revised and organized French law, and under his leadership, France developed a strong, efficient central government.

Napoleon was a military genius driven by ambition. By 1812, French forces under his leadership had conquered most of western and central Europe. But the new territory was too large for Napoleon to govern efficiently. He was forced to give up the throne in 1814. His final defeat came at Waterloo in 1815.

After 1815, the monarchy returned to France and tried to go back to the old ways of absolute rule. The attempt failed and revolution erupted in 1830 and again in 1848. Napoleon's nephew, Louis Napoleon, crowned himself Emperor Napoleon III in 1852.

After France lost Alsace and part of Lorraine to Germany in the Franco-Prussian War of 1870–1871, the people overthrew Napoleon III and declared France a republic once again.

France after World War I

Heavy losses during World War I (1914–1918) weakened France. During the 1930's, a worldwide economic depression and the rise of fascist leader Adolf Hitler in Germany caused political unrest. When Hitler invaded France on May 12, 1940, the northern part of the country was soon overrun and on June 22, France agreed to an *armistice* (truce) with Germany.

The northern two-thirds of France, including Paris, were occupied by German military forces. Meanwhile, General Charles de Gaulle headed a French government-in-exile in London. Four years later—on Aug. 25, 1944—the Allies defeated the German troops and entered Paris. De Gaulle soon formed a provisional government and became its president. Under his leadership, political and economic stability was restored. France once again became one of Europe's leading powers.

Paris

When Alexandre Gustave Eiffel designed his tower for the World's Fair of 1889, he could not have dreamed that it would become a symbol for one of the world's greatest cities—Paris, the capital and largest city of France. Today, the Eiffel Tower, rising 984 feet (300 meters), stands in the Champs de Mars, a park that was once a military training ground. From the top of the Eiffel Tower, visitors can enjoy spectacular views of Paris in all its beauty and charm, with its broad, treelined boulevards, historic buildings, and famous parks and gardens.

Paris is located in north-central France, in the heart of the Paris Basin. The Seine River flows through the city from east to west. The section north of the river is called the Right Bank. Busy offices, small factories, and fashionable shops are on the Right Bank. South of the river lies the Left Bank—the center of student life and a gathering place for artists.

Many historic monuments, churches, and palaces can be found throughout Paris. Each represents a different part of the fascinating, colorful story of this famous City of Light.

A vendor sells postcards and prints near the Cathedral of Notre Dame. The cathedral stands on the Île de la Cité, an island in the Seine. Paris was founded on this island more than 2,000 years ago.

A universal symbol of Paris, the Eiffel Tower rises over the city. Built as a part of the World's Fair in 1889, the tower now has restaurants, a weather station, and spaces for experiments.

Customers at a sidewalk cafe on the Champs Élysées, view the Arc de Triomphe while dining. The most famous avenue in Paris, the Champs Élysées is lined with beautiful trees and gardens.

Historic sights

At the western end of the Champs Élysées (Elysian Fields), Paris' best-known avenue, stands the Arc de Triomphe (Arch of Triumph). This huge stone arch was begun by Napoleon I in 1806 as a monument to his troops. The inner walls bear the names of 386 of his generals and 96 of his victories. After World War I (1914–1918), France's Unknown Soldier was buried beneath the arch.

At the eastern end of the Champs Élysées stands the Place de la Concorde (Square of Peace). It has eight statues, two fountains, and the Obelisk of Luxor, a stone pillar from the temple of Luxor in Egypt.

Built during the 1700's, the Place de la Concorde played a central part in a bloody chapter of French history. In this square, hundreds of people, including King Louis XVI and his wife, Marie Antoinette, were beheaded by the *guillotine.*

Across from the Place de la Concorde are the Tuileries Gardens—one of France's finest formal gardens and a favorite spot for Parisian children. They sail toy boats in the round lagoon fountains and enjoy Punch and Judy puppet shows in the park.

The 40-acre (16-hectare) Louvre, one of the largest art museums in the world, overlooks the Tuileries Gardens. Philip II originally built the Louvre as a fortress in about 1200. Today, it has about 8 miles (13 kilometers) of galleries and more than a million works of art.

The Louvre's six departments cover the art of every period from the Egyptians and Babylonians to the end of the 1800's. Its most famous works include the Greek sculptures *Venus de Milo* and *Winged Victory,* as well as Leonardo da Vinci's *Mona Lisa.*

The city's many historic churches include the magnificent Cathedral of Notre Dame. A fine example of Gothic architecture, the

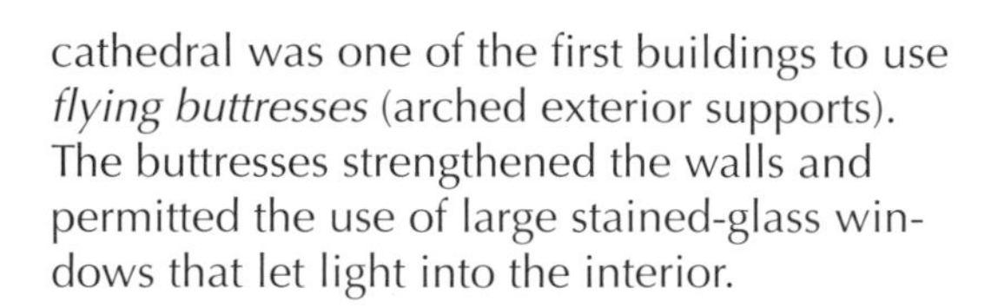

cathedral was one of the first buildings to use *flying buttresses* (arched exterior supports). The buttresses strengthened the walls and permitted the use of large stained-glass windows that let light into the interior.

Economy

Paris is the chief financial, marketing, and distribution center of France. Many company headquarters and financial institutions operate in the city. Over half the nation's business is done in Paris, and jobs provided by national and local governments contribute greatly to the city's economy.

Paris is also France's transportation center. The city is served by three major airports, and it is the hub of a national railroad network.

A fire-eater, *above,* entertains a crowd at the modern Pompidou Center. The building houses artwork, a library, and music and industrial design centers.

Paris, *right,* is rich in beautiful and historic sites that are visited each year by more than two million tourists. Many visitors come for the city's cultural attractions. Its museums and art galleries display some of the world's most precious works of art. Paris is also famous for its many restaurants, cafes, theaters, and nightclubs.

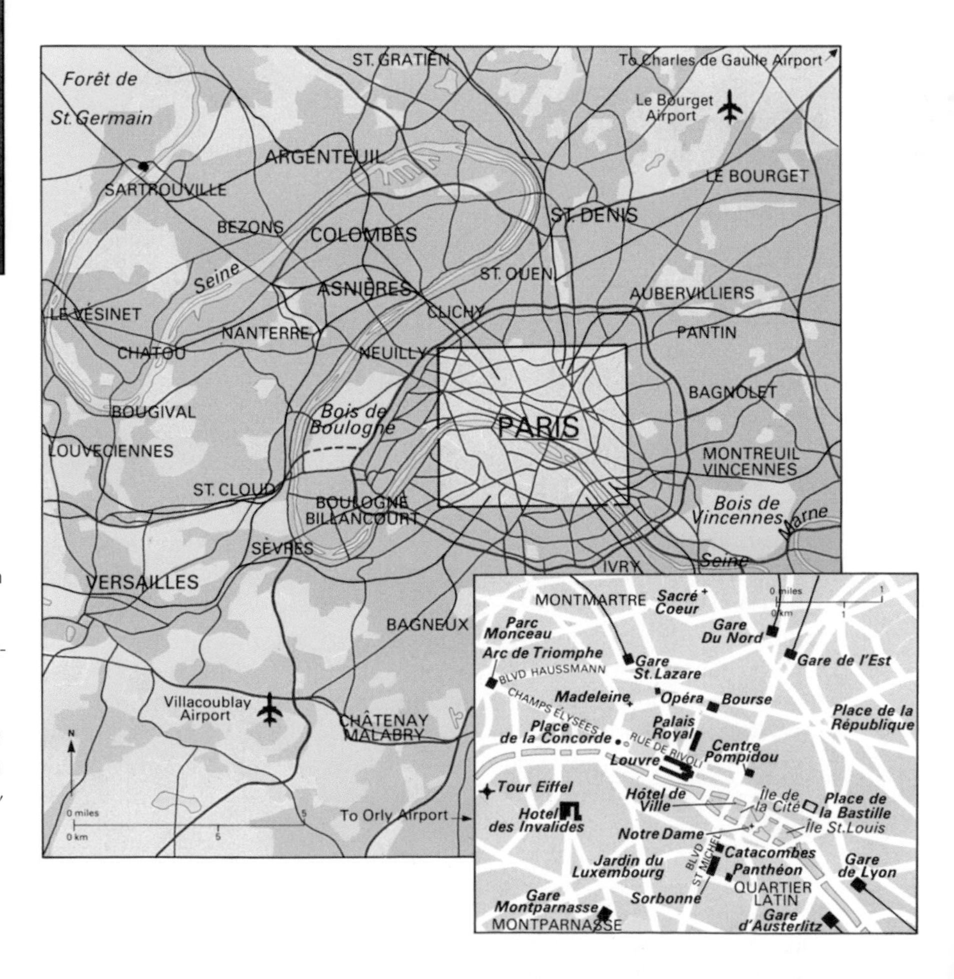

Agriculture

The crops that grow on more than a third of France's land are an important part of the nation's economy. Today, France is one of the world's leading exporters of farm products.

Crops of France

Wheat is the country's leading single crop. It is grown on large farms scattered throughout the northern region and Paris Basin.

Wine ranks as one of France's oldest and most famous agricultural products. About 2,000 years ago, the Romans cultivated vineyards in the southern region they called *Provincia* (provence). Today, grapes are grown throughout France.

Grapes used to make high-quality wine come from several regions, including Alsace, Bordeaux, Burgundy, Champagne, and the Loire Valley. The Mediterranean region produces grapes used for cheaper wines, and grapes from southwestern France are used in brandy.

Other major crops grown by French farmers include beans, carrots, cauliflower, cherries, flowers, peaches, pears, peas, potatoes, sugarbeets, sunflower seeds, and tomatoes.

Raising livestock

About a fourth of France's land area consists of grassland used for grazing. The chief areas for raising livestock are the north and northwest, the Central Highlands *(Massif Central),* and the Alps. Brittany is now a major center for the large-scale production of pigs and chickens.

About two-thirds of French farm income comes from meat and dairy products. Cattle, sheep, and lambs are the chief meat animals.

France produces more than 370 different kinds of cheese. Each region has its own specialties. Among the most famous of these cheeses are Gruyère from the French Alps, Camembert from Normandy, and Roquefort from south of the Massif Central.

The new French farm

In the early 1900's, about half of France's workers labored on farms. Today, only about 8 per cent of the people work in farming. Between 1950 and 1980, about 500,000 French farms went out of business—mainly village farms that were too small to be profitable in today's markets.

Gruyère de Comté cheese is one of three varieties of Gruyère, a popular cheese from France. Gruyère is made in the regions of Haute-Savoie and Franche-Comté. It requires very large amounts of milk and must cure for six months.

Rolling wheat fields, *right,* in western and northern France produce most of France's grain harvest. Wheat is by far the country's leading grain crop. French farmers also grow barley, corn, oats, and rapeseed for livestock feed.

Lavender fields in Provence, *right,* yield a sweet-smelling harvest. The lavender bush grows wild in the warm climate of the Mediterranean, but large fields are now cultivated. Its flowers and leaves are used to add fragrance to perfumes and soaps.

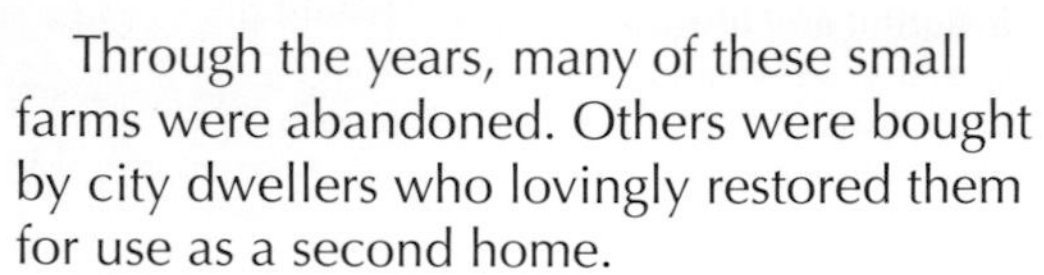

Through the years, many of these small farms were abandoned. Others were bought by city dwellers who lovingly restored them for use as a second home.

The life of a French farmer today differs greatly from that of a village peasant in the past. Modern farms are much larger and more efficient. The people who operate them use good business sense as well as the latest techniques in production and marketing.

Garlic, strung in garlands, is a strongly flavored plant used to season foods. It is an important part of French cooking.

Fertile soil makes farming possible in almost every region of France, *below.* Regional specialties include cider and calvados brandy made from apples grown in Normandy and Brittany; poultry in Bresse; and lamb grazed on the salty meadows near the bay of Mont-Saint-Michel.

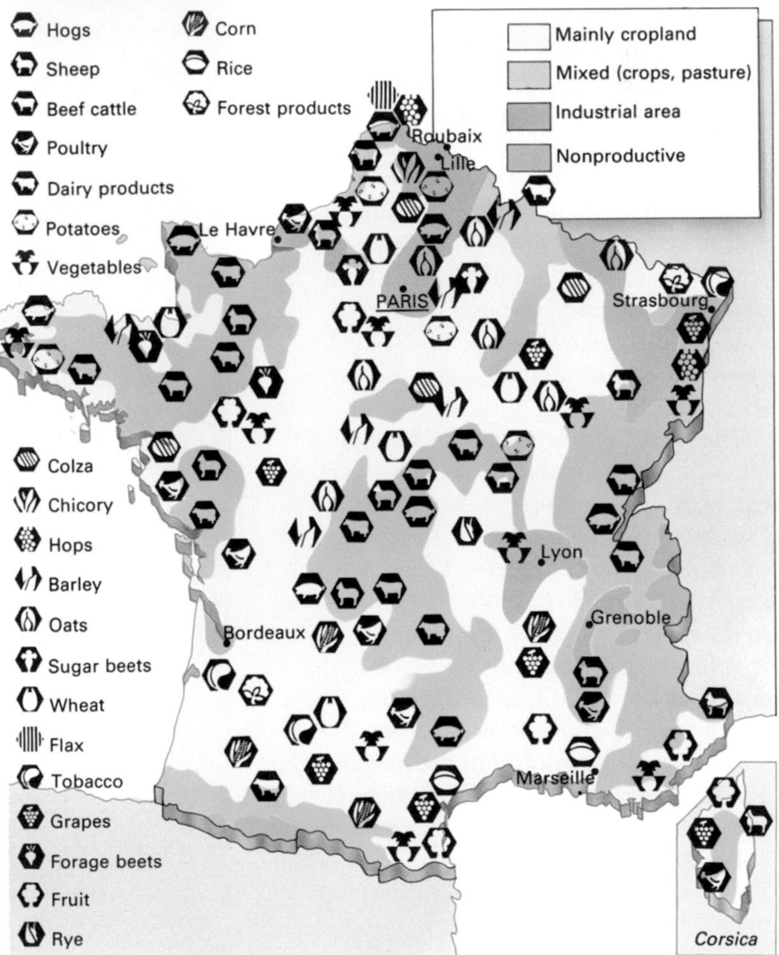

Industry

France ranks as one of the world's leading industrial nations, with about 22 per cent of its workers employed in manufacturing. The nation's many automobile, chemical, and steel plants use modern equipment and production methods.

France is also involved in the manufacture of many other products, including sophisticated military and commercial airplanes, industrial machinery, and textiles. In addition, the country's fast-growing electronics industry produces computers, radios, televisions, and telephone equipment.

The role of government

Most of France's industrial growth has occurred since the end of World War II (1939–1945). After the war, the French economy still depended chiefly on small farms and business firms. In 1945, however, the government introduced a program of industrial expansion and modernization.

High taxes and large government subsidies provided the necessary investment money. The French government also issued guidelines for private businesses. These reforms were made through a series of national plans, and they helped to make France an industrial leader in a relatively short time.

The French government has long owned all or part of several major businesses. For example, it has complete ownership of France's three largest banks. The government is also the sole owner of Renault, the largest automobile manufacturer in France.

Whether or not the government owns businesses often depends on whether the Socialists or the Conservatives are in control of the French Parliament. When the Socialists gained control of the presidency and the Parliament in 1981, they increased government ownership of businesses. But in 1986, the Conservatives gained control of Parliament and began to sell government-owned businesses.

The Paris Basin

Paris and its surrounding area is France's chief manufacturing, marketing, and distribution center. More than half of the nation's business is done in Paris. Industries in the Paris area include publishing and the

France's high-speed train

France's Train à Grande Vitesse (high-speed train), called the TGV, *right,* is the world's fastest train. Developed by France's national railroad company in the 1970's, the first TGV began operating between Paris and Lyon in 1981. In 1989, an even faster TGV began running between Paris and cities in western France. It operates at a top speed of 186 miles (300 kilometers) per hour. The TGV runs mainly on straight tracks, which allows it to keep up high speeds. The TGV was designed to cope with increasing traffic on France's main railroad lines.

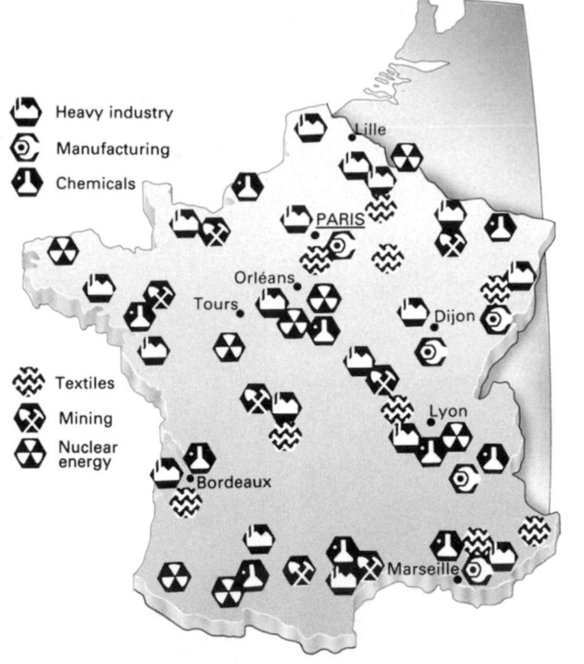

France, ***left,*** **is one of the world's leading industrial nations.** The country produces a wide range of products, including automobiles, chemicals, industrial machinery, perfume, steel, and textiles. France is the fourth largest manufacturer of automobiles in the world. Much of its success in industrial development is due to government planning. The state sets national economic goals and invests in research and development.

A nuclear power plant stands at Chinon on the Vienne River. Nuclear power plants provide more than half of France's electricity. The country is a world leader in nuclear energy technology and in the production of nuclear fuels.

A worker at the French mint, *left,* at Pessac begins the process that turns molten metal into coins. Service and high-technology industries now employ more workers than heavy industries such as steel and shipbuilding.

manufacture of automobiles, chemicals, clothing, electronics, leather goods, and transportation equipment.

Industrial centers

Although Paris is France's major industrial center, there are factories in cities and towns throughout the country. For example, the silk weavers of Lyon produce some of France's most beautiful fabrics. The Lille-Tourcoing-Roubaix area also produces a wide range of textiles.

Major construction programs have recently brought industry to previously underdeveloped areas. Huge smelting plants have been built at Dunkerque, in northern France, and at Fos, near Marseille. Many plants serving the automobile industry have relocated to areas such as Dijon.

Industry in the Rhône-Alps region, with its main towns of Lyon, St.-Étienne, and Grenoble, has also greatly expanded. New businesses in the area include chemical, petro-chemical, electrical, and electronic industries.

European Union membership

France is a member of the European Union (EU), a group of western European nations working to form a single market for their economic resources. The members of the EU have abolished all tariffs affecting trade among themselves. They have also set up a common tariff on goods imported from other countries.

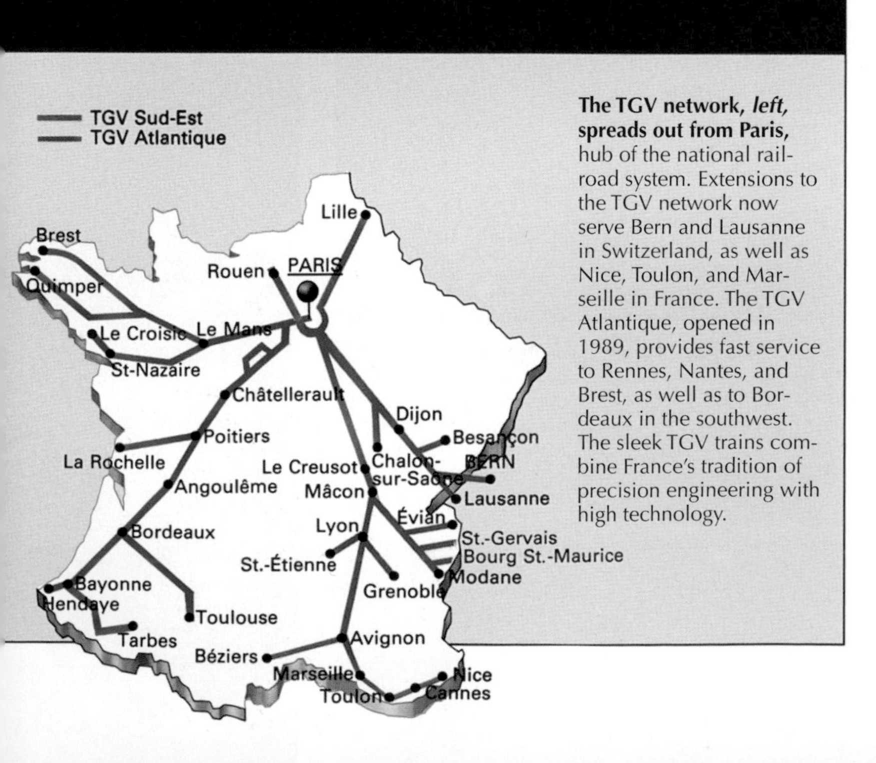

The TGV network, ***left,*** **spreads out from Paris,** hub of the national railroad system. Extensions to the TGV network now serve Bern and Lausanne in Switzerland, as well as Nice, Toulon, and Marseille in France. The TGV Atlantique, opened in 1989, provides fast service to Rennes, Nantes, and Brest, as well as to Bordeaux in the southwest. The sleek TGV trains combine France's tradition of precision engineering with high technology.

The Arts

Throughout the ages, France has given the world some of its most brilliant artists, composers, and writers. That tradition continues today. French culture is a source of national pride and a symbol of the quality of life. Artistic expression is encouraged by the government through massive funding projects.

In recent years, the French government has sponsored the *Maisons de la Culture* (Cultural Centers). These multipurpose art centers, located in 15 cities throughout France, are designed to bring culture to the provinces.

Early art movements

During the Middle Ages, France's major contributions to the arts were the magnificent Gothic cathedrals built from about 1150 to 1300. The most important examples of these huge churches include the Cathedral of Notre Dame in Paris, as well as cathedrals in Amiens, Chartres, Reims, and Rouen.

Great triumphs in French architecture continued through the Renaissance, which reached its height in the 1400's and 1500's. The most impressive châteaux of the Loire Valley were built during this time. Literature also flourished during the French Renaissance. The writer Michel de Montaigne, for example, is considered by many to be the creator of the personal essay as a literary form in the late 1500's.

Versailles and after

Baroque and rococo art developed in France during the 1600's and 1700's. Baroque art was large in scale and dramatic, while rococo art was smaller and more delicate.

France's greatest monument to baroque art is the magnificent Palace of Versailles and its beautiful gardens. Versailles, the royal residence for 100 years, is now a national museum. The palace has about 1,300 rooms, and the grounds cover some 250 acres (101 hectares).

French baroque music found its greatest expression in the operas of Jean Baptiste Lully and Jean Philippe Rameau. François Couperin composed music for the *harpsichord,* a stringed instrument similar to a piano. Classical art and drama also flourished during this period. Molière was the greatest writer of French comedy. Jean Racine and Pierre Corneille wrote tragedies.

The Hall of Mirrors in the Palace of Versailles, *right,* exemplifies the drama and splendor of baroque art. The graceful design of Versailles and its geometrically arranged gardens also reflect the classical themes of balance and order. Built by Louis XIV during the 1600's, Versailles lies about 11 miles (18 kilometers) southwest of Paris. Many of the rooms have been restored to look as they did when the palace was a royal residence. Visitors may also enjoy the many paintings and sculptures by famous European artists.

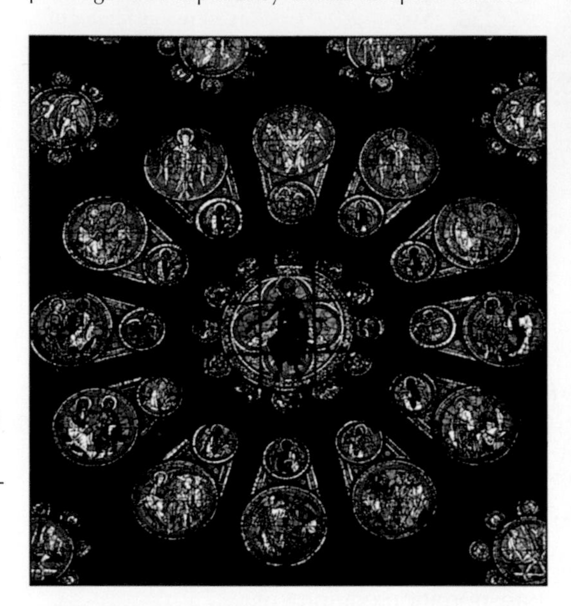

The Rose Windows, ***top,*** **at Chartres Cathedral** in north-central France display the beauty of stained glass. The cathedral, a masterpiece of Gothic architecture, also features hundreds of sculptured religious figures and two bell towers.

The Palace of the Popes, ***above,*** **in Avignon** was built in the 1300's, when Pope Clement V moved the pope's court from Rome to Avignon. Massive towers and walls that are 13 feet (4 meters) thick give the palace the look of a feudal castle.

The Age of Reason, also known as the Enlightenment, was a period of intellectual achievement in the late 1600's and 1700's. Writers such as Voltaire and Jean Jacques Rousseau stressed reason and observation as the best way to learn truth. The Age of Romanticism, from the late 1700's to the middle 1800's, was a reaction against the Enlightenment. The romantics stressed emotions and the imagination.

The great French novelists during the romantic period were Victor Hugo, Honoré de Balzac, Stendhal, and George Sand. The paintings of the romantics were colorful and dramatic. Hector Berlioz was the greatest French romantic composer. He gained fame for his large-scale orchestral works.

The middle and late 1800's brought a movement of realism and naturalism to France. Like many writers of the time, the novelist Gustave Flaubert tried to portray life accurately and objectively. Naturalism, an extreme form of realism, became popular in the vivid writing of novelist Émile Zola.

French painting soared to new heights in the late 1800's and early 1900's during the Age of Impressionism. Impressionist painters tried to capture the immediate impression of an object or event. Leaders of this movement were Edouard Manet, Camille Pissarro, Edgar Degas, Claude Monet, and Pierre Renoir.

The contribution of French artists, writers, and philosophers has continued into this century. Outstanding literary figures during the early and mid-1900's included Paul Claudel, André Gide, Jean-Paul Sartre, and Albert Camus. Georges Braque and Pablo Picasso (who was born in Spain) helped shape modern art. Pierre Boulez and Olivier Messiaen were leaders in experimental music.

An open-air sculpture, *left,* at the Pompidou Center intrigues a young visitor. The center provides a unique setting for exploring the vast collection of the National Museum of Modern Art.

Claude Monet's garden at Giverny, *far left,* in northern France inspired many of his impressionist works. *Water Lilies,* one of a series of paintings based on this garden, greatly influenced later abstract painters.

Corsica

About 100 miles (160 kilometers) off the southeast coast of France lies the island of Corsica—best known as the birthplace of Napoleon Bonaparte. Its mountainous landscape, colorful villages, and ancient ruins have made this Mediterranean island a popular tourist attraction. Corsica is a French island and makes up two of the country's *departments* (administrative districts).

Although Corsica has been a part of France since the 1700's, the islanders do not feel a strong sense of French identity. Most Corsicans speak an Italian dialect rather than French, and the language differences further emphasize the separation between Corsica and mainland France.

Some Corsicans favor independence from France. Many others would prefer greater local control over the island's government. In 1982, the French Parliament created a Corsican regional assembly, elected by island voters. The regional assembly controls local spending, as well as the development of Corsica's economy, education, and culture.

Early history

Tombs, *menhirs* (standing stones), and stone sculptures scattered throughout the island are all that remain of Corsica's prehistoric inhabitants. Beginning in about 560 B.C., the Phoenicians settled in Corsica, naming their new island home *Cyrnos.*

Since these earliest times, Corsica's location in the Mediterranean Sea has made it a place where many different peoples and cultures mixed. The island was conquered in turn by Etruscans, Carthaginians, and Romans. At one time, Corsica was controlled by Charlemagne, the most famous ruler of the Middle Ages.

About 1377, Corsica came under the control of the Italian city of Genoa. The Genoese sold the island to the French in 1768, who lost it to the British in 1794. Two years later, Napoleon sent an expedition to Corsica to reestablish French control. France has held the island ever since, except for a brief occupation by the British in 1814, and occupation by Italians and Germans during World War II (1939–1945).

Land and economy

Corsica is the fourth largest island in the Mediterranean. Its western side consists of barren mountains ending in steep cliffs that plunge into the sea. There are few natural harbors.

Corsica's rocky interior is covered with *macchia* (scrub). A variety of crops grow in its narrow, fertile valleys. Farmers raise olives, grapes and other fruits, grains, vegetables, and tobacco. The many olive trees, orange groves, vineyards, and flowers led Napoleon to once remark about his native island, "I would recognize Corsica with my eyes closed from its perfume alone."

Along the coast, people fish for sardines and hunt for coral. Since World War II, Corsica's main exports have been wool and cheese.

The French island of Corsica, *below,* lies in the Mediterranean Sea between the southeast coast of France and the northwest coast of Italy. Ajaccio, the capital and largest city, is located on the island's western side.

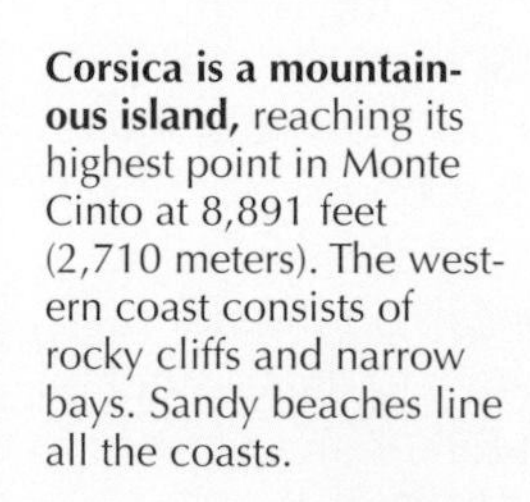

Corsica is a mountainous island, reaching its highest point in Monte Cinto at 8,891 feet (2,710 meters). The western coast consists of rocky cliffs and narrow bays. Sandy beaches line all the coasts.

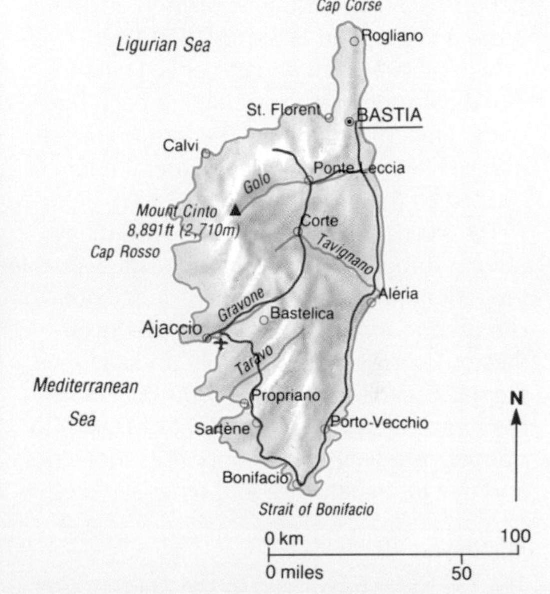

The mountain village of Corte, *above,* in the center of Corsica, was the capital of the island from 1755 to 1769. A fortress built in the 1400's stands high above the town on a rocky ridge. Today, the fortress is occupied by the French Foreign Legion, one of the world's most colorful and gallant fighting forces.

The narrow alleys and cobbled streets of Venzolasca, *right,* are a reminder of Corsica's ancient past. Today, the island attracts many tourists with its mild climate and rugged landscape surrounded by the Mediterranean Sea.

French Guiana

A land of many rain forests and rivers, French Guiana lies on the northeast coast of South America and ranks as the smallest territory on the continent. It is bordered by Suriname on the west, Brazil on the south and east, and the Atlantic Ocean on the north. Unlike its neighbors, French Guiana is not an independent nation, but rather an overseas *department* (administrative district) of France. It has a population of about 187,000.

Like other French departments, French Guiana has one representative in each house of the French Parliament. The territory is administered by a 16-member general council elected by the people.

The interior of French Guiana has many natural resources, including rich soil, valuable timberland, and large deposits of bauxite, an ore used in making aluminum. However, these resources remain largely undeveloped. French Guiana relies heavily on financial assistance from the government of France to operate its government, support its industries, and pay for health care and other services.

History

The French, who arrived in the 1600's, were the first Europeans to settle in what is now French Guiana. The territory became a French colony in 1667 and has been under French control ever since, except for a brief period in the early 1800's when it was governed by British and Portuguese military forces.

France began to send political prisoners to French Guiana in the 1790's, during the French Revolution, and in 1854, Napoleon III established a formal prison system in the colony. Political prisoners were kept on Devils Island, an offshore isle, while other convicts were kept in prison camps in the towns of Kourou and Saint-Laurent. About 70,000 people were held in French Guiana's penal colony between 1852 and 1945, when France closed the prisons, which had become notorious for their cruelty. In the 1960s, France turned the camp at Kourou into a space research center.

People

A strong movement for independence from France developed in the 1980's. But most citizens of French Guiana wish the territory to remain a French department. Today, the French government is helping the department develop its economy and improve the life of its people.

About 90 per cent of the people in French Guiana are blacks or Creoles, and most are descendants of slaves who were brought to the territory during the 1600's and 1700's. In French Guiana, unlike other South American countries, the word *Creole* is used to describe a person of mixed black and white ancestry.

Many other groups, including American Indians, Chinese, Europeans, Haitians, Indochinese, Lebanese, and Syrians, also live in French Guiana. Although the Indians were living in the territory long before the first Europeans arrived, only a few descendants of these people remain. Today, most Indians live in the rain forests of the interior. A few thousand Maroons, who are descendants of Black Africans who escaped from slavery, also live in the rain forests and follow African tribal customs.

French is the official language of the department, but many Creoles also speak a dialect that is a mixture of French and English.

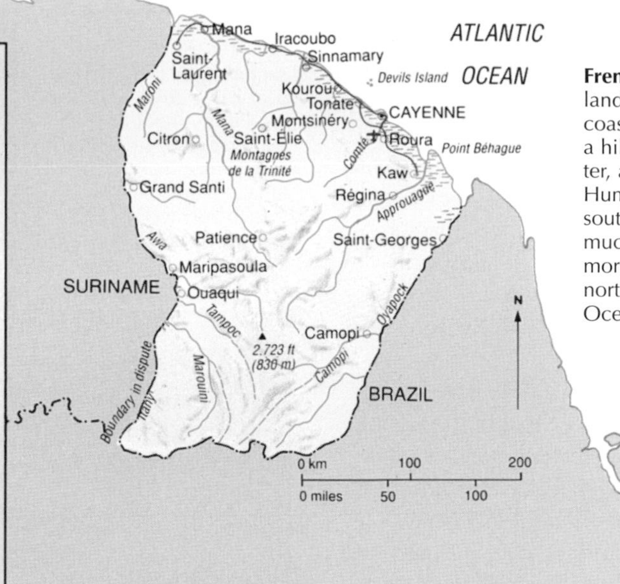

French Guiana's three land regions consist of the coastal plain in the north, a hilly plateau in the center, and the Tumuc-Humac Mountains in the south. Rain forests cover much of the land, and more than 20 rivers flow northward to the Atlantic Ocean.

Factory workers process the products of the palm tree, *far left,* which grows in French Guiana's rain forests. In addition to fruits, timber, and fibers for making rope, the palm tree yields oils and starches used in food production.

A prison barracks still stands on the shores of Devils Island, *left,* which served as a penal colony between 1854 and 1945. The prison camps were known for their cruelty to prisoners.

Buildings in the French country style line a street in Cayenne, the capital and largest city of French Guiana. Founded by the French in 1643, the city is home to about 38,000 people.

French Polynesia

The French overseas territory called French Polynesia lies about 2,800 miles (4,500 kilometers) south of Hawaii. The territory is made up of about 120 Pacific islands scattered over an area about the size of Western Europe. The islands are divided into the Austral, Gambier, Marquesas, Society, and Tuamotu island groups. Papeete, on Tahiti—one of the Society Islands—is the territory's capital.

French Polynesia has a population of about 270,000 people. Most of the people are Polynesians. French Polynesians elect representatives to the French Parliament and vote in France's presidential elections. Tourism is the territory's major industry, but agriculture and fishing are also important economic activities. The chief products include coconuts, pearls, and tropical fruits.

The Marquesas

About 10 volcanic islands in the French territory make up the Marquesas, which lie about 900 miles (1,400 kilometers) northeast of Tahiti. The total area of the Marquesas is 492 square miles (1,274 square kilometers). Most of the islands have steep mountains that drop sharply to the sea and fertile valleys with many streams and waterfalls.

About 8,700 people, mainly Polynesians, live on the Marquesas. Most provide their own food by farming or fishing. The chief crops are bananas, breadfruit, coconut, sweet potatoes, and taro. Copra (dried coconut meat) is the main export.

The Society Islands

This group of 14 islands lies southwest of the Marquesas. The island group's ancient volcanoes form many high peaks, making the land rough and mountainous. Some of the islands are low atolls used as fishing centers. The Society Islands cover an area of 613 square miles (1,587 square kilometers), and have a population of about 215,000.

Tahiti and Raiatea are the largest islands of the group, and Tahiti is the most populous. The capital of the Society Islands is the busy seaport of Papeete on Tahiti.

An outrigger canoe floats over living coral and colorful fish in Bora-Bora's coral lagoon. Agriculture and fishing, mainly for tuna, provide most of French Polynesia's exports.

Stilt houses going up in Bora-Bora's lagoon are part of a tourist resort. Tourism is the territory's major industry.

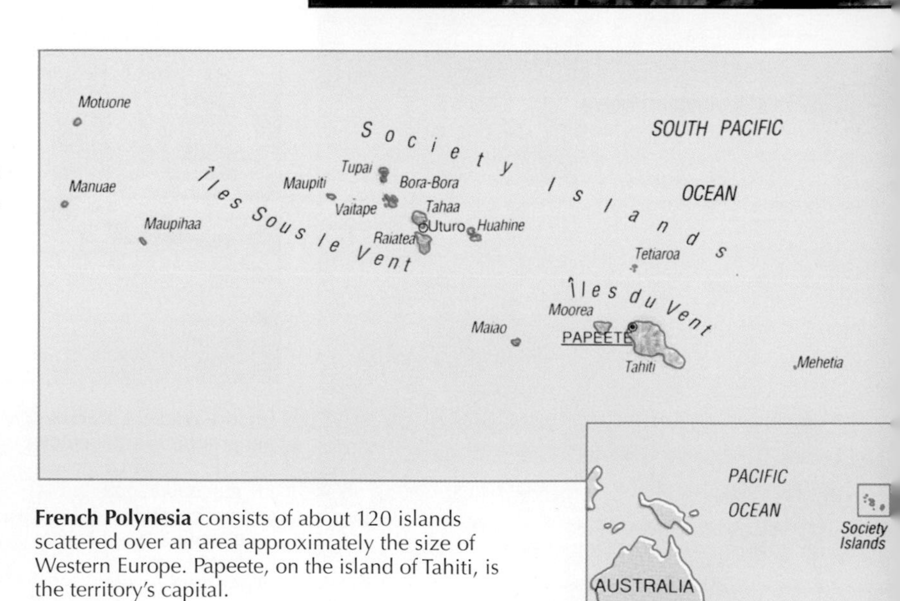

French Polynesia consists of about 120 islands scattered over an area approximately the size of Western Europe. Papeete, on the island of Tahiti, is the territory's capital.

Storm clouds encircle craggy peaks, *above,* and rain forest covers steep slopes on Mooréa, a volcanic island northwest of Tahiti.

Polynesians, whose ancestors migrated from Asia hundreds of years ago, make up the Society Islands' population. Today, many French military personnel also live in the territory.

One of the most picturesque islands of this group is Bora-Bora. It lies about 170 miles (270 kilometers) northwest of Tahiti. A barrier reef with a number of low islets encircles it.

Most of the people are Polynesians or have mixed Polynesian and European ancestry. A Chinese population of several thousand controls much of the retail and shipping trade on Tahiti. There and on the other Society Islands, the tourist industry is the major source of income. Many rural people make their living by farming, fishing, and diving for pearls.

The Tuamotu Islands

The Tuamotu group, made up of 75 reef islands and atolls, stretches across almost 1,000 miles (1,600 kilometers) of the Pacific Ocean. The islands themselves cover about 300 square miles (775 square kilometers), and have a population of about 16,000 Polynesian people. Pearl and copra are the islanders' chief sources of income.

The French government began testing nuclear weapons in Muroroa, one of Tuamotu's atolls, in 1965. Between 1975 and 1989, France performed 110 underground nuclear tests in the area. Scientific inspections in 1983 and 1988 revealed that the explosions caused environmental damage. As a result, many Pacific nations and environmental groups have protested the testing.

Pitcairn Island

Pitcairn, the main island of a British dependency called the Pitcairn Islands group, was the home of the infamous mutineers from the British naval ship *Bounty.* In 1790, 9 British sailors from that ship settled on Pitcairn with 19 Polynesians—6 men, 12 women, and a young girl. About 50 people—mainly descendants of the *Bounty* mutineers and their Polynesian wives—live on the island.

The island, which covers only about 2 square miles (5 square kilometers), rises sharply from the sea. Its interior is rugged, but Pitcairn has fertile soil and most of the people farm and fish for a living. Adamstown is the island's only settlement.

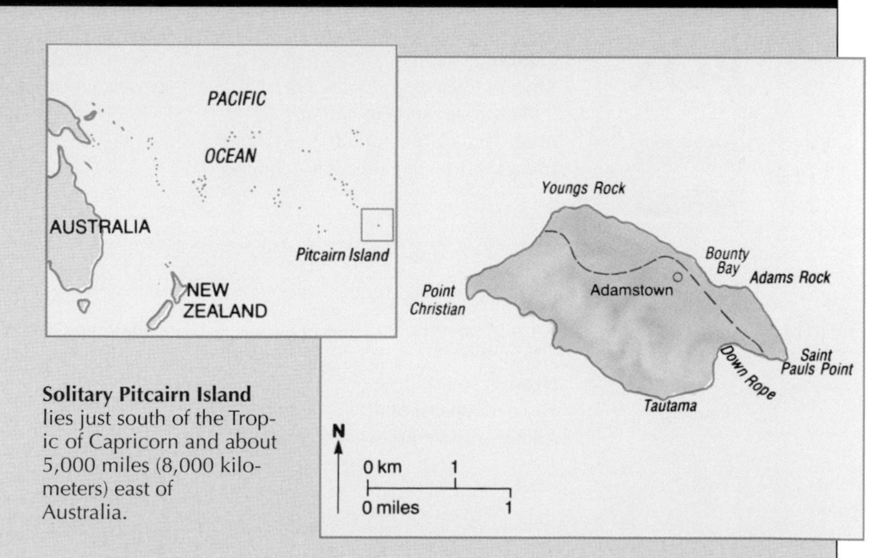

Solitary Pitcairn Island lies just south of the Tropic of Capricorn and about 5,000 miles (8,000 kilometers) east of Australia.

Gabon

The small, forested country of Gabon is one of the most thinly populated nations in Africa. Most of Gabon's people are farmers who live in small villages on the Atlantic coast, along the Ogooué River or one of its branches, or in the less forested north. One of the inland river towns, called Lambaréné, became known worldwide as the home of Albert Schweitzer—an acclaimed Alsatian doctor and missionary who built a hospital and leper colony near there in the early 1900's.

Clad in ceremonial dress, a musician sounds a fanfare on a holiday celebrating Gabon's independence.

Most of the people of Gabon once lived in houses with walls made of mud-covered branches and thatched roofs. Today, many houses have metal roofs, and Gabonese families try to save enough money to build homes with cement walls. Each Gabonese village usually has a meeting place where the older men gather to visit and discuss community affairs.

The people of Gabon represent many black African ethnic groups. The most important group, the Fang, live in the north and were once fierce warriors, feared by other African ethnic groups as well as by Europeans. Today, the Fang dominate national politics.

A small but important group of related peoples are the Omyéné, who live along the coast of Gabon and were the first to meet and deal with European traders and missionaries. These activities gave the Omyéné an early advantage in trade and education in the country.

Small groups of Pygmies hunt and trap animals for food in the southern forests of Gabon. The Pygmies were one of the first peoples to live in what is now Gabon.

The first Europeans to reach the region were Portuguese sailors who landed on the coast sometime in the 1470's. Europeans carried on a slave trade with the Omyéné people who lived along the coast, and in 1839 France established a naval and trading post there.

Soon, missionaries arrived and opened schools. In 1849, a group of former slaves landed at the trading post, which was then named *Libreville* (Free Town) and is now the capital of Gabon.

The French also explored the inland rivers and developed a trade in lumber. In 1910, Gabon became part of French Equatorial Africa, and French companies acquired a great deal of land—as well as complete control over Gabon's foreign trade and forest products.

After World War II (1939–1945), Gabon began to move toward independence, and on

FACT BOX

COUNTRY

Official name: Republique Gabonaise (Gabonese Republic)
Capital: Libreville
Terrain: Narrow coastal plain; hilly interior; savanna in east and south
Area: 103,347 sq. mi. (267,668 km^2)
Climate: Tropical; always hot, humid
Main river: Ogooué and its branches
Highest elevation: Mont Iboundji, 5,167 ft. (1,575 m)
Lowest elevation: Atlantic Ocean, sea level

GOVERNMENT

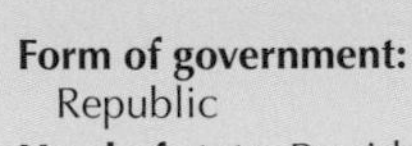

Form of government: Republic
Head of state: President
Head of government: Prime minister
Administrative areas: 9 provinces
Legislature: Senate with 91 members serving six-year terms and the Assemblee Nationale (National Assembly) with 120 members serving five-year terms
Court system: Cour Supreme (Supreme Court), Constitutional Court, Courts of Appeal, lower courts
Armed forces: 4,700 troops

PEOPLE

Estimated 2008 population: 1,457,000
Population density: 14 persons per sq. mi. (5 per km^2)
Population distribution: 84% urban, 16% rural
Life expectancy in years:
Male: 56
Female: 58
Doctors per 1,000 people: 0.3
Percentage of age-appropriate population enrolled in the following educational levels:
Primary: 132*
Secondary: 51
Further: N/A

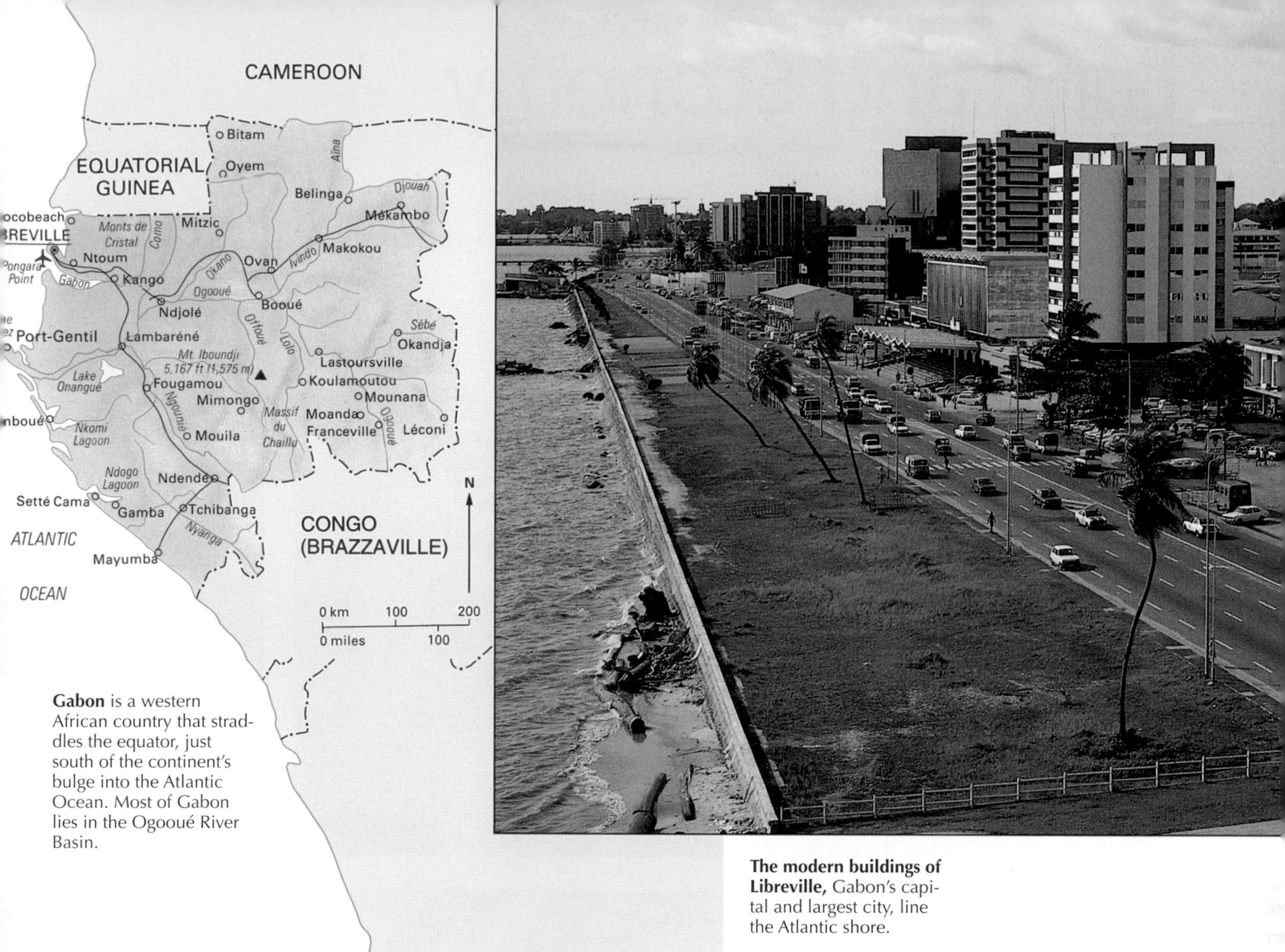

Gabon is a western African country that straddles the equator, just south of the continent's bulge into the Atlantic Ocean. Most of Gabon lies in the Ogooué River Basin.

The modern buildings of Libreville, Gabon's capital and largest city, line the Atlantic shore.

Languages spoken: French (official), Fang, Myene, Bateke, Bapounou/Eschira, Bandjabi

Religions:
Christian 55%-75%
Muslim less than 1%
Animist

*Enrollment ratios compare the number of students enrolled to the population which, by age, should be enrolled. A ratio higher than 100 indicates that students older or younger than the typical age range are also enrolled.

TECHNOLOGY

Radios per 1,000 people: 488

Televisions per 1,000 people: 308

Computers per 1,000 people: 22.4

ECONOMY

Currency: Cooperation Financiere en Afrique centrale franc

Gross domestic product (GDP) in 2004: $7.966 billion U.S.

Real annual growth rate (2003–2004): 1.9%

GDP per capita (2003–2004): $5,900 U.S.

Goods exported: Crude oil 75%, timber, manganese, uranium

Goods imported: Machinery and equipment, foodstuffs, chemicals, petroleum products, construction materials

Trading partners: United States, France, China, Cameroon

Aug. 17, 1960, it became a free nation. Leon Mba, who had headed the government since 1957, became president.

In January 1964, Mba dissolved the national legislature; he was arrested by army officers a month later in an attempt to overthrow his government. But French troops came to Mba's aid, crushed the revolt, and restored Mba to power. Mba was reelected in 1967, but he died that same year. He was succeeded by Vice President Bernard-Albert Bongo, who was elected in 1973 and again in 1979, 1986, 1993, and 1998. During that time, Bongo changed his name to El Hadj Omar Bongo.

Today, the people elect the president for a seven-year term, and the president appoints a council of ministers to help run the government. The president also appoints a governor for each of the country's nine provinces.

Land and Economy

Gabon lies directly on the equator, and much of its land is covered with dense forests. About 3,000 plant species, including 300 species of trees, thrive in the hot, humid tropical climate.

The average annual temperature in Gabon is about 79° F. (26° C). Rainfall is heavy throughout Gabon, especially along its northern coast on the Atlantic. There, the cold Benguela Current from the south meets the warm Guinea Current from the north, condensing moisture in the air into rain.

Much of the coast is lined with palm-fringed beaches, swamps, and lagoons, and large areas are covered with tall papyrus grass. Inland, the land gradually rises to forested rolling hills and low mountain ranges. Most of Gabon lies in the basin of the Ogooué River, which has cut many valleys through these highlands.

Agriculture and forestry

The majority of the Gabonese people live along the coast or the rivers and make their living by farming. They clear the forest around their villages and plant bananas, cassava, and yams—their main food crops. They also grow mangoes, oranges, and pineapples. Some farmers raise livestock, and many catch fish in the rivers or hunt animals in the forests. On the fertile land of northwestern Gabon, many farmers grow cacao and coffee for export.

Heavily forested Gabon is one of the richest countries in Africa in terms of resources. Its forests have long been its chief source of wealth, providing high-quality lumber for more than 100 years.

The lumber trade was developed in the mid-1800's when the French paddled up the Ogooué River into the interior. French companies eventually acquired control of the Gabonese forestry industry, and even now, much of Gabon's economy is controlled by French companies.

Today, lumber is the country's main export. Wood from the huge okoumé trees that grow in Gabon is used to make plywood, and the lush forests also produce ebony and mahogany, valuable dark hardwoods.

A Gabonese man loads cassava onto a small boat at Port-Gentil. Cassava is one of the country's major food crops, but very little is grown for export.

The Trans-Gabon Railway, *right,* was built to help the development of the mining industry in Gabon. It connects Franceville in the interior with the port of Owendo.

Port-Gentil, a major Gabonese port, lies on the Gulf of Guinea, an arm of the Atlantic Ocean, near the mouth of the Ogooué River. The river is still one of the chief means of transportation in the country.

Thick forests hug the beach in many coastal areas of Gabon, *right.* Such hardwoods as ebony and mahogany thrive in the hot, rainy equatorial climate.

Mining

Gabon has some of the richest deposits of iron and manganese in the world, and mining is becoming increasingly important to the Gabonese economy. In addition to iron and manganese, uranium and petroleum are now exported. Gabon is a member of the Organization of Petroleum Exporting Countries (OPEC).

An improving standard of living

In the past, most of the Gabonese people were poor, and their standard of living was correspondingly low. Since the mid-1900's, however, the development of their country's mineral resources has helped raise living standards by creating jobs, and the income from mining has helped pay for social services throughout the country.

Unlike many African nations, the great majority of children in Gabon go to elementary school, and the number attending secondary school is increasing rapidly. Since the early 1950's, the number of schools run by the government as well as by churches has also increased considerably, including a technical school in Libreville and an agricultural school in Oyem. Today, about two-thirds of the Gabonese people can read and write.

Roads built during the mid-1900's link various parts of Gabon, but one of the chief means of transportation is still the Ogooué River. In the 1970's and 1980's, the Trans-Gabon Railroad was built, linking the port of Owendo with the interior. The railroad will eventually help Gabon mine rich mineral deposits in remote areas. At present, some of the outlying regions in the north, east, and southwest must export their products through the neighboring countries of Cameroon and Congo.

Libreville, the capital city, is also a major port and the center of the country's commerce and culture. Since the 1970's, a convention center and many other buildings have been erected in the city. And, although manufacturing in Gabon is limited, factory workers in Libreville produce food products, furniture, lumber, and textiles.

Felled by a forester, one of the huge tropical trees that grow in Gabon crashes to the ground, *above*. Lumber has been the country's major export and its chief source of wealth for more than a century.

Gambia

Gambia is one of the smallest nations in Africa. The country lies on a narrow strip of land only 15 to 30 miles (24 to 48 kilometers) wide and 180 miles (290 kilometers) long on either side of the Gambia River. Except for its short Atlantic coast, Gambia is entirely surrounded by Senegal.

Government and economy

Gambia is a republic and a member of the Commonwealth of Nations. The people elect a president to a five-year term. They also elect 45 members of the National Assembly. The president appoints an additional 4. The president also appoints members of the Cabinet.

Gambia is a poor country with no valuable mineral deposits and little fertile soil. Mangrove swamps line the banks of the Gambia River from the coast to the center of the country. Beyond these swamps lie the *banto faros,* areas of ground that are firm during the dry season but turn into swamps when it rains. Near the ocean, the river water mixes with salt water and floods the banto faros, ruining the soil for crops. But farther inland, areas that the river floods with fresh water are used to raise rice.

Beyond the banto faros lie sandy plateaus, where farmers raise rice and peanuts. Rice is grown mainly by Gambian women as food for their families. Most Gambian farmers earn their living by raising peanuts, and peanut processing is one of Gambia's most important industries. Fishing and tourism also make important contributions to the economy.

People

Almost all Gambians are black Africans and Muslims. They belong mainly to five ethnic groups: the Mandinka, the Fula (or Fulani), the Wolof, the Serahuli, and the Jola.

The Mandinka live throughout Gambia and work as traders and peanut farmers. Most of the Fula live in eastern Gambia and raise cattle. Many of the Fula are nomadic people. The Wolof are mainly northern farmers or residents of Banjul, the capital. Most of the Serahuli are farmers in the east, where the soil is poor and agriculture is difficult.

The Jola, once bitter enemies of the Mandinka, have lived in Gambia longer than the other ethnic groups. The Jola are southeastern farmers who live in small villages surrounded by earthen walls.

FACT BOX

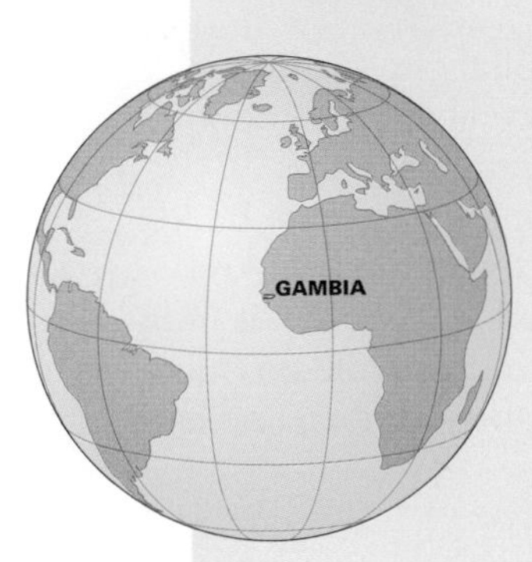

COUNTRY

Official name: Republic of the Gambia
Capital: Banjul
Terrain: Flood plain of the Gambia River flanked by some low hills
Area: 4,361 sq. mi. (11,295 km^2)
Climate: Tropical; hot, rainy season (June to November); cooler, dry season (November to May)
Main river: Gambia
Highest elevation: Unnamed location, 174 ft. (53 m)
Lowest elevation: Atlantic Ocean, sea level

GOVERNMENT

Form of government: Republic
Head of state: President
Head of government: President
Administrative areas: 5 divisions, 1 city
Legislature: National Assembly with 49 members serving five-year terms
Court system: Supreme Court
Armed forces: 800 troops

PEOPLE

Estimated 2008 population: 1,582,000
Population density: 363 persons per sq. mi. (140 per km^2)
Population distribution: 50% rural, 50% urban
Life expectancy in years:
Male: 52
Female: 56
Doctors per 1,000 people: Less than 0.05
Percentage of age-appropriate population enrolled in the following educational levels:
Primary: 85
Secondary: 34
Further: N/A

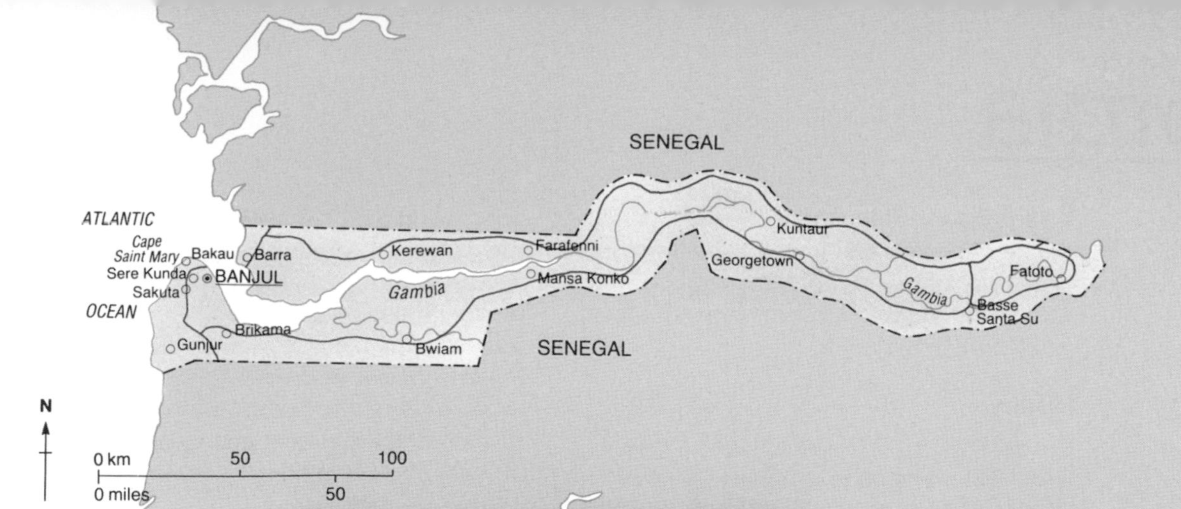

Gambia, *map above,* is a small, independent nation in west Africa. It lies along the Gambia River, and except for its short Atlantic coastline, it is surrounded entirely by Senegal. Most of Gambia's land is flat and sandy, with mangrove and scrub forests along the coast and river.

Passengers board a steamer that carries people and goods on the Gambia River. Small oceangoing ships can sail up the Gambia as far as Kuntaur.

In the shade on a quiet street in Banjul, *above right,* the capital of Gambia and its only large town, Gambians take a break from the midday heat. The shop sign in English is an indication of Gambia's official language, but many Gambians speak Wolof or other African languages.

Languages spoken:
English (official)
Mandinka
Wolof
Fula
Other indigenous vernaculars

Religions:
Muslim 90%
Christian 9%
Indigenous beliefs 1%

TECHNOLOGY

Radios per 1,000 people: 394
Televisions per 1,000 people: 15
Computers per 1,000 people: 13.8

ECONOMY

Currency: Dalasi
Gross domestic product (GDP) in 2004: $2.799 billion U.S.
Real annual growth rate (2003–2004): 6%
GDP per capita (2004): $1,800 U.S.
Goods exported: Peanuts and peanut products, fish, cotton lint, palm kernels
Goods imported: Foodstuffs, manufactures, fuel, machinery and transport equipment
Trading partners: Benelux, United Kingdom, Hong Kong, Japan

History

The area that is now Gambia was once part of the powerful west African Mali Empire. The English and Portuguese set up a slave trade in the area during the 1500's. All or part of the region was controlled by the English from the 1660's until 1965, when Gambia became independent.

On April 24, 1970, Gambia became a republic. David (now Sir Dawda, or Dauda) Jawara was elected president and reelected until a group of military officers overthrew him in 1994. They began ruling the country through a military council.

In 1996, Gambia adopted a new constitution, and Yahya Jammeh won presidential elections later that year. Soon after the elections, Jammeh dissolved the military council. In 1997, Gambia elected a National Assembly. Jammeh was reelected in 2001.

Georgia

Georgia, one of the former Soviet republics, is a mountainous nation on the eastern shore of the Black Sea. It is bordered by Turkey and Armenia in the south, Azerbaijan in the east, and Russia in the north.

History

Wars and invasions mark much of Georgia's long history. The first Christian state was established in Georgia in the A.D. 300's. In the 1200's, the country was conquered by the Mongols, and in the 1300's by the fierce Tatar warrior called Tamerlane. From the 1500's to the 1700's, the Ottoman Empire and Iran struggled for control of the territory.

Between 1801 and 1829, Russia acquired all of Georgia. In 1918, after the October Revolution, the Georgian Menshevik Party proclaimed Georgia's independence. Although the Soviet government in Moscow recognized the declaration in 1920, the Red Army invaded Georgia in 1921, and it became a Soviet republic. In 1922, Georgia and the Soviet republics of Armenia and Azerbaijan formed the Transcaucasian Federation. The federation was one of the four original republics that joined to form the Soviet Union in 1922. In 1936, Georgia became a separate Soviet republic.

In the late 1980's, a strong independence movement emerged in Georgia. In April 1991, the new parliament declared Georgia independent and called for a gradual separation from the Soviet Union. Following the political upheaval in the Soviet central government in 1991, the Soviet Union was dissolved in December and Georgia became an independent nation.

Land and people

Georgia's varied landscape ranges from snow-capped mountains and dense forests to fertile valleys and warm, humid coastlands along the Black Sea. The Great Caucasus Mountains extend across northern Georgia, and the Little Caucasus Mountains cover the south. In the central lowlands, vineyards produce grapes for fine wines. Tea and citrus fruits are grown along the coastlands.

While food processing is Georgia's chief industry, engineering and chemical industries developed rapidly in the early 1980's. Scientific research also flourished in the area.

About 70 per cent of the people are ethnic Georgians, and the remaining population is mainly Armenian, Abkhazian, Azerbaijani, Ossetian, and Russian.

FACT BOX

COUNTRY

Official name: Sak'art'velo (Georgia)
Capital: T'bilisi
Terrain: Largely mountainous; lowlands open to the Black Sea in the west; good soils in river valley flood plains and the foothills of Kolkhida Lowland
Area: 26,911 sq. mi. (69,700 km²)
Climate: Warm and pleasant; Mediterranean-like on Black Sea coast
Main rivers: Kur, Rioni
Highest elevation: Mount Shkhara 17,163 ft. (5,201 m)
Lowest elevation: Black Sea, sea level

GOVERNMENT

Form of government: Republic
Head of state: President
Head of government: Prime minister
Administrative areas: 53 rayons, 9 k'alak'ebi (cities), 2 avtomnoy respubliki (autonomous republics)
Legislature: Umaghiesi Sabcho (Supreme Council or Parliament) with 235 members serving four-year terms
Court system: Supreme Court, Constitutional Court
Armed forces: 17,770 troops

PEOPLE

Estimated 2008 population: 4,421,000
Population density: 164 persons per sq. mi. (63 per km²)
Population distribution: 52% urban, 48% rural
Life expectancy in years:
Male: 68
Female: 75
Doctors per 1,000 people: 3.9
Percentage of age-appropriate population enrolled in the following educational levels:
Primary: 90
Secondary: 80
Further: 38

Political problems

In 1991, Georgia elected Zviad K. Gamsakhurdia president. In January 1992, opposition leaders formed an alternate government, and Gamsakhurdia fled the country. Eduard A. Shevardnadze became president.

Conflicts between the Georgian government and rebels in South Ossetia erupted in the early 1990's. South Ossetia is a self-governing district of Georgia which had declared independence in 1990. Both sides agreed to a cease-fire in 1992.

Also in 1992, secessionist rebels in the autonomous Black Sea region of Abkhazia staged a revolt when the Georgian government sent troops to crush demands for greater sovereignty. A cease-fire was declared in 1993 but disagreement continued over Abkhazia's status. Fighting erupted again in the early 2000's.

In October 1993, Shevardnadze signed a decree ratifying Georgia's entry into the Commonwealth of Independent States (CIS). Abkhazia and South Ossetia both held elections in 1996; UN and Georgian officials called the elections invalid. In 1995, a new constitution went into effect. Voters elected Shevardnadze president and reelected him in 2000. Protests sparked by rigged elections forced him to resign in 2003. In 2004, Mikhail Saakashvili, who had led the oposition to Shevardnadze, was elected president.

A sea of poppies carpets the plains of Georgia. Farmers raise tobacco and grapes in the inland valleys, which are much drier than the coastlands.

Languages spoken:
Georgian 71% (official)
Armenian 8%
Russian 6%
Azeri 6%
Ossetian 3%
Abkhaz 2%

Religions:
Georgian Orthodox 65%
Muslim 11%
Russian Orthodox 10%
Armenian Apostolic 8%

TECHNOLOGY

Radios per 1,000 people: 568

Televisions per 1,000 people: 357

Computers per 1,000 people: 31.6

ECONOMY

Currency: Lari

Gross national income (GNI) in 2000: $3.2 billion U.S.

Real annual growth rate (1999–2000): 1.9%

GNI per capita (2000): $630 U.S.

Balance of payments (2000): -$162 million U.S.

Goods exported: Citrus fruits, tea, wine, other agricultural products; diverse types of machinery and metals; chemicals; fuel reexports; textiles

Goods imported: Fuel, grain and other foods, machinery and parts, transport equipment

Trading partners: Russia, Turkey, Azerbaijan, European Union

Georgia lies mostly in southwestern Asia, but part of northern Georgia is located in Europe.

Germany

Germany's geographical location in central Europe has drawn many different peoples throughout history. The nation has no natural barriers to the east or west, and migrating groups often crossed the seas to the north. As a result, Germany has long served as a meeting place for the exchange of social, economic, and intellectual ideas. The country has also witnessed a long history of political conflicts.

From 1949 to 1990, Germany was divided into the Federal Republic of Germany (West Germany) and the German Democratic Republic (East Germany). West Germany became a parliamentary democracy with strong ties to Western Europe and the United States. East Germany became a Communist dictatorship closely associated with the Soviet Union.

In 1989, however, reform movements calling for increased political and economic freedom swept through the Communist nations of Europe. In response to these protests, the Berlin Wall, a symbol of the East German government's control of its citizens, was opened in November 1989. Also, non-Communist political parties were permitted to organize and form their own policies in late 1989. Free parliamentary elections were held in March 1990, and non-Communists gained control of the East German government. In July 1990, East and West Germany united their economies into a single system, and on Oct. 3, 1990, the two countries were unified into a single nation. Berlin was made the capital of the united Federal Republic of Germany.

Unification brought many changes to the German people. For example, the East German government had restricted travel between East and West Germany, and erected a heavily guarded border between the two countries, separating many relatives and friends. Today, all Germans share the same freedoms, including freedom of speech and press and the freedom to travel throughout a single Germany undivided by borders and guards.

The German people have a reputation for being hardworking and disciplined, but they are also known for their love of music, dancing, good food, and good fellowship. People from all over the world come to Germany to participate in the country's lively festivals. Many Germans also enjoy visiting the nation's scenic mountain areas and such popular tourist attractions as Neuschwanstein Castle.

The German people have made many important contributions to culture. Germany has produced such famous composers as Johann Sebastian Bach and Ludwig van Beethoven, and such literary giants as Johann Wolfgang von Goethe and Thomas Mann.
In addition, German scientists have made significant breakthroughs in chemistry, medicine, and physics.

Germany Today

With the merging of the two countries' economic, legal, and social systems, the unification of East Germany and West Germany was officially completed on Oct. 3, 1990. Unity Day was welcomed with a mixture of joy and apprehension by the German people. At midnight, an estimated 1 million people gathered in Berlin to celebrate the unification. Some people, however, concerned about the costs of modernizing East Germany's crumbling economy, believed the occasion called for more solemn contemplation.

International reaction was also mixed. While many world leaders declared their hope for a more united Europe, others recalled Germany's violent history and expressed anxiety over the unification, fearing the future military strength of a united Germany. Aware of such concerns, German leaders from both the East and West pledged that Germany would serve the world in a "peace-loving, freedom-loving" role.

Political unity

On Dec. 2, 1990, the first national elections after unification brought an easy victory to Helmut Kohl, who had led the drive for unity, and his party, the Christian Democratic Union. Kohl had served as chancellor of West Germany from 1982 to 1990 and had continued to serve as chancellor after East Germany had reunited with the West in October. Free elections had not been held in a united Germany since 1932, when Adolf Hitler came to power.

A unified economy

One of the first steps taken toward unification was the union of the two nations' economies. Beginning on July 1, 1990, East Germans began trading their money for West German *Deutsche marks.* The Deutsche mark became the unit of currency throughout Germany, and East Germany began to operate under a free enterprise system.

Economic unification created some problems in Germany, particularly for East Germans. Because the East German government no longer controlled prices, the cost of many goods rose sharply. In addition, many East German businesses could not operate profitably without the government's support. Some of these businesses closed and others reduced their hours, which increased unemployment.

Before unification, West Germany was one of the world's leading industrial nations. Many economists believe the economy of the united Germany will remain strong.

FACT BOX

COUNTRY

Official name: Bundesrepublik Deutschland (Federal Republic of Germany)
Capital: Berlin
Terrain: Lowlands in north, uplands in center, Bavarian Alps in south
Area: 137,847 sq. mi. (357,022 km^2)
Climate: Temperate and marine; cool, cloudy, wet winters and summers; occasional warm foehn wind
Main rivers: Rhine, Danube, Elbe, Weser, Oder
Highest elevation: Zugspitze, 9,721 ft. (2,963 m)
Lowest elevation: Sea level along the coast

GOVERNMENT

Form of government: Federal republic
Head of state: President
Head of government: Chancellor
Administrative areas: 16 Laender (states)
Legislature: Parlament (Parliament) consisting of the Bundestag (Federal Assembly) with 669 members serving four-year terms and the Bundesrat (Federal Council) with up to 68 members
Court system: Bundesverfassungsgericht (Federal Constitutional Court)
Armed forces: 285,000 troops

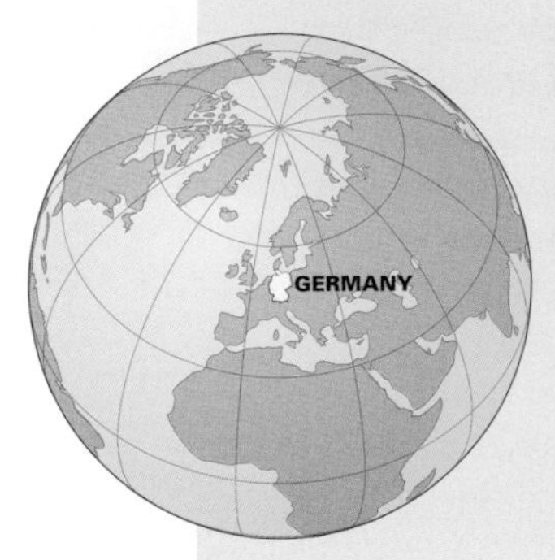

PEOPLE

Estimated 2008 population: 82,414,000
Population density: 598 persons per sq. mi. (231 per km^2)
Population distribution: 88% urban, 12% rural
Life expectancy in years:
Male: 75
Female: 81
Doctors per 1,000 people: 3.6
Percentage of age-appropriate population enrolled in the following educational levels:
Primary: 100*
Secondary: 100
Further: 49

The government of united Germany, officially called the Federal Republic of Germany, was established after unification and based on the democratic government system of West Germany. East Germany also adopted West Germany's economic system. In spite of a difficult transition period, many economists believe that membership in the European Union (EU) will strengthen united Germany's economy.

ECONOMY

Currency: Euro

Gross national income (GNI) in 2000: $2,063.7 billion U.S.

Real annual growth rate (1999–2000): 3.0%

GNI per capita (2000): $25,120 U.S.

Balance of payments (2000): -$18,707 million U.S.

Goods exported: Machinery, vehicles, chemicals, metals and manufactures, foodstuffs, textiles

Goods imported: Machinery, vehicles, chemicals, foodstuffs, textiles, metals

Trading partners: European Union, United States, Japan

Language spoken: German

Religions:
Protestant 34%
Roman Catholic 33%
Muslim 4%

*Enrollment ratios compare the number of students enrolled to the population which, by age, should be enrolled. A ratio higher than 100 indicates that students older or younger than the typical age range are also enrolled.

TECHNOLOGY

Radios per 1,000 people: 570

Televisions per 1,000 people: 675

Computers per 1,000 people: 484.7

History: 300 to 1945

About 1000 B.C., warlike tribes began to migrate from northern Europe into what is now Germany. In the A.D. 400's, these Germanic tribes had pushed farther south and plundered the city of Rome. They eventually carved up the western portion of the Roman Empire into tribal kingdoms, of which the kingdom of the Franks became the largest and most important. In 800, the most famous Frankish ruler, Charlemagne, was crowned emperor of the Romans.

In 919, the Saxon *dynasty* (a series of rulers from the same family) came to power over the part of Charlemagne's empire that lay east of the Rhine River. It ruled until 1024. The third Saxon ruler, Otto I, extended the borders of his kingdom and, in 962, was crowned emperor in Rome. This marked the beginning of the Holy Roman Empire.

Although Frederick I—called *Barbarossa* or *Red Beard*—had further expanded the empire's borders by 1152, a string of generally weak rulers after the Saxons reduced the emperor's power, and by the 1300's the emperor was almost powerless. Beginning in 1438, the Habsburgs—one of Europe's most famous royal families—ruled the Holy Roman Empire almost continuously until 1806. The center of Habsburg power lay in what is now Austria.

1500–1648

In 1517, the writings of a German monk named Martin Luther sparked the Protestant Reformation, which quickly spread through Europe. Luther protested some practices of the Roman Catholic Church and called for reforms. By 1600, the German lands were divided by political and religious rivalries.

In 1618, a Protestant revolt set off a series of wars that lasted for 30 years. The Peace of Westphalia ended the Thirty Years' War in 1648. Under this treaty, Germany became a collection of free cities and hundreds of states.

1648–1945

In the mid-1600's, Frederick William of the Hohenzollern family began to add lands to his state in the Berlin area. In 1701, his son, Frederick I, was given the title king of Prussia.

800 Charlemagne is crowned emperor of the Romans.
919 Saxon dynasty comes to power.
962 Beginning of the Holy Roman Empire.
1033 Burgundy unites with Holy Roman Empire.
1152 Frederick I—also called *Barbarossa* or *Red Beard*—expands the empire's borders.
1517 Martin Luther posts his Ninety-Five Theses in Wittenberg. Protestant Reformation begins in Germany.
1618–1648 Thirty Years' War devastates much of Germany.
1701 Frederick I is crowned king of Prussia.
1763 Prussia under Frederick II (the Great) is recognized as a great power.
1815 Congress of Vienna sets up the German Confederation.
1848–1849 German revolution is defeated.
1862 Otto von Bismarck becomes Prussia's prime minister.
1870–1871 Franco-Prussian War results in Prussian victory and unification of Germany.
1914–1918 The Allies defeat Germany in World War I, and the German Empire ends.
1919 Weimar Republic is established.
1933 Adolf Hitler is appointed chancellor and establishes Nazi dictatorship.
1939–1945 The Allies defeat Germany in World War II.

Adolf Hitler, *right,* shown conferring with Albert Speer, led the National Socialist (Nazi) Party. After the Nazis gained power in 1933, Hitler set up a brutal dictatorship and led Germany to defeat in World War II.

Martin Luther, *left* (1483–1546)

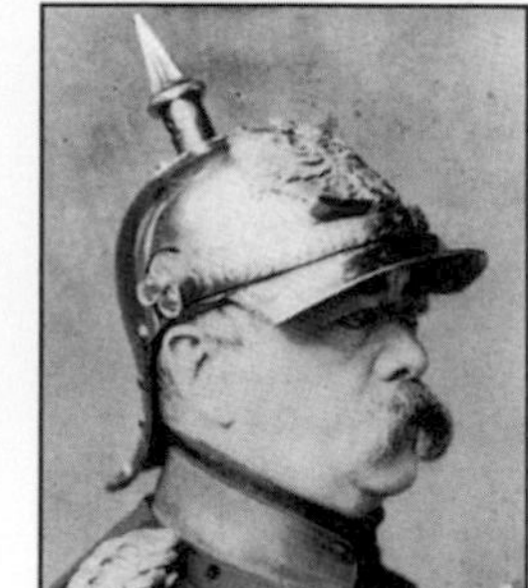

Frederick II, the Great, *far left* (1712–1786)

Otto von Bismarck, *left* (1815–1898)

Wilhelm I was proclaimed the first *kaiser* (emperor) of the German Empire in 1871 in the Hall of Mirrors at the Palace of Versailles. Germany had just defeated France in the Franco-Prussian War and was newly united, under the leadership of the Prussian prime minister, Otto von Bismarck, who became Germany's first chancellor in 1871.

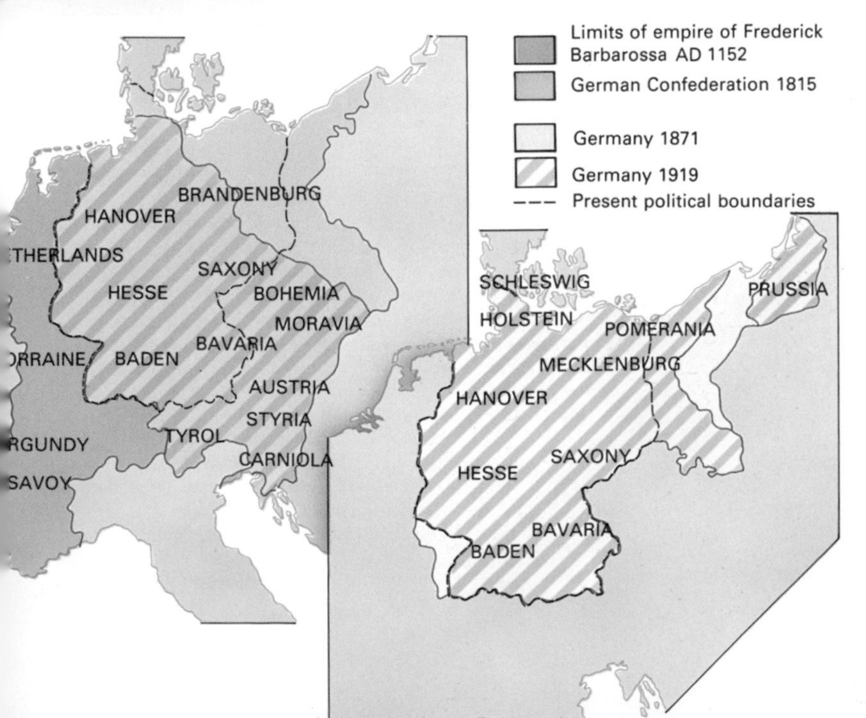

Germany's shifting boundaries from the time of Frederick I, also known as Barbarossa or Red Beard, in 1152 to the 1990's are shown on these maps.

By 1763, Prussia was recognized as a great power.

The French Revolution and the Napoleonic wars caused many changes throughout Europe from 1789 to 1815. In 1815, the powers that had defeated Napoleon met in the Congress of Vienna to restore order. Members of the Congress, including Prussia, divided Napoleon's lands among themselves. They also set up the German Confederation, a union of 39 independent states. Economic and political unrest led many Germans to revolt against the Confederation, but the Revolution of 1848 was defeated, and the German Confederation was reinstated in 1849.

In 1862, Otto von Bismarck became Prussia's prime minister, and in 1871, following the Franco-Prussian War, Bismarck succeeded in uniting Germany. Wilhelm I was crowned the first *kaiser* (emperor) of the new German Empire in 1871, and Bismarck was made chancellor and head of government.

World War I began in 1914. Although Germany and its allies seemed close to victory in the early years, the Allies defeated them in 1918. After the war, under the Treaty of Versailles, Germany lost its colonies and some of its European territory. Rejecting the government that had led them into the war, Germans created the Weimar Republic in 1919, but it was a weak state from the start.

During the political confusion of the 1920's and the early 1930's, the infamous Nazi Party rose to power. In 1933, Adolf Hitler was appointed chancellor of Germany and proceeded to build a dictatorship, which he called the *Third Reich,* or Third Empire. Hitler's determination to assert German superiority over Jews and other non-Germanic peoples, and to expand Germany's territory resulted in World War II (1939–1945).

During the war, the Nazis murdered about 6 million Jews and millions of other conquered people. However, the tide turned against Germany in 1943, and Hitler committed suicide on April 30, 1945, just before Germany's surrender to the Allies on May 7.

History: 1945 to Present

World War II ended in Europe with Germany's unconditional surrender on May 7, 1945. In June, the Allied Big Four—France, Great Britain, the Soviet Union, and the United States—officially took control in Germany, dividing the country into four zones with each power occupying one zone. Berlin, located deep in the Soviet zone, was also divided into four sectors. Some lands in eastern Germany were lost to Poland and the Soviet Union. Although the four powers had agreed to rebuild Germany as a democracy, it soon became clear that the Soviet Union intended to establish a Communist government in its zone. The tensions that subsequently arose between the Western powers and the Soviet Union came to be known as the Cold War.

France, Great Britain, and the United States combined the economies of their zones and prepared to unite them politically, but the Soviet Union kept its zone apart. In June 1948, the Soviets attempted to drive the Western powers from Berlin by blocking all highway, rail, and water routes to the city. The Allies retaliated by setting up a massive airlift that flew tons of supplies to Berlin every day. The Soviets finally lifted the blockade in May 1949, realizing it had failed, and the airlift ended in September.

West and East divided

As a result of the political division between the Eastern and Western zones, Germany was divided into two states in 1949. The three Western zones became the Federal Republic of Germany (West Germany), with Bonn as its capital. The Soviet zone became the German Democratic Republic (East Germany), with East Berlin as its capital. Both states became officially independent from their occupying powers in 1955.

By that time, the amazing economic recovery of West Germany had helped the country achieve both prosperity and political stability. East Germany's economy also recovered under the leadership of Communist Walter Ulbricht, but its stan-

The dramatic opening of the Berlin Wall in 1989 represented the collapse of Communism in Eastern Europe.

May 7, 1945 Germany surrenders, thereby ending World War II in Europe.

June 1945 The Allied powers divide Germany and the city of Berlin into four zones of military occupation.

July-August 1945 Leaders of Great Britain, the Soviet Union, and the United States meet in Potsdam, Germany, where they agree to govern Germany together and to rebuild it as a democracy.

June 1948 The Western Allies begin to rebuild the economy in their occupation zones. Under the Marshall Plan, U.S. aid pours into the Western zone.

1948–1949 A Soviet blockade fails to force the Western Allies out of Berlin.

1949 East Germany and West Germany are established.

1953 The Soviet Union crushes an East German revolt.

1955 East and West Germany are declared independent. West Germany joins the North Atlantic Treaty Organization (NATO), a Western military alliance, and East Germany joins the Warsaw Pact, an Eastern European military alliance.

1961 The East German government erects the Berlin Wall to prevent East Germans from escaping to West Berlin.

1971 West German Chancellor Willy Brandt wins the Nobel Peace Prize for his efforts to reduce tensions between Communist and non-Communist nations. He works to normalize relations between East and West Germany.

1973 East and West Germany ratify a treaty calling for closer relations between the two nations. Both nations join the United Nations (UN).

1989 East German leader Erich Honecker is forced to resign. East Germany opens the Berlin Wall and other border barriers, allowing its citizens to travel freely to West Germany for the first time since World War II.

March 1990 East Germany holds free elections, resulting in the end of Communist rule there.

July 1990 The economies of East and West Germany are united.

October 1990 East and West Germany are unified and become a single nation.

December 1990 Elections held in Germany make Helmut Kohl chancellor of the united country.

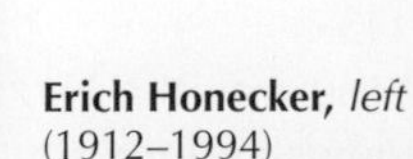

Erich Honecker, *left* (1912–1994)

Willy Brandt, *far left* (1913–1992)

Helmut Kohl, *left* (1930-)

The Brandenburg Gate, in the heart of Berlin, is a symbol of the city. Thousands of people gathered at the gate to celebrate the opening of the Berlin Wall in November 1989 and to welcome the unification of Germany in October 1990.

After World War II, Germany was divided into zones occupied by the Allied powers. Other parts of Germany were lost to Poland and the Soviet Union. In September 1990, the Allies gave up their remaining rights in Germany to help make reunification possible.

Konrad Adenauer, West German chancellor from 1949 to 1963, is seen here at the right of French President Charles de Gaulle in 1963. The two leaders met to sign a Franco-German Treaty of Cooperation.

dard of living remained much lower than West Germany's. Every week, thousands fled to West Germany. Almost 3 million East Germans left, and the country's labor force fell dramatically.

Most refugees fled through Berlin, because the Communists had sealed off the East-West border in the 1950's. In August 1961, the East German government built a wall between East and West Berlin to prevent people from leaving East Germany. The Berlin Wall became a hated symbol of divided Germany.

German reunification

In 1989, in response to public protests, Communist Hungary removed the barriers along its borders with non-Communist Austria. Immediately, thousands of East Germans fled through Hungary to West Germany, and citizens throughout East Germany protested for more freedom. In a dramatic policy change, the East German government responded by opening its own borders on November 9.

In March 1990, free elections took place in East Germany for the first time. Most East Germans voted for non-Communist candidates who favored rapid unification. Most West Germans also supported unification. In July, the economies of the two German states were united, and talks about unification were held between May and September. On August 31, East and West Germany signed a unification treaty that took effect on Oct. 3, 1990.

In November 1990, the German government charged former East German leader Erich Honecker with manslaughter for ordering East German guards to shoot people escaping to West Germany. Honecker's trial was halted due to his ill health. He died in 1994.

After unification, unemployment rose in the eastern states. For Germans who had jobs, taxes rose to pay for reunification. Neo-Nazis and other right-wing Germans attacked foreigners and immigrants, leading to a number of deaths. Despite these developments, German Chancellor Helmut Kohl was narrowly reelected to a fourth term in October 1994.

Environment

Germany is a land of scenic beauty with a varied landscape. The country's five main land regions are the North German Plain, the Central Highlands, the South German Hills, the Black Forest, and the Bavarian Alps.

Northern Germany

The North German Plain—the largest land region—covers all of northern Germany. This low, flat plain slopes down toward the coastal areas of the North and Baltic seas. Broad, slow-moving rivers that flow northward into these seas drain the region. Many of these rivers, such as the Elbe, Ems, Oder, Rhine, and Weser, are important commercial waterways.

The wide river valleys of the North German Plain have soft, fertile soil, but large areas between the valleys are covered with sand and gravel deposited by glaciers thousands of years ago. However, the soil on the southern edge of the plain is highly fertile, and this area is heavily cultivated and densely populated.

Central and southern Germany

The Central Highlands, which includes a mixture of landforms ranging from nearly flat land to mountainous areas, are covered with rock and poor soil. Peaks in the Harz Mountains and the Thuringian Forest rise more than 3,000 feet (910 meters).

The many rivers that flow through the Central Highlands have carved steep, narrow valleys in the region. These rugged gorges, especially that of the Rhine River, are among the most beautiful areas in Germany.

The South German Hills include a series of long, parallel ridges that extend from southwest to northeast. The lowlands between these ridges have fertile clay soil that provides some of the best farmland in Germany. Much of this region is drained by the Rhine River and two of its branches—the Main and Neckar rivers. The Danube River, which drains the southern part of the hills, is the only major river in Germany that flows eastward.

The Black Forest, the setting for many old German fairy tales, is a rugged, mountainous region. Its name comes from the thick

The Oker River, *above,* has its source in the Harz Mountains on the northern edge of the Central Highlands. The climate in this region is marked by biting mountain winds, cool summers, and snowy winters.

The mountainous Black Forest region, *left,* gets its name from the thick forests that cover its granite and sandstone hills. The Black Forest is known for its mineral springs and health resorts.

The varied landscape of Germany, *map left,* includes the flatlands of the North German Plain, the rugged hills of the Central Highlands, and the rocky ridges of the South German Hills. The country's major land regions also include the Black Forest, a mountainous region of southwestern Germany; and the Bavarian Alps, which form part of Germany's southern border. The North German Plain covers the largest land region in the country. Many of Germany's oldest cities, including Bonn and Cologne, are located in this area.

The large island of Rügen, *above,* with its steep Stubbenkammer chalk cliffs, lies in the Baltic Sea off the North German Plain. Rügen is a popular resort area.

Lüneburger Heide, on the North German Plain, *below,* is one of the largest heathland areas in Germany, and a portion of it was made Germany's first nature reserve in 1921. Largely infertile, the area supports sheep, forestry, and some farming.

forests of dark fir and spruce trees that cover the mountainsides. The Feldberg, the highest peak in the Black Forest, rises 4,898 feet (1,493 meters). The region is also known for its mineral springs and health resorts.

The majestic, snow-capped Bavarian Alps, part of the largest mountain system in Europe, include the 9,271-foot (2,963-meter) Zugspitze, the highest peak in Germany. This beautiful mountain region, with its sheer rock faces and crystal-clear, glacial lakes, is a favorite, year-round tourist spot.

Germany's mild climate is largely due to west winds from the sea that help warm the country in winter and cool it in summer. Southern areas located away from the sea have colder winters and warmer summers. Most of Germany receives moderate rainfall the year around, but deep snow covers some mountainous areas throughout the winter. In December 1993 and January 1995, regions along the Rhine were hit by severe floods. In 1995, thousands of people were evacuated from their homes and several died. The German government offered about $20 million in low-interest loans to flood victims.

South-Central States

1 Bavaria
2 Baden-Württemberg
3 Hesse

Picturesque scenery—the Alps, the Black Forest, the Rhine River—and such major cities as Munich, Frankfurt, and Stuttgart characterize Germany's south-central area. The area includes the states of Bavaria, Baden-Württemberg, and Hesse.

Bavaria

Bavaria is the largest of Germany's federal states, covering the eastern half of southern Germany. Thousands of tourists come to visit Bavaria's beautiful mountains and lakes each year.

Munich, Bavaria's state capital and largest city, is often called "Germany's secret capital." The city's manufactured goods include electronic products and optical instruments, and Munich is also an important publishing center.

Nuremberg, which lies north of the Danube River, is Bavaria's second largest city—long famous for its toys and gingerbread. Today, it is an important industrial center. Other important Bavarian cities include Regensburg, the cathedral city of Würzburg, and Bayreuth, famous for its annual festival of German composer Richard Wagner's operas.

Many of Bavaria's industries are located in the northern part of the state. Processing and refining plants predominate, but electrical engineering, mechanical engineering, and the manufacture of chemicals, textiles, and motor vehicles are also important. Also, hundreds of breweries produce the world-famous Bavarian beer. In other parts of the state, many people work in agriculture and forestry.

The Bavarian people are proud of their state's long history. The beautiful state of Bavaria is also known as the "Free State of Bavaria"—and the only German state that marks its borders.

Baden-Württemberg

Baden-Württemberg lies in southwest Germany, where the Rhine River forms much of the state's border, separating it from France and Switzerland. To the south, Baden-Württemberg extends to Lake Constance. Stuttgart, which lies along the Neckar River, is the capital of Baden-Württemberg and the state's economic, political, and cultural center.

A border post, *above,* marks the border between Austria and the state of Bavaria in the Bavarian Alps.

The Zugspitze—Germany's highest peak at 9,721 feet (2,963 meters)—towers over the town of Garmisch-Partenkirchen in Bavaria.

Picturesque farmhouses dot the Black Forest in Baden-Württemberg.

A beer hall at Andechs Monastery near Munich in Bavaria features beer that is both brewed and sold by monks.

One of the state's most popular regions is the Black Forest, an area of dense woods and mineral springs. The people of the Black Forest have traditionally been known for manufacturing fine toys, cuckoo clocks, and musical instruments.

Baden-Württemberg is an important industrial state. Mercedes and Porsche cars, Pforzheim jewelry, and Black Forest clocks are among the state's best-known products. Other industries include electronics, precision engineering, and the production of chemical and optical equipment. Many of the state's farmers specialize in cattle raising, but cereals, vegetables, tobacco, and wine grapes are also grown.

Baden-Württemberg is the only German state that owes its existence to a popular referendum. At the end of World War II, the Allied occupying forces divided Baden and Württemberg into three new states, but in 1951, a large majority of the people voted to merge the three states into the single state of Baden-Württemberg.

Hesse

Hesse lies in west-central Germany. Agriculture dominates in the northern part of the state, while the area around Frankfurt, in the south, is highly industrialized.

Wiesbaden, the state capital, lies in a valley on the southern slope of the Taunus Mountains. Many visitors are drawn to the mineral springs around Wiesbaden and to the scenic upper valley of the Lahn River, where the university town of Marburg lies.

The state's other principal towns include Frankfurt, Kassel, and Darmstadt. Frankfurt is the hub of Germany's transport network, with motorways, railroads, and shipping and air routes all converging in the city. Its main industries include the production of chemicals, rubber and leather goods, machinery, and cars. Frankfurt also hosts an international annual book fair.

Kassel is the economic center of northern Hesse. Major industries include mechanical engineering and motor vehicle manufacturing.

The fertile soil north of the Main River yields cereals, sugar beets, and vegetables. Some of Germany's finest wines come from vineyards south of Darmstadt and in the Rheingau region.

Munich

Munich is the capital of Bavaria and the third largest city in Germany, ranking in population after Berlin and Hamburg. The city is less than 100 miles (160 kilometers) from Brenner Pass, which straddles the border of Austria and Italy in the Alps. Its location, near where northern and southern Europe meet, has made Munich a major transportation link.

Munich was founded in 1158 by Duke Henry the Lion. In 1255, the city became the seat of a family called the House of Wittelsbach. This family ruled Munich and the rest of Bavaria until 1918.

Munich in the 1900's

A small political group founded in Munich in 1919 became the Nazi Party in 1920. In 1923, the leader of this party, Adolf Hitler, proclaimed a Nazi revolution, or *putsch,* at a rally in a beer hall in Munich. The following day, Hitler tried to seize control of the Bavarian government, but he failed. This unsuccessful revolt became known as the "Beer Hall Putsch." However, Adolf Hitler became chancellor of Germany in 1933 and soon established a dictatorship through legal means. In 1938, in an attempt to avoid war with Nazi Germany, Great Britain, Italy, and France signed an agreement at Munich to give Czechoslovakia's Sudetenland to Germany.

Today, Munich is one of Germany's most important centers of economic activity. Electronics, publishing, and the production of chemicals are among the city's major industries. Munich is also well-known for its breweries.

Historic and popular attractions

Although much of the city was destroyed during World War II, Munich's buildings, unlike those of Frankfurt, were restored in their former style. As a result, many beautiful palaces, churches, and public buildings can be found throughout Munich. The Palace, the former residence of Bavarian monarchs, attracts visitors throughout the year. The German Museum houses one of the world's most famous exhibits of science and technology. And the twin towers of Munich's magnificent Cathedral symbolize the city.

The twin towers of Munich's Frauenkirche, *right,* the Roman Catholic cathedral, rise above the rooftops and spires of the city's old section. Heavily damaged during World War II, this area has been carefully restored. The city's German name—München, or Place of the Monks—points to its origins as a monastic settlement established in the 700's. The city was founded in 1158 by Duke Henry the Lion.

Munich's Oktoberfest, *below,* begins with a procession of innkeepers followed by the opening of the first barrel of beer by Munich's mayor. The festival has attracted as many as 7 million visitors who consumed nearly 1-1/2 million gallons (5 million liters) of beer.

The old center of Munich, *right,* is concentrated along the banks of the Isar River. At its heart, the Town Hall stands in the busy Marienplatz, the city's main square, and the Cathedral and National Theater are nearby. These sites are ringed by famous museums—the Old and New Pinakotheks, with art collections; the Glyptothek, with ancient sculpture; the German Museum, with technological collections; the Bavarian National Museum; and the Residence, a complex of theaters and museums.

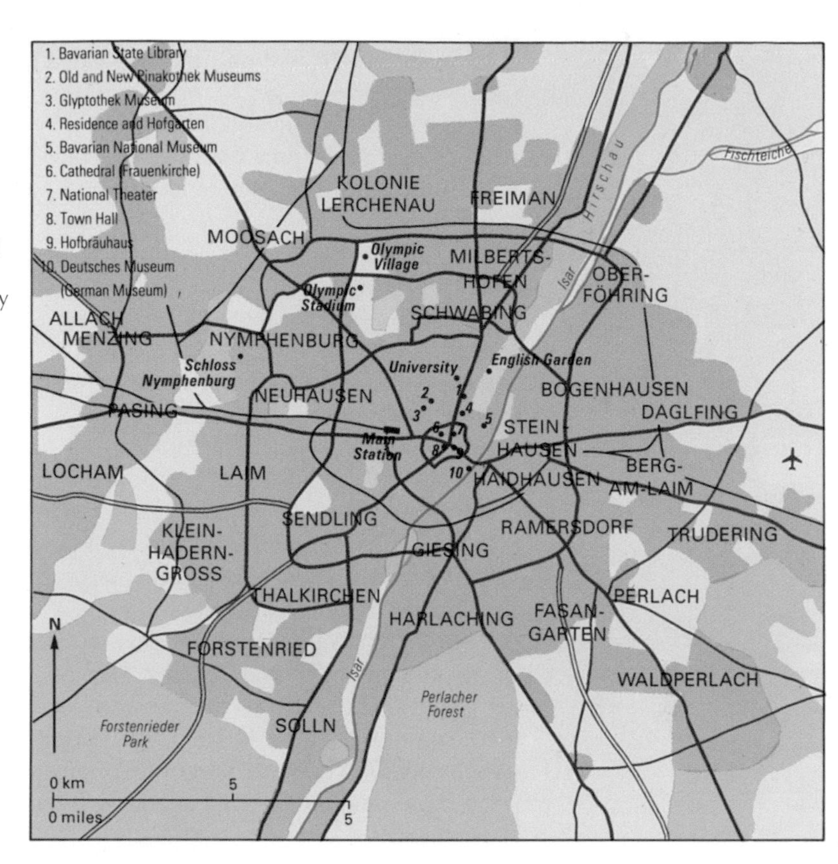

Munich's Olympic Village, *left,* located northwest of the city center, hosted the 1972 Summer Olympics. The Olympic stadium, swimming pool, and sports hall can accommodate over 100,000 people.

Three famous museums—the old Pinakothek, the new Pinakothek, and the Glyptothek—were restored after being severely damaged during the war. Some of the valuable paintings and sculptures in these museums were saved and are exhibited there today.

Munich is also a city of festivals and pleasure. In the summer, people get together in open-air beer gardens, or stroll in the many parks on the banks of the Isar River. Around the city's town hall in the heart of Munich, colorful street markets offer a variety of goods. And the Oktoberfest, the most famous beer festival in the world, is held in Munich from the middle of September to the beginning of October.

Western States

1 North Rhine-Westphalia
2 Rhineland-Palatinate
3 Saar

The forested hills of the Rhine and Moselle river valleys and the heavy industry of the Ruhr and Saar are located in Germany's western states. The area includes the states of Saar, Rhineland-Palatinate, and North Rhine-Westphalia.

Saar

Saar, the smallest of Germany's federal states—apart from the city-states—lies in southwest Germany and borders on France and Luxembourg.

The industrial towns of Neunkirchen and Saarlouis rely on Saar's many productive coal mines to fuel their iron and steel plants. Saar's other industries include metal processing, mechanical engineering, and the production of chemicals and glass.

After World War I (1914–1918), the Treaty of Versailles gave France the use of the Saar coal mines, and the land was governed by the League of Nations. In 1935, the people of Saar voted to become part of Germany once more, and the region became a federal state in 1957.

Rhineland-Palatinate

The federal state of Rhineland-Palatinate borders on Belgium, Luxembourg, and France. Located in southwestern Germany, Rhineland-Palatinate is a beautiful land of woods and castle-topped hills. The mountainous Eifel region in the north is dotted with volcanic lakes that are almost perfectly circular. The many ruins of picture-book castles that still stand along the Rhine Valley make this area the typical German landscape for many tourists.

The state's largest towns include Mainz, its capital, as well as Koblenz and Trier. Industries include mechanical engineering, shoemaking, jewelry manufacture, and chemical production. Rhineland-Palatinate is also a major wine-producing region, with two-thirds of all German vineyards located there.

The state of Rhineland-Palatinate was created in 1946 from parts of Bavaria, Hesse, and Prussia.

North Rhine-Westphalia

North Rhine-Westphalia borders on Belgium and the Netherlands. Located in the western part of the country, it is Germany's most densely populated state, with more than 20 per cent of the country's population. The capital of North Rhine-Westphalia is Düsseldorf.

Industry dominates much of the state, and its industrial core, the Ruhr region, is one of the world's richest coal-mining areas. For many years, the Ruhr was characterized by its mining and the manufacture of iron and steel. But today, many new industries have been developed there, including electronics and vehicle manufacture.

Northeast of the Ruhr region, however, agriculture is the major activity. The richly forested, hilly regions in the southern and eastern parts of the state attract thousands of visitors each year.

North Rhine-Westphalia's flourishing industries are served by a complex transport network that links its large cities, including Cologne and Essen. Bonn, where most of Germany's government offices are located, lies on the Rhine south of Cologne.

North Rhine-Westphalia was created in 1946 from territories that had belonged to Prussia in the 1800's.

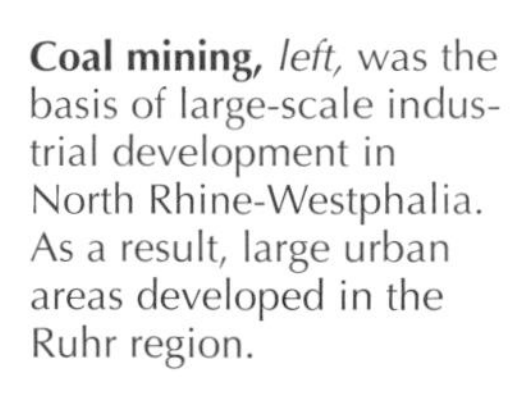

Coal mining, *left,* was the basis of large-scale industrial development in North Rhine-Westphalia. As a result, large urban areas developed in the Ruhr region.

Cologne Cathedral in North Rhine-Westphalia is a masterpiece of Gothic architecture. Nearby, the modern complex housing the Wallraf-Richartz and Ludwig museums provides a striking contrast.

Fertile vineyards overlook the Moselle River, *left,* as it curves through the state of Rhineland-Palatinate.

A sculpture by Henry Moore, *above,* stands before the Federal Chancellor's Office in Bonn in North Rhine-Westphalia.

Frankfurt

Before World War II, Frankfurt was Germany's best-preserved medieval city. During the war, however, Allied bombers leveled nearly half of the city and, today, gleaming skyscrapers have replaced some of Frankfurt's Gothic spires. In 1798, the Rothschild family opened its first bank in Frankfurt, and the city is still a world center of commerce and banking.

Frankfurt also hosts international trade fairs every year. Publishers and booksellers from around the world meet every autumn at the Frankfurt Book Fair, and consumer goods fairs are held there twice a year.

As its full name—Frankfurt am Main—implies, Frankfurt stands on the banks of the Main River. A network of railroads and highways links the city with all parts of Western Europe, and Frankfurt's airport is one of the largest and busiest on the continent. The city also ranks as one of Germany's busiest inland ports. Frankfurt's role as the transportation hub of western Germany dates back to the time of the Roman Empire. The shallow part of the Main River where Frankfurt now stands provided the easiest north-south river crossing in all Germany, and was widely used by travelers from all over Europe. The Franks who forded the river gave the city its name, which means *ford of the Franks.*

Historic sites

Although many of Frankfurt's buildings are modern, part of the old city still remains. The Römer—the city's ancient town hall—stands in the area surrounding Frankfurt Cathedral on the Main River, the historic center of the old town. Dating from the 1400's, this building contains the *Kaisersaal* (imperial room), the scene of the glittering celebrations and huge banquets that marked the coronations of the Holy Roman emperors. Nearby is the Paulskirche, a church where leaders of the unsuccessful Revolution of 1848 met to draft a German national constitution. The great German writer Johann Wolfgang von Goethe was born in Frankfurt in 1749. His home has been reconstructed and is now a museum.

Popular attractions

Although Frankfurt is one of western Germany's most lively cities, it retains a degree of small-town character. The Sachsenhausen district, for example, offers modern entertainment, but the narrow lanes that surround this area delight visitors with their Old World charm. Sachsenhausen is also the home of Germany's largest street market. The Schaumainkai, a riverside walkway, offers a fine view of the old town. Eight museums—Frankfurt's "museum mile"—line the walkway.

The city's traditional drink is a strong, bitter apple cider that goes especially well with frankfurters—the popular sausages to which the city lends its name. Experts believe that frankfurters were first made in Germany during the Middle Ages. In America, these sandwiches are called hot dogs. (Another favorite American food—the hamburger—is named after the German city of Hamburg.) Pork ribs with sauerkraut is also a hearty specialty of Frankfurt.

The Frankfurt Opera House was reopened in the early 1980's—after extensive reconstruction—as a center for concerts and meetings.

Frankfurt's city center was reduced to rubble in World War II bombing, *right.* Some of Frankfurt's medieval buildings have been reconstructed. But modern skyscrapers have replaced other old buildings.

The characteristic gabled roofs of the Römer, Frankfurt's ancient town hall, *left,* contrast with the surrounding modern buildings. In the background, the headquarters of the BfG, one of Frankfurt's banking corporations, towers over the Römer.

The Frankfurt Book Fair, *above,* is one of many international trade fairs held in the city each year.

Most of Frankfurt's major landmarks, *right,* are north of the Main River. The Festhalle is the site of trade fair exhibitions, sports events, and popular music festivals. St. Paul's Church was the site where leaders of the unsuccessful Revolution of 1848 met to draw up a constitution for a united Germany. The Frankfurt Zoo breeds many endangered species and has a building that displays nocturnal animals active in dim light.

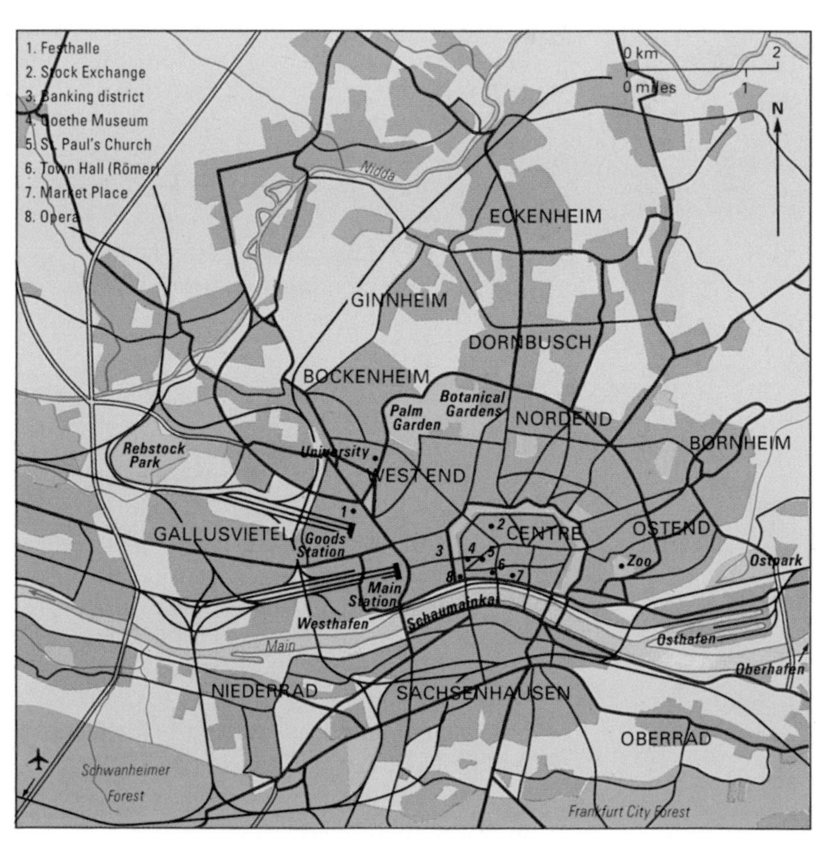

Northern States

1 Lower Saxony
2 Schleswig-Holstein
3 Mecklenburg-Western Pomerania

Northern Germany consists mostly of a plain dotted with marshes and lakes. The area borders the North and Baltic seas. It includes the states of Lower Saxony, Schleswig-Holstein, and Mecklenburg-Western Pomerania.

Lower Saxony

The state of Lower Saxony in northwestern Germany extends southward from the East Frisian Islands and the North Sea coast to the Central Highlands, and westward from the Harz Mountains to the Ems River on the Netherlands border. Lower Saxony has some of the richest farmland in Germany.

The coastal area is an important industrial region. Oil and natural gas are extracted from the German sector of the North Sea and from the Ems River Valley. The inland city of Hanover, the capital of Lower Saxony, is internationally renowned for its industrial and trade fairs, while nearby Wolfsburg is the home of the Volkswagen motor company. The East Frisian Islands, the Harz Mountains, and the Lüneburger Heide are popular tourist spots.

Lower Saxony covers most of the old Duchy of Saxony, a powerful state whose territory was broken up in the 1100's. Later, members of the royal family of Saxony established power in a central German region that was part of East Germany from 1949 to 1990. The old region of Saxony adopted the name *Lower Saxony* to distinguish it from the East German region. In 1946, the Prussian province of Hanover merged with other areas to form the present state of Lower Saxony.

Schleswig-Holstein

The most northerly of the German federal states, Schleswig-Holstein lies between the North and Baltic seas and shares its northern border with Denmark. The state includes the North Frisian Islands and the island of Helgoland.

The narrow inlets along the Baltic coastline provide natural harbors for thriving port cities, such as Kiel—the state's capital—as well as Flensburg and Lübeck. Ferries link these Baltic ports with Denmark, Sweden, and Finland. The Kiel Canal, one of the busiest waterways in the world, connects the North and Baltic seas.

Schwerin, *right,* which lies on Schwerin Lake in the state of Mecklenburg-Western Pomerania, was formerly the seat of the dukes of Mecklenburg. The Gothic cathedral in the background was built of red brick because of a lack of local stone.

An East Frisian fisherman, a citizen of Lower Saxony, plies his trade on the stormy waters of the North Sea.

While Schleswig-Holstein is primarily an agricultural region, shipbuilding is a major industry in the Baltic ports, and other industries have been built up in the southern part of the state in recent years. In addition, resorts on the North Sea and Baltic coasts attract many tourists.

For 1,000 years, Germany and Denmark fought over the Schleswig-Holstein area, which includes part of present-day Denmark. Denmark lost the territory in 1864 as a result of a war with Austria and Prussia, and in 1866 it became a Prussian province. However, most of the people of northern Schleswig were Danish, and in 1920, they voted to secede from Germany and join Denmark. Since 1920, northern Schleswig has remained a part of Denmark and southern Schleswig has been part of Germany. In 1946, Schleswig-Holstein became a German state.

Mecklenburg-Western Pomerania

The state of Mecklenburg-Western Pomerania in northeastern Germany extends southward from the Baltic Sea to the Mecklenburg Lake Plateau. Rostock, in the northern part of the state, is an important industrial center and seaport. The city lies on the Baltic Sea at the mouth of the Warnow River and serves as a major port of entry for much of Germany's petroleum supplies. Rostock's industries produce cargo ships, container ships, machinery, and motors.

For many centuries, the sparsely populated region around Schwerin, the state's capital city, has provided farmland, while cattle are raised in the lowlands area. East of Schwerin, the Mecklenburg Lake Plateau is a popular vacation spot. The hilly region east of the plateau contains many forests and lakes.

Mecklenburg was created in 1934, when the former grand duchies of Mecklenburg-Schwerin and Mecklenburg-Strelitz were merged. After World War II, Mecklenburg was combined with part of the old Prussian province of Pomerania. The state was called Mecklenburg-Western Pomerania until 1952, when the East German government split the region into several districts. The state's former name was readopted in 1990, after unification.

The Kiel Canal, *left,* one of the world's busiest waterways, runs through the state of Schleswig-Holstein for a distance of 61.3 miles (98.7 kilometers), providing a short cut for ships sailing between the North Sea and the Baltic Sea.

The age-old charm of Hörnum, a North Frisian village on the island of Sylt in Schleswig-Holstein, coexists with the island's modern resorts and discos.

Hamburg and Bremen

1 Hamburg
2 Bremen

In the late 1100's, the port cities of Hamburg and Bremen were important members of the Hanseatic League, a confederation of north German cities that grew out of trade associations. The Hanseatic League, whose members also included Lübeck and Cologne, controlled trade in the North and Baltic seas. Hamburg and Bremen, which both developed as a result of shipping and trade, remain thriving commercial and industrial centers today.

Hamburg

Hamburg—also known as "Germany's gateway to the world"—is located on the Elbe River, about 68 miles (110 kilometers) from the North Sea. This city-state ranks as Germany's most important port and the country's second largest city after Berlin. Hamburg is also an important political and cultural center.

Hamburg's industries include chemical plants, iron and steel works, sawmills, and shipbuilding. The harbor, which stretches along the Elbe, is a center for foreign and inland shipping, and the hub of Hamburg's economic activities. Many of Germany's industrial products, including automobiles, machinery, and optical goods, are exported from Hamburg.

The city is also one of Germany's leading railroad centers and a major center for national publications. The news magazine *Der Spiegel* and the newspapers *Die Zeit* and *Die Welt* are published in Hamburg. In addition, the city's many foreign consulates enable businesses in Hamburg to maintain contacts throughout the world.

Hamburg began to develop into an important trade center in the late 1100's, and the city's leadership in the Hanseatic League in the 1200's considerably strengthened its position. In the late 1800's and early 1900's, Hamburg was a state in the German Empire and the Weimar Republic. The city-state's present constitution went into effect in 1952.

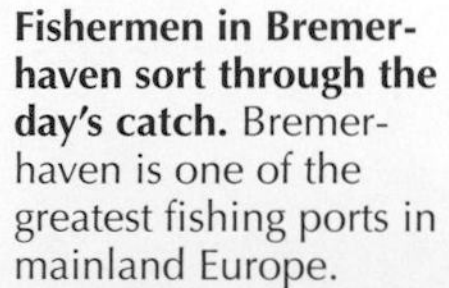

Fishermen in Bremerhaven sort through the day's catch. Bremerhaven is one of the greatest fishing ports in mainland Europe.

The magnificent facade of Bremen's town hall dates from the early 1400's. Bremen grew rich through its membership in the Hanseatic League, an association of trading cities.

Hamburg harbor can accommodate large ocean-going vessels. The harbor connects Germany with other major trading nations throughout the world.

More than half of Hamburg's buildings were destroyed during World War II, along with large parts of the city's port and commercial areas. Since then, the city has been extensively rebuilt, and the *Rathaus* (town hall), the new opera house, and many other attractive buildings give Hamburg a modern appearance today.

Bremen

The city-state of Bremen, the smallest of Germany's 16 federal states, consists of the cities of Bremen and Bremerhaven. Bremen, the state capital, lies along both banks of the Weser River, about 45 miles (72 kilometers) south of the North Sea.

While Bremen's economy is based chiefly on shipping and trade, other industries include food processing, oil refining, shipbuilding, and the production of automobiles, aircraft, electrical equipment, and textiles. Bremerhaven ranks as one of mainland Europe's largest fishing ports.

Bremen was founded sometime before A.D. 787, when it became a seat of bishops. The city gained economic importance in the 1300's through membership in the Hanseatic League, and like Hamburg, became a state of the German Empire in the late 1800's. Bremen was also badly damaged during World War II, though the damaged areas were quickly rebuilt. Today, Bremen's landmarks include the Romanesque-Gothic Cathedral of St. Peter, begun in the 1000's, and the Gothic Rathaus, which dates from the early 1400's.

On the embankment of the Weser River in Bremen, *above,* stands old St. Martin's Church. The nearby Schnoor area of the city includes streets and buildings in the style of the 1400's and 1500's.

Berlin

Berlin—now Germany's official capital and largest city—was also the capital of Germany before World War II. However, in 1949, four years after the war, Berlin—like Germany itself—was divided into two sections: West Berlin, which was allied with West Germany, and East Berlin, which became the capital of East Germany. The city of Bonn was established as West Germany's capital.

In 1961, the Communists built a 26-mile- (42-kilometer-) long wall to stop East German refugees from fleeing to West Berlin. In 1989, after a movement for freedom, the Berlin Wall and other borders separating East and West Germany were opened and in 1990, East Germany and West Germany were unified. Although Bonn remained the seat of German government, a united Berlin became the official capital of the new Germany. It is also one of the country's 16 states.

The city

A new city grew up in Berlin after World War II. In West Berlin, areas devastated by the war were replaced with towering skyscrapers, wide streets, and large parks. Much of East Berlin was also rebuilt. Today, Berlin is one of Europe's most beautiful cities.

The Brandenburg Gate lies at the heart of downtown Berlin. The central section of the huge stone structure was built between 1788 and 1791. Today, the Brandenburg Gate is a famous symbol of the city. Unter den Linden, the grandest of prewar Berlin's avenues, runs east from the gate. Many cultural buildings line this street, including the State Opera House and Humboldt University. Marx-Engels Platz, once the site of mass rallies and demonstrations in East Berlin, lies at the end of Unter den Linden.

The Tiergarten, a huge park that extends to the west of the Brandenburg Gate, includes the city's Zoological Gardens and Aquarium. Northwest of the Tiergarten stands the Hansa Quarter, which was designed by leading architects from 14 nations. Southwest of the Tiergarten, fashionable stores and theaters line the Kurfürstendamm, one of Berlin's most famous boulevards. At its east end, the bomb-scarred tower of Kaiser Wilhelm Memorial Church stands as a warning against war.

The Grunewald, an area of lakes and forests, stretches along the Havel River on the city's west side. The 100,000-seat Olympic Stadium, built for the 1936 Olympic Games, stands just north of the Grunewald.

History until 1945

The village of Berlin grew up on the northeast bank of the Spree River in the 1200's, and was later united with the village of Kölin (or Cölin), founded on an island in the Spree. By the 1400's, Berlin was an important town in the province of Brandenburg, and in 1470, the town became the official home of the Hohenzollern family, who ruled the province.

From 1640 to 1688, Berlin flourished under the rule of Frederick William of Hohenzollern. His son, Frederick I, became the first king of Prussia in 1701 and made Berlin his capital. In 1709, Berlin, Kölin, and three nearby settlements united as the city of Berlin.

During the 1700's, Berlin became a thriving trading and industrial center, and the arts and sciences also flourished. Immigrants from all over Europe flocked to the city. When the German Empire was formed in 1871, Berlin became its capital. The empire collapsed with Germany's defeat at the end of World War I in 1918, and Berlin became the capital of the weak Weimar Republic.

Berlin was especially hard hit by the worldwide economic depression of the 1930's. Hunger, unemployment, and widespread discontent paved the way for Adolf Hitler and his National Socialist Party. In 1933, Hitler seized control of Germany, and in 1939, the dictator initiated World War II.

By the end of the war in 1945, one-third of Berlin had been destroyed by bombings, and the city's population had dropped dramatically. On May 2, 1945, Berlin surrendered to the Allied Soviet army.

Cafes on the Kurfürstendamm, *above,* provide refreshment for Berlin's shoppers. The ruined tower of Kaiser Wilhelm Memorial Church, at the avenue's east end, stands as a permanent reminder of the devastation of war.

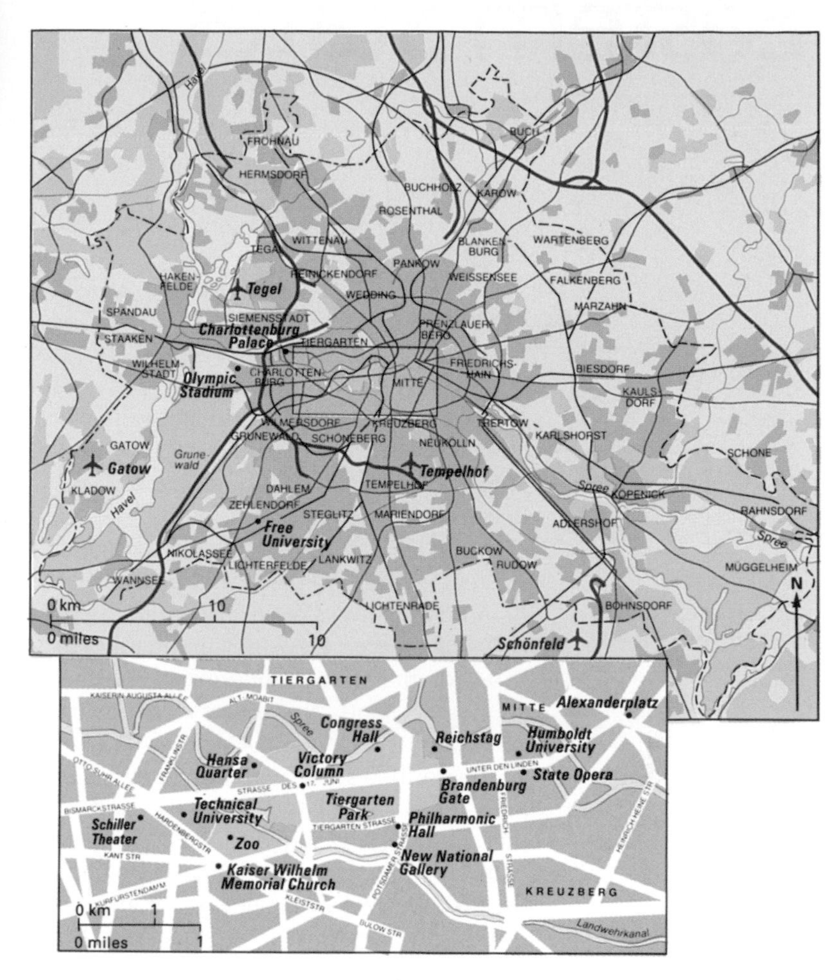

The Berlin Wall, a forbidding barrier of concrete slabs and barbed wire, once split the city in two, *below left.* The Brandenburg Gate stood just inside the wall in East Berlin. The 28-year-old Berlin Wall was opened in 1989 and later removed.

The village of Berlin, *below,* grew up on the northeast bank of the Spree River in the 1200's, and later united with Kölln (or Cölln), an island in the Spree. In 1709, the townships merged with three other settlements to form the city of Berlin.

Berlin, *above,* located near the junction of the Spree and Havel rivers, is one of Germany's 16 federal states as well as the country's capital. Berlin contains many historical and cultural landmarks, and with more than 3 million people, it is Germany's largest city.

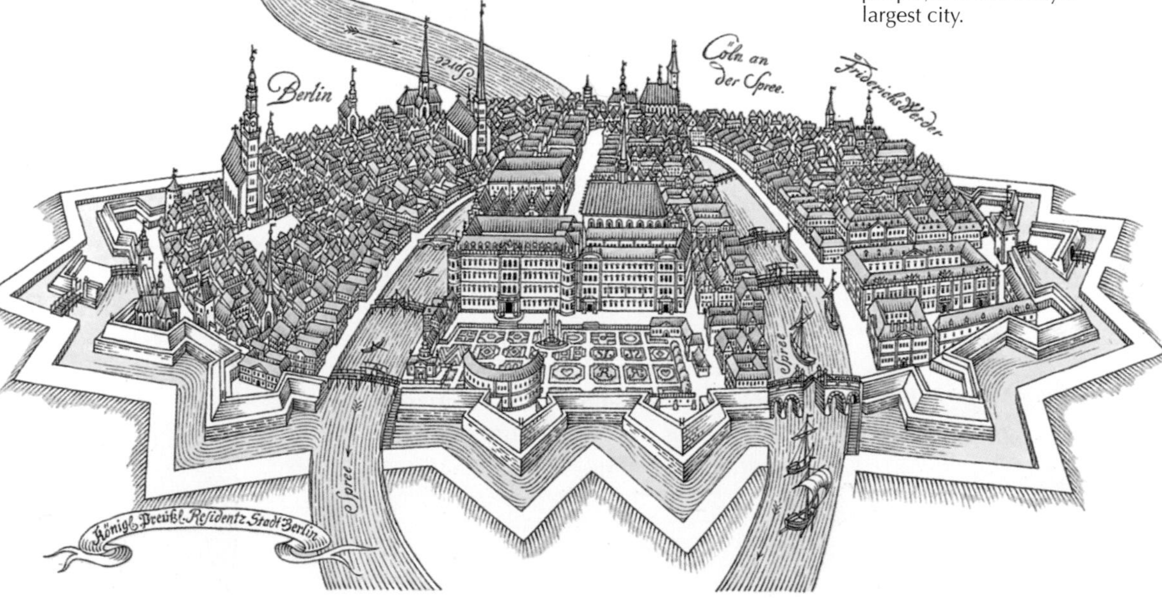

Eastern States

1 Brandenburg
2 Saxony
3 Saxony-Anhalt
4 Thuringia

The eastern states are flat in the north, but hilly in the south. They surround but do not include the city of Berlin. They consist of Brandenburg, Saxony, Saxony-Anhalt, and Thuringia.

Brandenburg

The federal state of Brandenburg is bounded on the west by the Elbe and Havel rivers and on the east by the Oder and Neisse rivers along the Polish border. Although Berlin lies at the center of Brandenburg, the city of Berlin makes up a separate state.

While agriculture and forestry dominate the region around Potsdam, the capital of Brandenburg, the state is also an important industrial center. Industries in the town of Schwedt include oil refineries and fertilizer and paper plants, and Lower Lusatia, near Cottbus, has huge deposits of brown coal. To the north lies the beautiful Spree Forest area, where sparkling waterways wind through thickly wooded islands.

In 1871, the ruler of Brandenburg became the first emperor of the new German Empire. After World War II, the part of historic Brandenburg east of the Oder and Neisse rivers became part of Poland. In 1952, East Germany divided the remainder into districts, including Cottbus, Frankfurt, and Potsdam. Following Germany's reunification in 1990, the districts were rejoined to form the state of Brandenburg.

Saxony

Saxony, which extends from the Ore Mountains across gently rolling hills to the edge of the North German Plain, has long been an industrial region, with coal mines, machine works, and textile factories. Smaller-scale industries in the Ore Mountains produce lace, musical instruments, and toys.

Saxony's major cities include Leipzig and the state's capital city of Dresden—both important industrial and cultural centers. Leipzig, the home of such composers as Johann Sebastian Bach, Robert Schumann, and Felix Mendelssohn, has played an important role in the history of German music. Dresden is one of the largest cities in Germany and a major European art center.

Wittenberg, *left,* where Martin Luther, a leader of the Protestant Reformation, began his call for church reforms, is located in what is now the state of Saxony-Anhalt. Luther posted his famous Ninety-Five Theses on the door of the town's Castle church (the tower in the center), where he is now buried. Factories in the background reflect Wittenberg's industrial importance.

Barbecued sausages, *above,* called *rostbratwürste* are a delicious specialty of Erfurt, the capital of Thuringia.

Acid rain has damaged many trees near Oberwiesenthal near Chemnitz (formerly Karl-Marx-Stadt) in Saxony, *left.*

The resort of Rathen, *top,* lies near the *Bastei* (bastion), a weathered rock outcrop and famous viewpoint above the Elbe River. Southeast of Dresden, the Elbe winds through a landscape of sandstone mountains called "Saxony's Switzerland."

During the 1700's, Saxony played an important part in European politics, and Dresden became a magnificent European capital. Saxony became part of the German Empire in 1871. In 1952, East Germany split the state into several smaller districts including Dresden, Karl-Marx-Stadt, and Leipzig. The districts were reunited to form Saxony once again in 1990.

Saxony-Anhalt

The landscape of Saxony-Anhalt is bounded by the Harz Mountains in the southern part of the state and the Havel River in the north. Halle is an important center of Germany's brown coal and chemical industries. Magdeburg, the state's capital, has fertile farmland and also serves as a major center for food processing and the production of heavy machines.

The state of Saxony-Anhalt was formed in 1945 by combining the state of Anhalt with the Prussian province of Saxony. In 1952, most of Saxony-Anhalt was divided, but it reassumed its former boundaries in 1990.

Thuringia

Thuringia is bounded by the Thuringian Forest in the south and the Harz Mountains in the north. The fertile Thuringian Basin between these regions grows such crops as barley, sugar beets, and wheat.

Erfurt, Thuringia's capital, is also the state's economic center. Both light and heavy engineering industries have been developed in Erfurt. Nearby lies Jena, the site of the famous Zeiss optical works, and Gera, an important center for uranium mining. Thuringia also contains the historical towns of Eisenach and Weimar. The state's popular tourist regions include the health resorts in the Thuringian Forest and artificial lakes in the Thuringian Hills.

As part of the German Empire of 1871, Thuringia consisted of many dukedoms and principalities. Most of these were united in 1920 to form the state of Thuringia. In 1952, the state was split up, mostly into the districts of Erfurt, Gera, and Suhl. However, like Brandenburg, Saxony, and Saxony-Anhalt, Thuringia was reunited in 1990.

Dresden

Dresden, which lies along both banks of the Elbe River in east-central Germany, is the capital of the state of Saxony. Before World War II, Dresden was one of the most beautiful cities in Europe, renowned for its magnificent baroque buildings and fabulous art collections. In February 1945, however, Allied bombing raids killed thousands of people in Dresden and destroyed most of its architectural monuments. Although much of the city has since been rebuilt, Dresden today has not regained its former splendor.

The reconstruction of Dresden's historic buildings began in the 1950's with the restoration of the Zwinger, a museum complex. Built during the 1700's, the Zwinger is an outstanding example of the decorative baroque style, and houses a magnificent collection of art treasures, including many porcelain artworks, jewels, and paintings by famous old masters. Other restored buildings from the same period include the Japanese Palace; the Hofkirche, a Catholic church; and nearby Pillnitz Castle. However, another church, the Frauenkirche, whose dome once dominated Dresden's skyline, has been left in ruins as a reminder of the devastation of war.

Much of the city has been rebuilt in a modern, functional style, and Dresden today has many broad streets lined with boxlike concrete buildings. Dresden's Central Institute for Nuclear Physics and the Dresden Technical University have made the city a center for scientific education and research.

The Zwinger, Dresden's great museum complex, was built by the rulers of Saxony in the 1700's. Now restored to its former splendor, its collections include porcelain, jewels, and paintings.

The Hofkirche, Dresden's Roman Catholic cathedral, stands in the restored city center. The church's ornamented style is an example of baroque architecture, popular in Germany in the 1700's. The Semper Opera House stands in the background.

Industry

The city's position on the Elbe River makes it an important German port, and Dresden's major products include pharmaceuticals; electronics equipment; furniture; optical and precision instruments; and machinery, including food-processing and food-packing machines.

Dresden china, a famous porcelain, is manufactured in nearby Meissen. The Meissen factory became the first producer of true porcelain in Europe. Augustus the Strong, Elector of Saxony and King of Poland, established the factory in 1710.

History

German settlers from Meissen founded Dresden in the early 1200's, and its location on the Elbe made Dresden an important link along the trade route that handled the silver mined in the Ore Mountains. In the 1400's, the city became the capital of Saxony. During the next 400 years, the Saxon rulers made the city an important cultural center and established and enlarged Dresden's art collection.

Dresden enjoyed its golden age under Frederick Augustus I during the late 1600's and early 1700's. Under his rule and that of his son, Frederick Augustus II, the city became a showcase of baroque architecture. By the 1800's, Dresden had also become a center of music. Several of Richard Strauss's operas were first performed in Dresden's Semper Opera House, built in the late

The ruined Frauenkirche, *below,* serves as a stark reminder of Dresden's suffering during World War II. In February 1945, Allied bombing attacks devastated the city, killing thousands of its inhabitants.

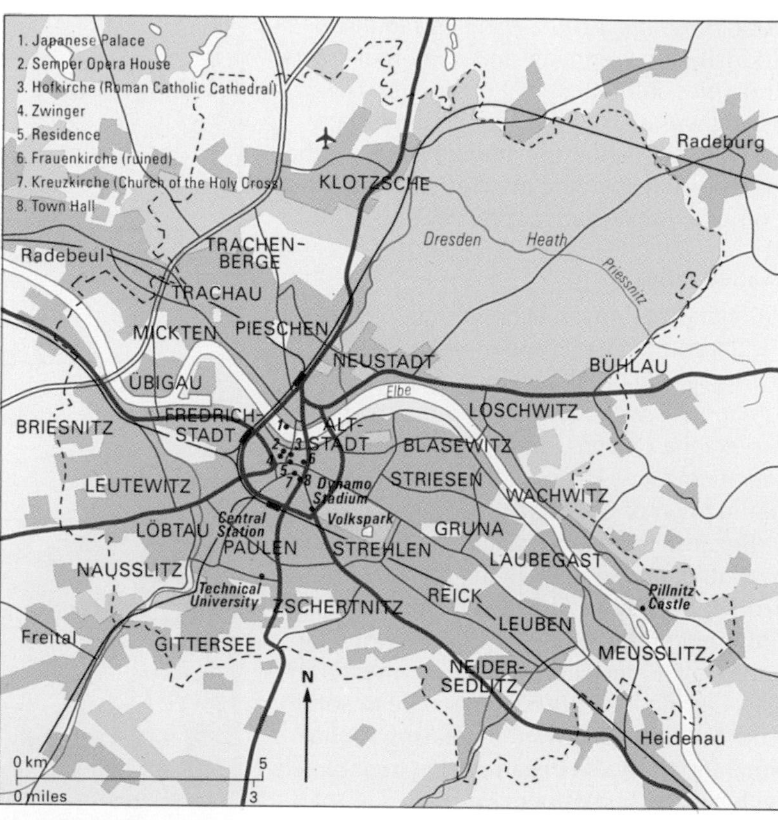

Dresden, *above,* the capital of Saxony, lies on the Elbe River close to the border of Czechoslovakia. Dresden was founded in the 1200's by settlers from Meissen to the northwest. World-famous Dresden porcelain is produced in Meissen.

The restored interior of the Semper Opera House is once again the setting for ballets, concerts, and operas. In the 1800's Dresden was a great music center and the birthplace of composer Richard Wagner (1813–1883).

1800's. After Saxony became part of the German Empire in 1871, Dresden gained importance as a commercial center.

During World War II, many of Dresden's art treasures were hidden outside the city for safekeeping. Soviet troops seized the collection in 1945, but most of the works were returned in the mid-1950's. Today, Dresden's restored museums and galleries once again display these works of art.

Today, the city faces a number of problems. Some of the city's older residential areas are neglected and decaying. Air pollution from coal-operated power stations has reached severe levels. In addition, the production facilities of many of Dresden's industries are outdated. It will be the task of the united Germany to modernize Dresden and perhaps restore it to its prewar brilliance.

Medieval Townscapes

During the Middle Ages, between the 1000's and the 1200's, trade and industry flourished in Germany, and numerous towns and cities still reflect that prosperity. This period was also an age of faith, and many magnificent medieval cathedrals built at that time are still standing in cities such as Cologne, Regensburg, and Trier.

Rothenburg, *above,* with its almost perfectly preserved walls and gates, offers an authentic glimpse of life in the Middle Ages.

Walled cities

The nation's thriving economy brought new wealth to merchants and craftworkers, and towns sprang up along the trade routes. Great walls were built around the towns to protect the citizens, and soldiers guarded the entrance gates. The old city gates still stand in the town of Breisach.

Because the walls limited the amount of available land, houses in the towns were crowded together. Streets were unpaved and filthy, because the people threw all of their garbage into the streets. Disease spread rapidly. During the 1200's, the people in some towns took steps toward increasing sanitation, and they also began paving their streets with rough cobblestones. In many German towns, these cobblestoned streets have been preserved.

As merchants and craftworkers settled in towns, they set up organizations called *guilds* to protect their interests. Guild meetings were held in town halls and churches built by the merchants themselves. Outstanding examples of these guildhalls are brick buildings still standing in Lübeck, Rostock, Stralsund, and Wismar, which all belonged to the Hanseatic League (the medieval confederation of north German cities). The towns in the league were economically powerful and largely self-governing.

After World War II

During World War II (1939–1945), bombing destroyed a number of medieval towns and cities in Germany. In some devastated cities, such as Nuremberg, many of the buildings were restored in their original style. In some smaller towns in southern Germany, the original structures survived intact. Rothenburg, for example, still has its ancient walls, gates, tall gabled houses, cobbled streets, guildhalls, and Gothic churches.

Many of the cities under Communist rule between 1949 and 1990 were rebuilt in modern style after the devastation of World War II. However, the East German government restored selected historic buildings, houses, and streets in some cities. In Leipzig, for example, the Old City Hall and many other buildings constructed during the Middle Ages were restored. The medieval cathedrals of Magdeburg and Naumburg were preserved, and buildings in the towns of Quedlinburg and Wernigerode in Saxony-Anhalt were restored in the medieval half-timbered style.

Quedlinburg, a town located at the edge of the Harz Mountains, was founded in 922 by King Henry I of Saxony, known as Henry the Fowler. Henry's widow Mathilde established an abbey there in 936, and the town was long a favorite royal residence.

Bautzen lies above the Spree River in the state of Saxony. At various times in the past, the town formed part of German, Bohemian, and Saxon domains. During the 1300's, Bautzen joined in a federation of Lusatian cities.

The half-timbered town hall in Michelstadt, *below,* reflects the prosperity of its medieval merchants.

Roots and Traditions

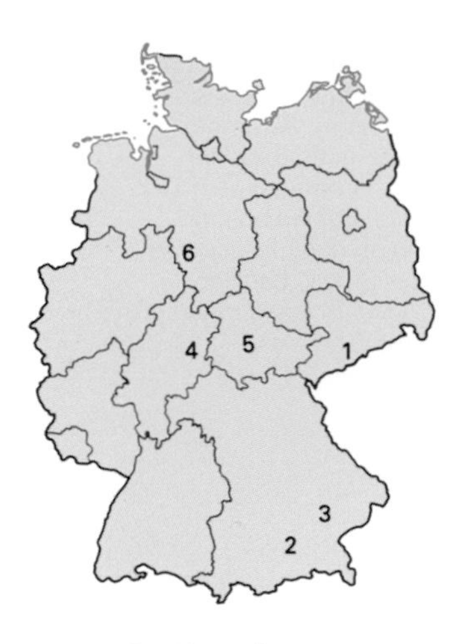

1 Ore Mountains
2 Munich
3 Landshut
4 Bad Hersfeld
5 Thuringia
6 Hameln

Tourists make special efforts to time their visits with traditional folk festivals, such as the Prince's Wedding in Landshut and the Lullus Festival in Bad Hersfeld.

The German people are the descendants of various warlike tribes that migrated from northern Europe to what is now Germany sometime after 1000 B.C. In the 100's B.C., the tribes moved south to the Rhine and Danube rivers, the northern frontiers of the Roman Empire. The Romans named the tribes' land *Germania* and called the people *Germani,* though that was the name of only one tribe. Other tribes included the Cimbri, Franks, Goths, and Teutons.

The German language

Each of the early Germanic tribes had its own dialect. For many years, however, these existed only as spoken languages. The people composed ballads and stories about their gods and heroes and passed these on by word of mouth from one generation to the next.

The oldest known record of written German is a Latin-German dictionary dating back to around A. D. 770, the time of the great Frankish ruler Charlemagne. Latin was the language used by the Romans, particularly among the nobles and the clergy. Charlemagne wanted his people to learn about the law and religion in a language similar to their own, and even encouraged German stories to be collected and written down. But few works from this period survive. The German language commonly used today—Standard German—has its roots in Old High German, one of the dialects used in Charlemagne's time.

In general, there are two principal forms of spoken German. High German is spoken in the mountainous regions of central and southern Germany, and Low German is spoken in the lowlands of northern Germany. Regional dialects, however, still survive, and a dialect spoken in one region may not be understood by someone from another area.

Folk traditions

Many German folk traditions also differ from region to region. The people in the old mining towns around the Ore Mountains, for example, celebrate their particular history with processions, songs, and colorful costumes. In the state of Thuringia, the Thuringian Fair is still celebrated in the traditional style, and ancient wedding rituals are still observed.

The maypole is a pagan symbol of May Day, a celebration of spring. In Bavaria, every village takes pride in its own "May tree."

A group of Bavarian women, *right,* dress in traditional costumes to celebrate a Christian feast day.

Some traditional folk festivals cross over regional boundaries, including many festivals originally held to honor the seasons and the harvest. The Shrove Tuesday festivals, for instance, were once rites of spring held in rural areas to greet the returning fertility of the earth. Seasonal traditions are also carried on in small villages and towns along the Rhine and Moselle rivers where wine festivals celebrate the September grape harvest. The famous Oktoberfest held in Munich honors the harvesting of hops and other grains used to make beer.

As Christianity spread throughout Europe in the early Middle Ages, people began to mix old pagan customs with the new religious festivals. Thus, Shrove Tuesday, or *Fastnacht* in German, became a celebration held before Lent, the 40-day period of spiritual renewal before Easter. Today, Shrove Tuesday is celebrated throughout Germany. The people of Cologne, Mainz, and other cities mark the occasion with wild merrymaking.

Participants in Shrove Tuesday festivals sometimes wear grotesque masks to "frighten away" the winter, *below.* To ensure a good harvest, masked dancers traditionally acted out the conflict between spring and winter.

A wedding procession, *bottom,* provides an occasion for Sorbs to display their decorative costumes. This ethnic minority lives along the Spree River in Upper and Lower Lusatia. The people maintain their own Slavic language and traditions.

The Pied Piper of Hamelin, *left,* leads a procession of children in an annual summer commemoration of the famous legend. According to the story, the Pied Piper agreed to rid Hamelin (now Hameln) of its rats for a sum of money, but the mayor refused to pay him. Angry, the piper led the town's children into a cave. The cave closed upon them, and they were never seen again.

Festivals

Germany hosts a variety of festivals, which feature concerts, operas, plays, and films. These festivals are held in large and small towns alike.

The oldest festival in Germany takes place in the Bavarian village of Oberammergau. Every 10 years, the town's residents perform the Passion Play, which portrays the suffering and death of Jesus Christ. The play was first performed in 1634 after the townspeople were delivered from a terrible plague that killed 1 out of every 10 citizens. The survivors pledged to perform the story of Christ's passion every 10 years. About half a million visitors attend each Passion Play season.

Berlin festivals

Festivals are held in Berlin almost all year round. The first organized event, initiated in 1951, was the Berlin International Film Festival, a competition of the latest films from all over the world. Other major events include the Berlin Festival, which offers concerts, operas, and ballets; the Berlin Jazz Festival, a top showcase for international jazz artists; and the Drama Festival, which presents plays from German-speaking countries.

Dramatic presentations are also offered in the ruins of a monastery in Bad Hersfeld and at the Ruhr Festival in Recklinghausen. The rococo theater in Schwetzingen provides a special setting and atmosphere for the plays presented there.

Music festivals

Musical events dominate the festival scene in Germany; some concentrate on a single composer. For instance, Bonn honors Ludwig van Beethoven; Wolfgang Amadeus Mozart is celebrated in Würzburg; and the eastern German town of Zwickau pays homage to Robert Schumann. During the International Bach Festival held in Leipzig, the famous St. Thomas Church Choir performs some of Johann Sebastian Bach's major oratorios.

Festivals devoted to a particular instrument or musical period include the Nuremberg Organ Festival and a festival in Herne called the Early Music Festival. Donaueschingen's Festival of Contemporary Music, on the other hand, offers the best in contemporary musical pieces.

Germany's best-known music festival is undoubtedly Bayreuth's Wagner Festival, Europe's oldest summer music festival. Richard Wagner himself built the Festival Opera House there and initiated the festival in 1876 with a performance of *The Ring of the Nibelung.*

More recently, the Schleswig-Holstein Music Festival got underway in 1986. Internationally renowned musicians participate in the event every year, and its concerts are performed throughout the state.

Festival calendar

The map on the right shows the location of some of the festivals that take place throughout Germany.

Ansbach: Bach Festival (*July/August*)

Augsburg: Mozart Festival (*May*)

Bad Hersfeld: Theater Festival (*July/August*)

Bad Segeberg: Karl-May Plays (*July*)

Bayreuth: Wagner Festival (*July/August*)

Berlin: International Film Festival (*February/March*)
Drama Festival (*May*)
Berlin Festival (*September/October*)
Jazz Festival (*November*)

Bonn: International Beethoven Festival (*September*)

Donaueschingen: Festival of Contemporary Music (*October*)

Hanover: Music and Drama at Herrenhausen (*May-September*)

Herne: Early Music Festival (*December*)

Hitzacker: Summer Music Festival (*July/August*)

Kassel: Music Festival (*September/October*)

Leipzig: International Bach Festival (*every 4 years, summer*)

Munich: Drama Festival (*May/June*)
Opera Festival (*July/August*)

Nuremberg: Organ Festival (*June/July*)

Oberammergau: Passion Play (*every 10 years, summer*)

Recklinghausan: Ruhr Festival (*May/June*)

Schleswig-Holstein: Music Festival (*June-August*)

Schwetzingen: Festival (*May*)

Witten: Festival for Modern Chamber Music (*April*)

Würzburg: Mozart Festival (*June*)

Zwickau: Schumann Festival (*every 4 years, summer*)

Thousands of visitors attend the Wagner Festival held at the Festival Opera House in Bayreuth every summer. Richard Wagner, who lived in Bayreuth and composed several operas there, intended his festival to serve as a model for other theaters. The composer is buried in Bayreuth, and his house is now a museum.

A fresco on a store in Oberammergau illustrates the passion of Jesus Christ. The fresco bears the date 1633, the year a plague killed 1 out of every 10 people in Oberammergau.

The Berlin Philharmonic Orchestra, *below,* seen here under the direction of Claudio Abbado in the city's Philharmonic Hall, is renowned world wide.

Economy

In 1945, at the end of World War II, Germany's economy lay almost in ruins. The controlling Allied powers rebuilt the country's economy, but the economic systems established in West Germany and East Germany were entirely different.

West Germany's economy was based on a social market economic system, which combines the economic freedom of the individual with the public good. Greatly aided by the funds that the United States began to send in 1948 under the Marshall Plan, the West German economy recovered at an amazing rate in the 1950's. By the mid-1960's, West Germany had one of the world's strongest economies.

In East Germany, the Soviet Union set up a strong Communist state where the government controlled the economy, including production, distribution, and pricing of almost all goods. Under this system, East Germany grew to be one of the wealthiest Communist countries, though it lagged well behind West Germany.

Effects of unification

When the two countries began economic unification on July 1, 1990, East Germany started to operate under West Germany's economic system. The East German government began selling government-owned businesses and breaking up the country's large government-controlled farms in order to sell land to individuals.

However, the transformation of East Germany's economy has posed a number of problems. Its industries and telecommunications systems must be modernized. In addition, many businesses could not compete without government support and were forced to close, causing high rates of unemployment. Many economists say it will take several years for living conditions in the east to catch up with the rest of the country.

Industry and agriculture

Manufacturing, Germany's fastest-growing industry, was the basis of West Germany's rapid economic recovery after World War II. Today, Germany is one of the world's leading producers of steel, much of which is used to make automobiles and trucks, industrial and agricultural machinery, ships, and tools.

Hanover's Exhibition Center hosts a major industrial fair each spring, *left.* The Hanover Fair, founded in 1947, ranks as the largest of its kind in the world, and draws more than 5,000 German and foreign firms.

A Turkish vendor in Berlin, *below,* finds plenty of customers for his oranges and bananas. Most non-Germans who live in the country moved there from Turkey, Yugoslavia, and Italy as *Gastarbeiter* (guest workers).

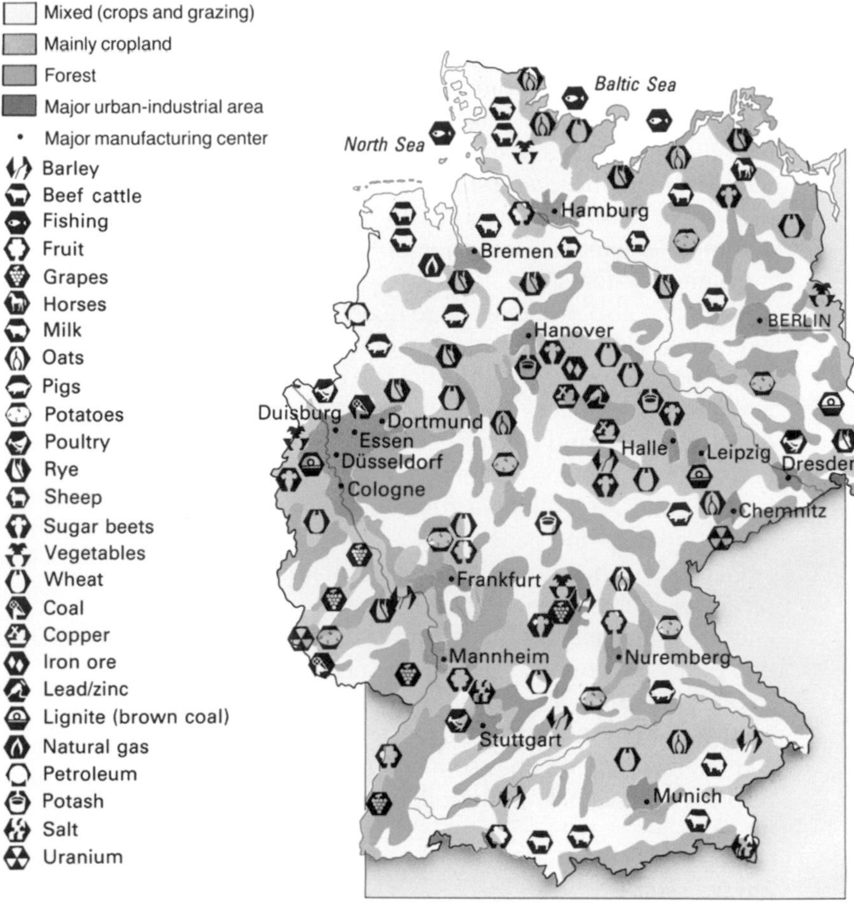

This map shows the economic uses of land in Germany. It also indicates the main farm and mineral products and identifies important manufacturing centers.

Germany has many major manufacturing regions, and there are factories throughout the nation. However, the Ruhr, in the western part of the country, is Germany's most important industrial region and one of the busiest in the world. Industries in the Ruhr produce chemicals, iron and steel, and textiles. Other important industrial regions include the areas around Berlin, Bremen, Chemnitz, Dresden, Frankfurt, Halle, Hamburg, Hanover, Leipzig, Mannheim, Munich, and Stuttgart.

Although about two-thirds of Germany's land can be used for some form of agriculture—cropland, grazing land, or forestry—German farmers produce only about two-thirds of the food consumed within the country. Germany is the world's largest importer of agricultural goods.

Potatoes are the only food produced in such plentiful quantities that they do not have to be imported. Other important crops are grains, sugar beets, and vegetables, as well as apples, grapes, and other fruits. Fine wines are made from grapes grown in vineyards along the Rhine and Moselle rivers.

Many German farms, particularly in western Germany, are small. Most of the farmers who operate these small farms have other jobs, too. Livestock and livestock products are important sources of farm income, and large numbers of farmers raise beef and dairy cattle, hogs, horses, poultry, and sheep.

Potatoes have formed the basis of the German diet since the 1700's, when Frederick the Great ordered their cultivation. Today, potatoes, cereals, and sugar beets are among Germany's most important crops.

Industry

Germany's economy depends largely on industry. Steel production is one of the country's leading industries. Most of the nation's steel is used to manufacture automobiles, machinery, and ships, and Germany now ranks as the world's third largest automobile manufacturer, after Japan and the United States.

Germany also produces large quantities of cement, clothing, electrical equipment, and processed foods and metals. The chemical industry produces large quantities of drugs, fertilizer, and plastics. Other important products include cameras, computers, leather goods, scientific instruments, toys, and wood pulp and paper.

Germany exports many of its industrial products, including automobiles, chemicals, iron and steel products, and machinery. While the country's exports exceed its imports, Germany imports large amounts of food as well as raw materials for industry. Although the country has some natural resources, it has only minor oil and natural gas supplies.

Energy sources

In the 1800's, coal deposits near the Ruhr River helped German industries grow. But by the 1970's, most of the high-quality deposits had been exhausted, and coal use in industry had declined. Oil, natural gas, and nuclear power had become more efficient energy sources.

Germany has modest supplies of natural gas, particularly in the western part of the country, and much of this gas is used in the chemical industry. Most of Germany's natural gas, however, is imported. The nation also depends on imported oil, mainly from the Middle East. Before unification, East German industries relied heavily on *lignite* (brown coal) as their main energy source. East Germany was one of the largest producers of brown coal in the world. Today, natural gas and oil are replacing coal.

Modern problems

In the early 1970's, an oil crisis resulted in greatly increased oil prices worldwide, compelling many nations, including Germany, to limit the use of imported oil. The oil crisis also caused a reduction in jobs and high unemployment.

Robot technology positions a car battery at the Mercedes-Benz factory at Sindelfingen, *above.* Automobiles built in Germany are shipped worldwide, making motor vehicles one of the nation's chief exports.

Chemical manufacturing, *above right,* represents one of Germany's most important export industries. New laws ensure safer production methods.

Industries based on new technology, *right,* are areas of major growth and innovation in the German economy. Southern Germany is home to developing industries such as computers, aeronautics, telecommunications, and space research.

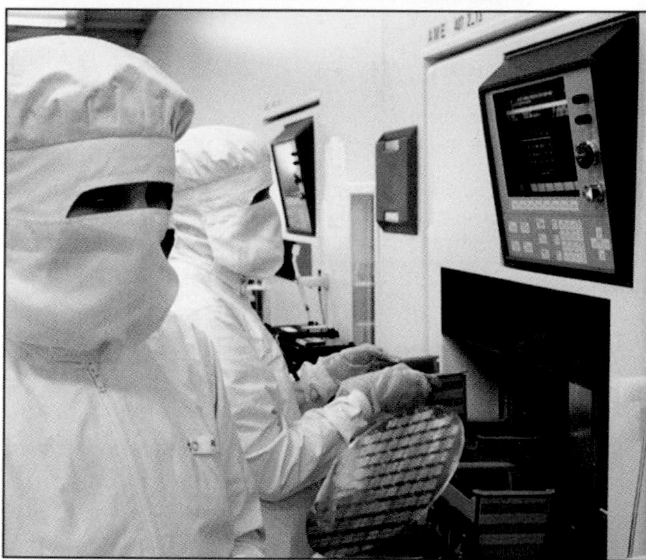

Specialized mining machinery strips away seams of *lignite* (brown coal) near Grevenbroich in North-Rhine Westphalia. Lignite still represents an important source of energy for German industry, but natural gas and oil are replacing coal today.

Mechanical engineering in the Ruhr area, *left,* ranks as one of the largest branches of German industry. German companies in this sector enjoy a worldwide reputation for their high-quality engineering products.

The central and northern parts of the country where mining, shipbuilding, and steel plants dominate were especially hard hit. In contrast, the southern regions, the home of new technology and machine and vehicle construction, have prospered. Unification also profoundly affected the labor situation. Many former East German companies were forced to close or reduce production, because they were out of date or unable to operate and compete without government support. Unemployment rose greatly, and still more workers were reduced to working shorter hours for less pay.

Pollution is another problem in Germany. More than half of the country's forests have been damaged by acid rain, and the nation's soil, rivers, and lakes are also threatened. As a result of its coal-burning industries, environmental conditions in some areas of eastern Germany are among the worst in Europe because of the lack of environmental protection under the former Communist government before 1990. Many rivers are too polluted even for industrial use.

Laws designed to limit this damage have already had desirable effects. Government bonuses, too, encourage the use of lead-free fuel. In addition, industries are currently seeking methods to dispose of wastes—particularly nuclear waste—that will not harm the environment.

Historic Sites

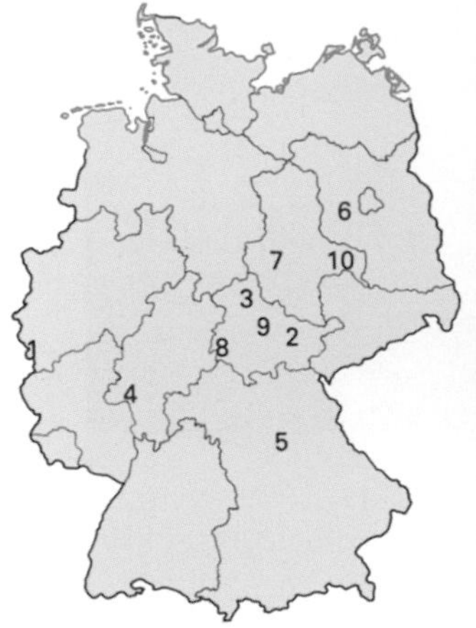

1 Aachen
2 Jena
3 Kyffhäuserburg
4 Mainz
5 Nuremberg
6 Potsdam
7 Quedlinburg
8 Wartburg
9 Weimar
10 Wittenberg

Locations of the historic, religious, and cultural sites discussed in this section are indicated on this map.

The Wartburg, a famous medieval castle, crowns a steep hill overlooking the town of Eisenach, *right.* During the Middle Ages, a famous contest between wandering minstrels and poets took place in the castle. In 1521 and 1522, Martin Luther worked here translating the New Testament into German, while under the protection of the Prince of Saxony.

The landmarks and artwork in many towns and cities throughout Germany reflect the country's long and rich heritage. Many of these historic sites were once the homes of great rulers. The western city of Aachen, for example, was the birthplace of Charlemagne, the most famous ruler of the Middle Ages. Charlemagne, who ruled in the late 700's and early 800's, made Aachen the capital of his western European empire. The city's magnificent cathedral was built during the emperor's rule and contains his tomb.

In 922, the founder of the Saxon dynasty, Henry I (the Fowler), built a fortress at Quedlinburg, at the foot of the Harz Mountains, as a stronghold against the invading Hungarians. Henry's son, Otto I, drove the Hungarians out of southern Germany in 955.

Nearby stand the Kyffhäuser Mountains. Frederick I—also known as *Barbarossa* or *Red Beard*—king of Germany from 1152 until 1190, is said to "sleep" beside a huge table in a cave beneath the mountains. According to legend, when his beard grows completely around the table, Barbarossa will arise and defeat Germany's enemies.

Wittenberg

Other cities in Germany are remembered for their religious significance. The Reformation, one of the most important religious movements in Europe, began in Wittenberg, a small university town in east-central Germany. The movement began when Martin Luther, a monk and professor of theology at the University of Wittenberg, attacked some of the church's practices.

On Oct. 31, 1517, Luther posted his famous Ninety-Five Theses on the door of Wittenberg's Castle church. As a result of this action, Luther was eventually expelled from the Catholic church and sentenced to death. However, Frederick the Wise, Prince of Saxony, protected Luther and concealed him at Wartburg, Frederick's castle in the Thuringian Forest.

Cities of culture

Many of Germany's cities and towns are also renowned cultural sites. Mainz, located at the junction of the Rhine and Main rivers, gained importance during the 700's when it became the seat of archbishops. However, Mainz is perhaps best known today as the home of Johannes Gutenberg, who invented the type mold that made printing from movable metallic type possible. Gutenberg used his invention to produce splendid books in Mainz during the 1400's.

During the late 1400's and early 1500's, a number of German artists lived in the Bavarian city of Nuremberg. Albrecht Dürer, the most famous painter and printmaker in the history of German art, was born in Nuremberg in 1471. One of Dürer's most famous oil paintings, *Four Apostles,* was painted for the Nuremberg city hall in 1526. The sculptor Veit Stoss

Nuremberg stands on the Pegnitz River, *right,* which divides the central part of the city in half. During the late Middle Ages, Nuremberg became a prosperous trade and cultural city.

The Chinese Pavilion stands in the grounds of the Palace of Sans Souci in Potsdam, *below.* The palace was built according to the plans of Frederick II, under whose leadership Prussia became a great power.

Charlemagne's marble throne sits in the cathedral in Aachen, *above,* the Frankish emperor's birthplace and his capital during the late 700's and early 800's. The cathedral also holds Charlemagne's tomb.

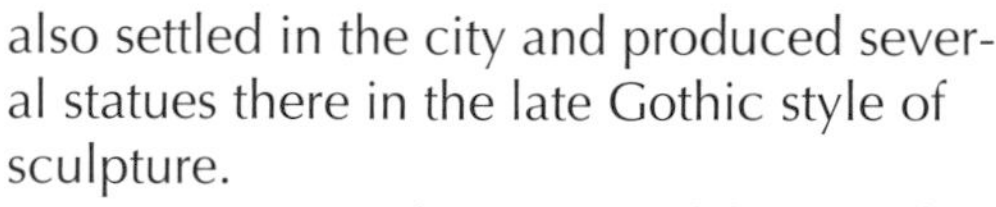

also settled in the city and produced several statues there in the late Gothic style of sculpture.

Weimar, once the capital of the grand duchies of Sachsen-Weimar-Eisenach, also has a long connection with the arts. In the 1770's, after Duchess Anna Amalia founded the "court of the Muses" in Weimar, the duchy began attracting such giants of German culture as Johann Wolfgang von Goethe and Johann Christoph Friedrich von Schiller. In 1842, the Hungarian composer Franz Liszt was appointed court music conductor and made Weimar the European headquarters for music. The city also became a center for architecture in 1919 when the Bauhaus, an influential school of design, was founded in Weimar by Walter Gropius.

A short distance from Weimar lies Jena, another center of German intellectual life. In the late 1700's, Jena witnessed the birth of German Romanticism, a style of art and literature that stressed emotion over reason.

Frederick II of Prussia, known as Frederick the Great, was another ruler who was a benefactor of the arts, particularly architecture. During his reign, from 1740 to 1786, many of the most beautiful buildings in the royal seat of Potsdam were erected. Frederick himself planned the most famous of them all, the Palace of Sans Souci.

Ghana

The tropical country of Ghana lies on the Gulf of Guinea, where the continent of Africa bulges westward into the Atlantic Ocean. From its heavily populated coastal plain along the Gulf of Guinea, Ghana rises to the Kwahu Plateau. The plateau runs from the northwest to the southeast across the center of the country.

The Kwahu Plateau helps form a divide between Ghana's rivers. In the south and west, the Tano, Ankobra, and Pra rivers flow south to the gulf. In the north and east, rivers such as the Black Volta and White Volta flow into Lake Volta in east-central Ghana. The main Volta River then flows from the lake to the gulf. Lake Volta, one of the world's largest artificially created lakes, was formed when the Akosombo Dam was built on the Volta River in 1965.

In general, Ghana has a tropical climate, with an average temperature of 80° F. (27° C) in the south and warmer temperatures in the north. Most of Ghana receives 40 to 60 inches (100 to 150 centimeters) of rain a year, but heavier rains fall in the southwest. Axim, on the southwestern coast, receives more than 80 inches (200 centimeters) annually. Northern and eastern Ghana have a *savanna climate,* with severe dry spells from November to March.

Variations in Ghana's climate lead to differences in vegetation throughout the country. Southwestern Ghana is heavily forested. Farther north, the land gradually becomes a *savanna* (thinly wooded grassland) that merges with grassy plains.

The forests of Ghana provide one valuable natural resource—tropical hardwoods such as mahogany. The country also has important mineral deposits of *bauxite* (used to make aluminum), diamonds, gold, and manganese. However, Ghana is largely an agricultural country, and its economy depends greatly on its farm products.

Cacao, a seed from which cocoa is made, is the most important crop and Ghana's chief export. Other valuable crops include coffee; coconuts; kola nuts, which are used to make soft drinks; and palm kernels, which are used to make oils.

Most of Ghana's factories are small plants that process agricultural products or the nation's timber. Manufactured goods include beverages, cement, and clothing.

Moshi-Dagomba women sort rice, part of an abundant harvest in northern Ghana. Women raise food crops, while men grow cacao and other export crops on the country's small farms.

Cacao workers prepare sacks of cacao beans for export in a Ghana warehouse. Cacao seeds are the country's most important crop and leading export.

The White Volta River flows through northern Ghana into Lake Volta, one of the world's largest artificial lakes. The waters of this huge reservoir help produce electricity for much of the country.

An aluminum smelter at Tema is the largest factory in the country. Hydroelectric power plants at the southern end of Lake Volta produce electricity for much of Ghana as well as the nearby countries of Togo and Benin.

After two decades of decline, the economy of Ghana began to improve in the late 1980's, largely because of economic reforms started in 1982 after the 1981 military overthrow of the government. The reforms included a plan to grow and market crops other than cacao. Ghana had long depended on cacao exports for much of its income, and the nation's economy was seriously hurt when prices for cacao dropped. The addition of other crops, such as avocados, pineapples, and papayas, made Ghana less dependent on cacao. At the same time, the shortage of basic foods in local markets was eased when farmers started growing corn, sorghum, yams, and cassava.

The government also encouraged Ghanaians to catch and sell lobster and shrimp. In addition, it promoted the development of light industry, such as furniture manufacture. Also, foreign companies were encouraged to prospect in Ghana for gold and other minerals. New deposits of gold were soon discovered, abandoned mines were rebuilt, and gold production increased more than 10 per cent.

An agricultural worker harvests the fruit of the oil palm, which is processed to produce palm oil. One of the most widely used vegetable oils in the world, palm oil is a basic ingredient in many kinds of soaps, ice creams, and margarines.

Ghana Today

When Portuguese explorers landed in what is now Ghana in 1471, they found so much gold there that they called it the Gold Coast. When the Gold Coast became an independent nation in 1957, it was named the Republic of Ghana after an ancient and powerful African kingdom.

Ghana was the first member of the British Commonwealth of Nations to be governed by black Africans. However, a series of military revolts since Ghana's independence often left the country in the hands of army officers.

Lieutenant Jerry Rawlings became Ghana's military head of state in 1981, when he overthrew the civilian government. In 1992, Ghana's voters approved a new Constitution. Political parties were made legal and multiparty elections were held. Rawlings was elected president. His party, the National Democratic Congress, won a majority in parliament. Rawlings was reelected in 1996. In 2000, the opposition New Patriotic Party came to power. John A. Kufuor became president. He was reelected in 2004.

Almost all the people of Ghana—99 per cent—are black Africans. They belong to about 100 different ethnic groups, including the Ashanti, the Fante, the Ewe, the Ga, and the Moshi-Dagomba. The Ashanti and the Fante, who make up a large percentage of Ghana's population, are closely related and belong to an even larger cultural group of African peoples called the Akan. Many Ghanaians speak the African language of their ethnic group, but a large number also speak English, the official language of the country.

FACT BOX

COUNTRY

Official name: Republic of Ghana
Capital: Accra
Terrain: Mostly low plains with dissected plateau in south-central area
Area: 92,098 sq. mi. (238,533 km^2)
Climate: Tropical; warm and comparatively dry along southeast coast; hot and humid in southwest; hot and dry in north
Main rivers: White Volta, Black Volta, Ankobra, Pra, Tano
Highest elevation: Mount Afadjato, 2,887 ft. (880 m)
Lowest elevation: Atlantic Ocean, sea level

GOVERNMENT

Form of government: Constitutional democracy
Head of state: President
Head of government: President
Administrative areas: 10 regions
Legislature: Parliament with 200 members serving four-year terms
Court system: Supreme Court
Armed forces: 7,000 troops

PEOPLE

Estimated 2008 population: 23,542,000
Population density: 256 persons per sq. mi. (99 per km^2)
Population distribution: 54% rural, 46% urban
Life expectancy in years: Male: 57 Female: 59
Doctors per 1,000 people: 0.1
Percentage of age-appropriate population enrolled in the following educational levels: Primary: 79 Secondary: 39 Further: 3

The Republic of Ghana lies on the Gulf of Guinea, a part of the Atlantic Ocean. Known in colonial times as the Gold Coast, it became an independent nation in 1957. Accra is the capital and largest city.

Fishing boats are beached near the town of Cape Coast, on Ghana's heavily populated coastal plain. The majority of the country's people live in rural villages.

About two-thirds of Ghana's people live in rural areas and farm for a living. Many women raise food crops for their families on plots of ground, while men raise cacao on small farms. The cacao is sold to make chocolate and cocoa. Life in rural Ghana centers mainly on the village marketplace, where people come to buy and sell goods.

Ghana's major food crops include cassava, yams, plantains, corn, rice, and peanuts. The country produces most of its own food but must import all of its wheat, half of its rice, and 80 per cent of its beef.

In central and southern Ghana, people live in rectangular houses with mud walls and thatched or tin roofs. Farther north, the mud houses are round with cone-shaped thatched roofs. Ghana's cities have many Western-style buildings, but large numbers of urban Ghanaians also live in traditional houses with mud-brick walls and tin roofs.

Traditional Ghanaian clothing is made from brightly colored cloth. Men wear it wrapped around their bodies, and women make it into blouses and skirts. But many Ghanaians wear Western-style clothing.

People in Ghana travel on crowded buses or flat-bed trucks, or walk—fewer than 1 per cent of the people own an automobile. There are about 70 radios for every 100 people in the country, but no village is without one. Only about 5 per cent of the people own a television set.

Languages spoken:
English (official)
African languages (including Akan, Moshi-Dagomba, Ewe, and Ga)

Religions:
Indigenous beliefs 40%
Muslim 20%
Christian 40%

TECHNOLOGY

Radios per 1,000 people: 695

Televisions per 1,000 people: 53

Computers per 1,000 people: 3.8

ECONOMY

Currency: New cedi

Gross national income (GNI) in 2000: $6.6 billion U.S.

Real annual growth rate (1999–2000): 3.7%

GNI per capita (2000): $340 U.S.

Balance of payments (2000): -$413 million U.S.

Goods exported: Gold, cocoa, timber, tuna, bauxite, aluminum, manganese ore, diamonds

Goods imported: Capital equipment, petroleum, foodstuffs

Trading partners: United Kingdom, Togo, Nigeria, United States, Italy

History

The earliest known inhabitants of the region that is now Ghana probably came from African kingdoms to the northwest during the 1200's. In the late 1600's, an Ashanti leader named Osei Tutu united his people and became the first Asantehene, or king, of this unified nation.

Tutu made the inland settlement of Kumasi the capital of his kingdom. During the 1700's, the Ashanti developed a powerful army and conquered surrounding territories. At its height in the early 1800's, the Ashanti Empire included much of modern-day Ghana, eastern Ivory Coast, and western Togo.

Meanwhile, Europeans had come to the coast of Ghana. As early as 1471, Portuguese explorers landed on the shore and named the region the Gold Coast. Later, the Dutch came to compete with the Portuguese for gold, and by 1642 the Dutch had seized all the Portuguese forts and ended Portuguese control of the coast.

During the 1600's, a large and profitable slave trade developed. The Dutch began competing with the English and the Danes for the trade. The buying and selling of slaves lasted for 200 years, ending in the 1860's. By 1872, Great Britain had gained control of the Dutch and Danish forts.

In 1874, Britain made the lands from the coast to the inland Ashanti Empire a British colony. The Ashanti and British fought each other for control of trade in western Africa. In 1901, the British defeated the Ashanti and made their lands a colony.

The cacao industry prospered in the British colony in the early 1900's. Britain extended roads and railways, built hospitals, developed schools, and gradually gave the Africans more power. In 1946, a majority of the members of parliament in the colony were Africans elected to represent the people. However, most of the power was still in the hands of the British governor and Cabinet.

Then in 1951, an African leader, Kwame Nkrumah, formed a Cabinet, and in 1952 he became prime minister. By 1954, the people were running their own government except for police, defense, and foreign affairs. Finally, in 1957, full independence was granted, and the new nation of Ghana was born.

The inner courtyard of a village in southern Ghana is surrounded by traditional, stoutly built mud-brick houses. In the late 1600's, the Ashanti people created an empire in this region that lasted until 1901.

A fortress on the coast of Ghana stands as a grim reminder of the African slave trade. From the 1600's to the 1800's, European slave traders built such forts as bases for their operations.

In 1960, the people of Ghana voted to become a republic, and elected Nkrumah president. Through the mid-1960's, however, Nkrumah worked to increase his personal power. Government debt and corruption, along with the falling price of cacao, began to weaken the economy.

In 1966, a military council overthrew Nkrumah, suspended the Constitution, and dismissed the legislature. The council named General Joseph Ankrah head of the government. Ankrah resigned in 1969 and was replaced with Brigadier Akwasi Amankwa Afrifa.

A series of military revolts took place in Ghana during the 1970's. In September 1979, a civilian government took over, but in 1981, it too was overthrown. Lieutenant

Jerry Rawlings then took control of the government and began economic reforms.

Before Rawlings seized power, many Ghanaians had moved to neighboring Nigeria to seek work. But Nigeria also began experiencing economic difficulties, and in 1983, it forced about 1 million people to return to Ghana. The return of these people created shortages of food, housing, water, and jobs in Ghana, and led to tension between Nigeria and Ghana.

From 1983 to 1988, Ghana's government took strong measures to improve the economy. It tripled prices paid to cacao growers, which led to a 20 per cent increase in production and reduced smuggling of the crop to neighboring countries. It also removed price controls. As a result of these moves, Ghana's economy experienced greater growth than any other African nation.

In 1992, Ghana's voters approved a new Constitution. Political parties, which had been banned since 1981, were legalized. In multiparty elections held later that year, Rawlings was elected president.

Dressed in traditional brightly colored robes, Ghanaian women gather for a ceremony. About 100 different ethnic groups live in Ghana today. Many are descendants of people from African kingdoms to the northwest of present-day Ghana who migrated to the region in the 1200's. The Ashanti probably are descended from people who lived in western Africa thousands of years ago.

Nkrumah

1909 Born in Nkroful, Ghana.
1935 Enters Lincoln University, United States.
1946 Publishes *Towards Colonial Freedom.*
1947 Returns to the Gold Coast.
1950 Imprisoned by British.
1951 Wins the Gold Coast's first general election.
1957 Sees birth of the nation of Ghana.
1958 Legalizes imprisonment without trial.
1961 Presides over widespread unrest.
1962 Survives assassination attempt.
1964 Becomes president for life of one-party state.
1966 Is overthrown by army.
1967 Exiled in Guinea.
1972 Dies in Conakry, Guinea.

The independence movement that swept through the countries of Africa after World War II (1939–1945) owes much to the work of Kwame Nkrumah. In the 1940's, Nkrumah sought self-government for his country, then called the Gold Coast. In 1951, the British, who controlled the country, asked him to form a Cabinet, and in 1952 he became prime minister. He continued to lead the Gold Coast's drive for full independence, and in 1957 it became the first black African colony to win its freedom. In 1960, Nkrumah was elected president. In an effort to develop Ghana's economy and improve living conditions, he promoted industry, introduced health and welfare programs, and expanded the school system. But he also began jailing his opponents, and his government became corrupt. The economy began to weaken, and in 1966, when Nkrumah was visiting China, army leaders took control of the country. Nkrumah went into exile in nearby Guinea, where Guinean President Sékou Touré made him honorary president. He died there in 1972.

Gibraltar

The tiny peninsula of Gibraltar, a British dependency on the southern tip of Spain, has a land area of only 2.3 square miles (6 square kilometers), but its landmark—the great Rock of Gibraltar—is recognized the world over. For many people, this huge limestone mass has come to symbolize strength and security.

The Rock of Gibraltar lies at the entrance to the Mediterranean Sea, rising 1,398 feet (426 meters) above sea level on the north and east side. From the top of the Rock on a clear day, visitors can see as far as Spain's Sierra Nevada and Morocco's Atlas and Rif mountains.

For centuries, the Rock of Gibraltar has been the subject of myth and legend. In Greek mythology, it was one of the Pillars of Hercules. According to legend, Hercules created the Rock by tearing a mountain in two, thus separating Europe and Africa and creating the Strait of Gibraltar.

Early conquests

The ancient Phoenicians were the first to establish a trading post on the Rock, realizing that the location of Gibraltar gave it great strategic importance. In A. D. 711, the Moors of North Africa settled in Gibraltar and held it for almost 600 years. Their leader, Tariq ibn Ziyad, ordered a fortress built high on the hillside overlooking the Bay of Gibraltar. Spain conquered Gibraltar in 1309 but lost it to the Moors again in 1333. Spain reclaimed ownership of Gibraltar in 1462 and held it until 1704 when it was captured by a British naval force, aided by the Dutch.

Great Britain's base

The Rock was promptly developed as Great Britain's major west Mediterranean base.

Britain's naval power on Gibraltar played an important role in stopping Napoleon from conquering all of Europe during the 1800's. Later, during World War II (1939–1945), the Allies launched an attack from Gibraltar against German and Italian forces in North Africa. After the war, Gibraltar's military importance gradually declined. In 1991, Great Britain withdrew its military forces from Gibraltar.

In 1964, Britain considered granting independence to Gibraltar. Spain strongly

A Barbary ape suns itself on an upper ledge of the Rock. The only wild monkeys in Europe, Barbary apes are about half the size of a large dog. Only about 5,000 are left in the wild, mostly in Morocco.

A ship awaits repair in dry dock at Gibraltar's main shipyard. Gibraltar has been an important British naval base since the early 1700's. Its extensive port facilities contribute much to the economy, but tourism is important as well.

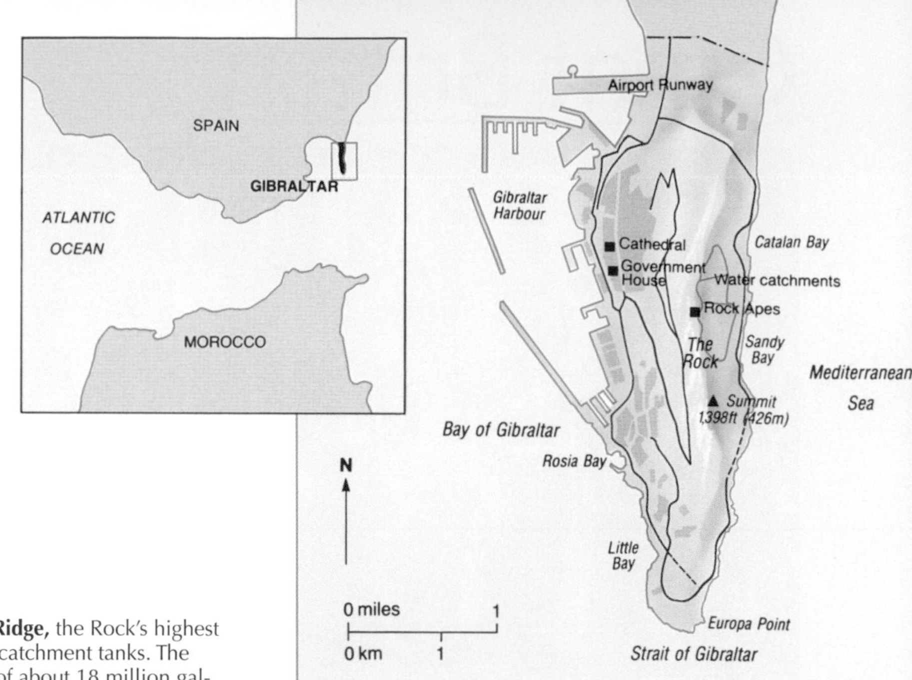

Gibraltar is a narrow peninsula at the entrance to the Mediterranean Sea. The Rock of Gibraltar covers most of its land area. Gibraltar has always had great military value, keeping enemy ships from entering or leaving the Mediterranean Sea. The Strait of Gibraltar is about 32 miles (51 kilometers) long and 8 to 23 miles (13 to 37 kilometers) wide.

Metal sheeting on Summit Ridge, the Rock's highest point, drains rainwater into catchment tanks. The tanks, with a total capacity of about 18 million gallons (68 million liters), serve all of Gibraltar, which has no natural water supply.

Fishermen prepare their nets on a beach in the shadow of the Rock of Gibraltar, *left.* The population of Gibraltar is largely made up of British military personnel and civilians descended from Italian, Maltese, Portuguese, and Spanish settlers. Many Moroccans now live and work on Gibraltar.

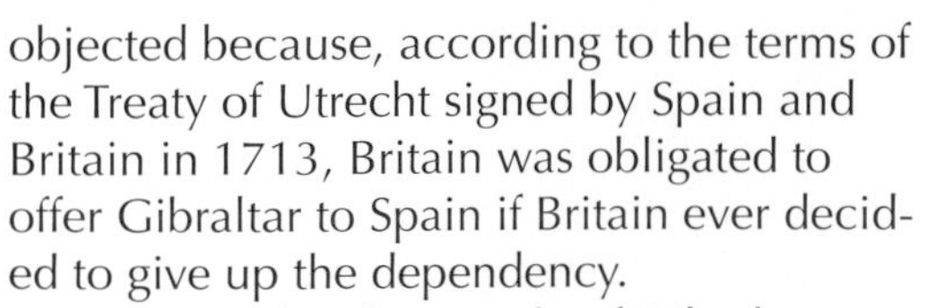

objected because, according to the terms of the Treaty of Utrecht signed by Spain and Britain in 1713, Britain was obligated to offer Gibraltar to Spain if Britain ever decided to give up the dependency.

In 1967, after the people of Gibraltar voted for continued British control, Britain decided to keep the dependency. The Spanish government responded by closing their border, thus restricting overland travel between Spain and Gibraltar. The border was not fully reopened until 1985.

Vacationers and residents

In addition to being an important naval and air base, Gibraltar has also developed a thriving tourist trade. Its duty-free status offers vacationers great bargains on European luxury goods and North African craft products.

The Rock's most famous natives may well be the shaggy-haired creatures known as Barbary apes. These animals are actually tailless monkeys related to the rhesus monkey of India, and not true apes.

Protected by the British government, the Barbary apes wander freely over the upper Rock. Legend has it that the monkeys once warned the British of a surprise attack on Gibraltar by Spain, and that the British will never lose control of the Rock as long as the apes live there.

The tradition was kept alive even during the darkest days of World War II, when British Prime Minister Winston Churchill sent an urgent message to his colonial secretary. It read: "The establishment of apes on Gibraltar should be 24 . . . every effort should be made to reach this number as soon as possible and maintain it thereafter."

Great Britain

Great Britain, officially known as the United Kingdom of Great Britain and Northern Ireland, is an island country in northwestern Europe slightly smaller than the state of Wyoming and has about 1 per cent of the world's total population. Yet for hundreds of years, Great Britain was one of the world's most important countries. The British started the Industrial Revolution and founded the largest empire in history. For centuries, Great Britain ruled the seas and ranked as the world's greatest trading nation.

However, by the end of World War II (1939–1945), the power of the United Kingdom had been much reduced. The crippling costs of two world wars and competition from other industrial countries led to one economic crisis after another. Today, although the United Kingdom is still a leading industrial and trading nation, it is no longer the world power it once was.

By the early 1950's, the United Kingdom's empire was declining rapidly, as many of its former colonies became independent nations. Today, most of these nations are linked with the United Kingdom and with one another through membership in the Commonwealth of Nations. This association of free countries and other political units once under British rule recognizes the British monarch as head of the Commonwealth. However, the monarch is mainly symbolic and has no governing power.

Although the United Kingdom is part of Europe, it is separated from mainland Europe by the North Sea on the east and by the English Channel on the south—a separation that has helped shape the independent character of the British people. Throughout its history, the United Kingdom has preferred to stay out of "European" affairs, and the channel has also helped protect the United Kingdom from invasion. In May 1994, a 31-mile (50-kilometer) railway beneath the English Channel opened. The Channel Tunnel, nicknamed "the chunnel," cost between $15 billion and $16 billion and was hailed as an outstanding engineering feat. It allows passengers to travel by rail from London to Paris in about 3 hours.

Still, although most British people do not think of themselves as Europeans, they do not generally consider themselves to be "British" either. The United Kingdom is made up of four political divisions united under one government. These political divisions are England, Northern Ireland, Scotland, and Wales. Although all British people share certain customs and traditions, each division has its own dialect, culture, history, and traditions, and most of the people have a strong sense of regional identity. The United Kingdom is sometimes called the *U.K., Great Britain,* or simply *Britain.* London is the capital and largest city of the United Kingdom.

Great Britain Today

Great Britain is a constitutional monarchy. The British Constitution is not a single document. It consists partly of *statutes* (laws passed by Parliament) and of documents such as the *Magna Carta* (a charter passed in 1215 to limit the monarch's power). It also includes *common law* (laws based on custom and supported in the courts). Much of the Constitution is not even written. These unwritten parts include many important ideas and practices that have developed over the years.

Number 10 Downing Street, guarded night and day by the London police, has been the official residence of the United Kingdom's prime ministers since 1732.

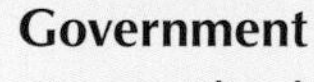

Government

Queen Elizabeth II is the United Kingdom's head of state. Her powers are largely ceremonial, however, and a Cabinet of government officials, called *ministers,* actually rules the country. The Cabinet is responsible to Parliament, which makes the laws of the United Kingdom.

Parliament consists of the monarch, the House of Commons, and the House of Lords. The queen must approve all bills passed by Parliament before they become laws, but no monarch has rejected a bill since the early 1700's.

The prime minister, who is usually the leader of the political party with the most seats in the House of Commons, serves as the head of government. The monarch appoints the prime minister after each general election. The prime minister selects about 100 ministers to head governmental offices and chooses the Cabinet members from among them.

People

The United Kingdom is a densely populated country, and almost 90 per cent of the people live in urban areas. English is the official language, but some people in Wales, Scotland, and Northern Ireland speak the traditional language of their areas.

Most of the British are descendants of the many early peoples who invaded Great Britain, including the Celts, Romans, Angles, Saxons, Jutes, Danes, and Normans. However, since World War II ended in 1945, many immigrants from Commonwealth countries in the West Indies, Asia, and Africa have settled in the United Kingdom. The United Kingdom has also offered asylum to refugees from around the world.

There are many divisions in British life. For example, Scotland and England have their own national churches, and there are separate legal and educational systems in England and Wales, Scotland, and Northern Ireland. For centuries, the British people were also separated by a rigid class system, but most of these class barriers were greatly reduced during World War II.

FACT BOX

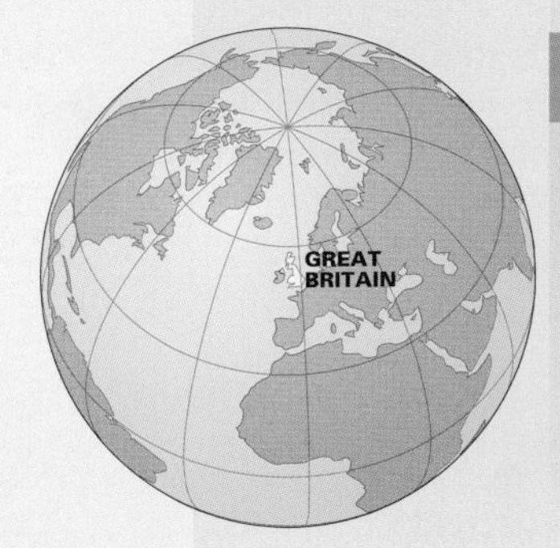

COUNTRY

Official name: United Kingdom of Great Britain and Northern Ireland
Capital: London
Terrain: Mostly rugged hills and low mountains; level to rolling plains in east and southeast
Area: 93,784 sq. mi. (242,900 km^2)
Climate: Temperate; moderated by prevailing southwest winds over the North Atlantic Current; more than one-half of the days are overcast
Main rivers: Thames, Severn, Mersey, Humber, Clyde, Forth
Highest elevation: Ben Nevis, 4,406 ft. (1,343 m)
Lowest elevation: Great Holme Fen, 9 ft. (2.7 m) below sea level

GOVERNMENT

Form of government: Constitutional monarchy
Head of state: Monarch
Head of government: Prime minister
Administrative areas: 47 counties, 7 metropolitan counties, 26 districts, 9 regions, 3 islands areas
Legislature: Parliament consisting of House of Lords with about 700 members and House of Commons with 646 members serving five-year terms
Court system: House of Lords
Armed forces: 200,000 troops

PEOPLE

Estimated 2008 population: 60,715,000
Population density: 647 persons per sq. mi. (250 per km^2)
Population distribution: 89% urban, 11% rural
Life expectancy in years:
Male: 76
Female: 80
Doctors per 1,000 people: 1.7
Percentage of age-appropriate population enrolled in the following educational levels:
Primary: 100*
Secondary: 178*†
Further: 64
Languages spoken:
English
Welsh
Scottish form of Gaelic

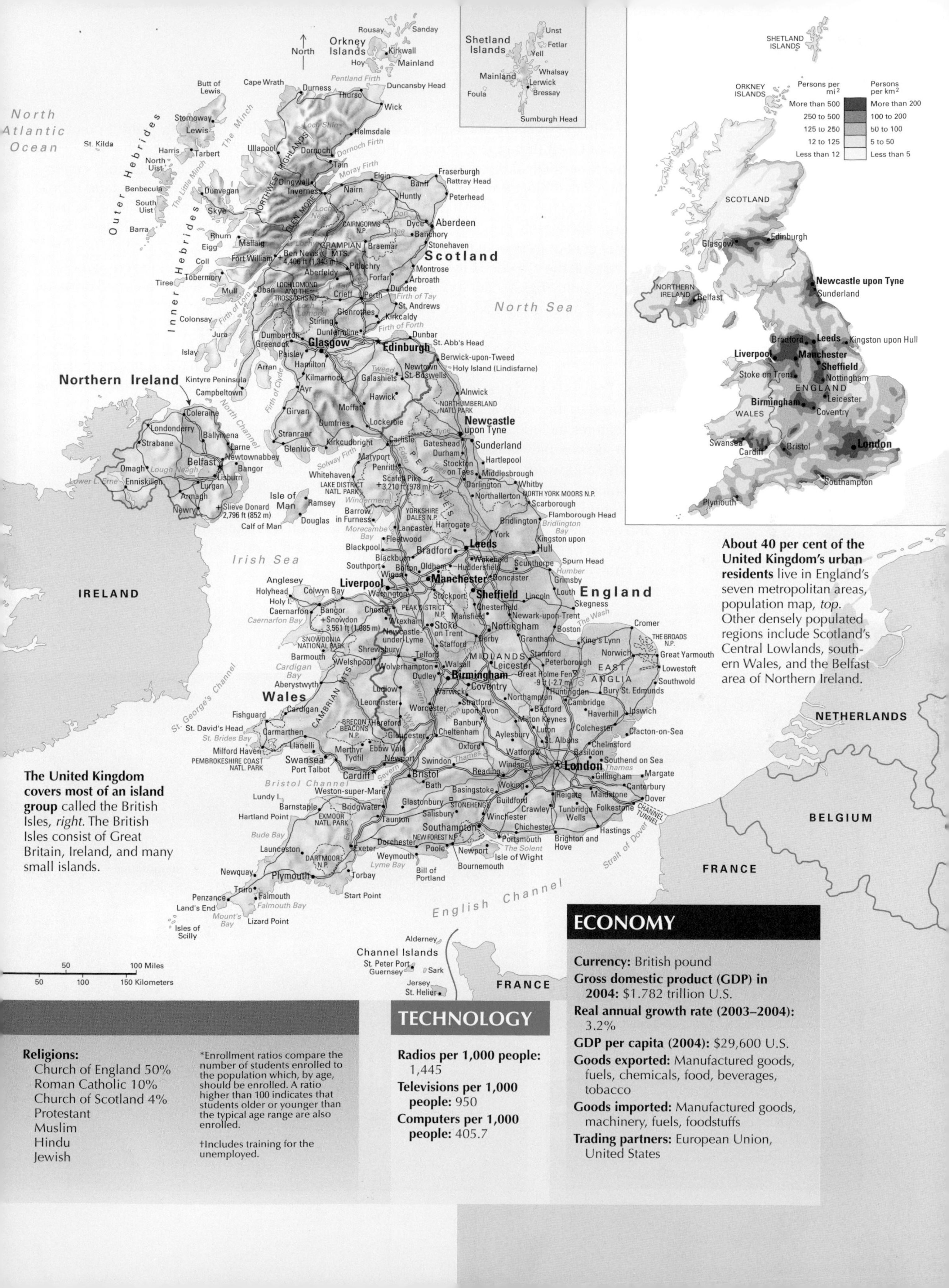

About 40 per cent of the United Kingdom's urban residents live in England's seven metropolitan areas, population map, *top*. Other densely populated regions include Scotland's Central Lowlands, southern Wales, and the Belfast area of Northern Ireland.

The United Kingdom covers most of an island group called the British Isles, *right*. The British Isles consist of Great Britain, Ireland, and many small islands.

ECONOMY

Currency: British pound

Gross domestic product (GDP) in 2004: $1.782 trillion U.S.

Real annual growth rate (2003–2004): 3.2%

GDP per capita (2004): $29,600 U.S.

Goods exported: Manufactured goods, fuels, chemicals, food, beverages, tobacco

Goods imported: Manufactured goods, machinery, fuels, foodstuffs

Trading partners: European Union, United States

TECHNOLOGY

Radios per 1,000 people: 1,445

Televisions per 1,000 people: 950

Computers per 1,000 people: 405.7

Religions:
Church of England 50%
Roman Catholic 10%
Church of Scotland 4%
Protestant
Muslim
Hindu
Jewish

*Enrollment ratios compare the number of students enrolled to the population which, by age, should be enrolled. A ratio higher than 100 indicates that students older or younger than the typical age range are also enrolled.

†Includes training for the unemployed.

England

England has a population of 50,803,000 in an area of 50,352 square miles (130,410 square kilometers). The capital city of London is home to more than 7 million people.

England lies in the southern and eastern part of the island of Great Britain, and ranks as the largest of the four political divisions that make up the United Kingdom. With more than 80 per cent of the total British population, England is a densely populated area. About 95 per cent of its people live in urban areas, but outside the crowded city centers stretches the scenic English countryside, with its charming villages, green pastures, and neat hedges.

Land

In general, England's land slopes from the north and west to the south and east. Characteristic features include *moors* (open grasslands), *downs* (hilly grasslands), *fens* (marshlands), and *wolds* (low, chalky hills).

The Pennines—England's major mountain system—extend from Scotland to central England, and are often called the "backbone of England." West of the Pennines lies the Lake District, known for its beautiful mountain scenery, including England's highest point—the 3,210-foot (978-meter) Scafell Pike.

A large plateau called the Midlands covers east-central England. Much of this land is broken by rolling hills and fertile valleys, drained by the Thames, Severn, Trent, and Ouse rivers. Along the North Sea coast, the Midlands are low and flat, particularly in the Fens, where much of the land has been reclaimed from the sea.

Southeastern England is crossed by ranges of hills consisting of layers of limestone and chalk. Along the English Channel, the hills drop sharply to form steep cliffs, including the famous white cliffs of Dover.

England's westernmost point—Land's End—and the southernmost point in the British Isles—Lizard Point—sit on the rugged Southwest Peninsula.

England's offshore islands include the Isle of Wight, off the southern coast, and the Scilly Islands, off Land's End.

Economy

Until the early 1800's, most English people lived in the countryside and worked on farms. However, during the Industrial Revolution, huge numbers of people moved to cities and towns to work in the new factories. Today, more than a third of the population lives in seven large metropolitan areas. Greater London, with more than 7 million people, is the largest metropolitan area in England and one of the largest in the world. The other metropolitan areas include Birmingham, Manchester, Leeds-Bradford, Liverpool, Sheffield, and Sunderland-Newcastle upon Tyne.

The shift in population from rural to urban areas reflected the shift from an agricultural to an industrial economy. In recent years, however, many of the factories built near coal fields, their source of power, have closed. Nuclear energy, oil, and gas are the modern energy sources. As a result, many new industries have developed around London and in the southeastern section of England, where there is little coal. The decline of the factories around the northern coal fields has led to a drop in that region's prosperity, and the new industries in the south have drawn even more people to an already crowded area.

Service industries are also important to England's economy. About 70 per cent of English workers are employed in service industries, particularly in social services, wholesale and retail trade, and financial services.

Other service industries are concerned with communication and leisure activities. Many English people have much more free time than their rural ancestors enjoyed. The leading leisure activities include gardening, sports, watching television, and attending movies, plays, and concerts. The neighborhood public house, or pub, is also a favorite social center for many people. In addition, hotels and restaurants serve not only English people but also many foreign tourists and business travelers.

Blackpool, *left,* has long been a popular seaside resort for industrial workers in northern England.

England's Lake District, *above,* where 16 lakes lie within a circle about 30 miles (48 kilometers) in diameter, is celebrated for its beauty.

A ship passes under the Tyne Bridge in Newcastle upon Tyne, *far left.* The city has long been a coalmining center.

London's Dockland, *below,* is home to some of the city's wealthiest people.

Wales

Wales has a population of 2,992,000 in an area of 8,015 square miles (20,758 square kilometers). The capital city of Cardiff has more than 305,000 residents.

Wales covers about 10 per cent of the island of Great Britain, but has only about 5 percent of the population of the United Kingdom. About 17 per cent of the people speak Welsh, an ancient Celtic language. English and Welsh are both official languages.

Wales has a history filled with poets and singers with a literature that dates back more than 1,000 years and an ancient choral music tradition. A festival called the *eisteddfod* (pronounced *ay STEHTH vahd*), featuring musicians, poets, and singers, is held once a year.

As part of the United Kingdom, Wales elects as least 35 of the 646 members of the House of Commons. In addition, in 1997, by a very small margin of 50.3 per cent, Welsh voters established a National Assembly for Wales. The National Assembly does not have the authority to pass its own primary legislation or to raise taxes. However, it is allowed to pass legislation in certain local matters. The National Assembly of Wales is composed of 60 members.

Land

The Cambrian Mountains cover about two-thirds of Wales. The highest peak in Wales, Snowdon (*Yr Wyddfa* in Welsh), reaches 3,561 feet (1,085 meters). Coastal plains and river valleys cover about a third of Wales.

The longest rivers are the Severn and Wye, which both empty into the Bristol Channel. The Isle of Anglesey (*Môn* in Welsh), a large island off the northwest coast of Wales, is separated from the mainland by the Menai Strait.

Wales has three national parks—Snowdonia, the Brecon Beacons, and the Pembrokeshire Coast. These and other protected areas provide a refuge for endangered species.

Economy

The largest urban areas in Wales are Cardiff, Swansea, and Newport, all located on the southern coast, and the majority of Wales's nearly 3 million people live in the southeast. A major increase in Wales's population occurred during the Industrial Revolution in the 1700's, when large numbers of people immigrated from England to work in the coal mines. Many Welsh people also left the farms to find work in the coal-mining towns of southern Wales, where most mining activities were concentrated. By the late 1900's, although a few mines were still operating, the economy of Wales had come to depend primarily on manufacturing and service industries.

Everyday life in industrial Welsh cities and towns is similar to life in the industrial areas elsewhere in the United Kingdom. After work, many people enjoy watching television or socializing at the local pub. Other popular pastimes include rugby football as well as soccer and cricket.

Coal-mining villages such as Abercynon, *left,* are still found in the valleys of southern Wales.

Queen Elizabeth named her eldest son, Charles, Prince of Wales in 1958, when he was 9 years old. In 1969, she presented him to the Welsh people in a ceremony at Caernarfon Castle in Wales. The title dates back to 1301.

The Pass of Llanberis, *left,* lies southeast of Llanberis, the starting point for the rail ascent of Snowdon, the highest peak in Wales. Nearby, Snowdonia National Park covers 845 square miles (2,190 square kilometers).

The Royal National Eisteddfod, *above,* a festival of music and the arts, takes place in August. It is held in various cities and towns, alternately in southern and northern Wales. Only Welsh is spoken during this event.

Scotland

Scotland has a population of 5,181,000 in an area of 29,767 square miles (77,097 square kilometers). The capital city of Edinburgh has a population of more than 448,624.

Scotland occupies the northern third of the island of Great Britain, but only about 9 per cent of the total population of the United Kingdom live there. Most of the population is concentrated in an industrial area located in the central part of Scotland.

Although the Scots have kept numerous symbols of their long and colorful history, industrialization has eliminated many of Scotland's old traditions and ways of life. The historic Scottish *clans* (groups of related families) have lost much of their importance, and kilts are usually worn only on special occasions.

However, their strong sense of identity has inspired a number of Scots to seek greater independence from Great Britain. In a referendum held in 1997, voters chose to establish a Scottish Parliament. The Parliament, which is composed of 129 members, has the power to impose some taxes and to control education, health, the environment, agriculture, and the arts in Scotland. Scotland also elects 59 members to the United Kingdom's House of Commons.

Land

Scotland has three main land regions: the Highlands, the Central Lowlands, and the Southern Uplands. The Highlands are a magnificent, rugged region that cover the northern two-thirds of Scotland. The area's two major mountain ranges, the Northwest Highlands and the Grampian Mountains, are divided by a deep valley called *Glen Mor,* or the *Great Glen,* and the Highlands are dotted with sparkling *lochs,* or lakes.

The Central Lowlands, an area of scenic green valleys, fertile fields, and scattered woodlands, has Scotland's richest farmland, most of its mineral resources, and about 75 per cent of its people. The region includes two of Scotland's leading cities—Glasgow, its largest city, in the west, and Edinburgh, its capital, in the east.

The Southern Uplands consist of rolling moors, which are broken in places by rocky cliffs. Sheep and cattle are raised on the rich pastureland that covers most of the lower slopes.

A bagpiper in traditional Highland costume plays at a gathering of the clans. The bagpipe, which dates back thousands of years, is one of the oldest musical instruments still in use.

Loch Awe, *right,* is one of the many beautiful lakes in the Scottish Highlands. On its banks stand the ruins of Kilchurn Castle, built in 1440 and once the scene of fierce battles between the clans.

Economy

Most of the Scottish people live in industrial cities and towns. However, after the decline of heavy manufacturing in the late 1900's, service industries became the main employers of Scotland's work force. About two-thirds of Scotland's people work in such areas as retail sales, finance and business services, tourism, transportation, and communication.

Oil and gas fields under the North Sea provide much of Scotland's energy. Nuclear power plants supply about 45 per cent of Scotland's power, coal provides about 30 per cent, and natural gas and hydroelectric power plants supply most of the rest. One of Scotland's most valuable resources is its excellent fishing grounds off the east and north coasts. Other important products include chemicals, electronic equipment, steel, textiles, and whisky.

In their leisure time, the Scots enjoy sports, including golf, which is believed to have originated in Scotland, and soccer. Highland Games, held throughout the Highlands during the spring, summer, and early fall, include field events, foot races, and dancing and bagpipe competitions.

The distillery at Tain on Dornoch Firth, along with more than 100 other distilleries in Scotland, exports some 950 million bottles of Scotch whisky yearly. The biggest market is the United States.

The annual Edinburgh International Festival of the Arts, *left and inset,* features musical and dramatic productions from around the world and a spectacular Military Tattoo, presented every evening of the festival on the Esplanade of Edinburgh Castle.

Mallaig, ***above,*** **on the northwest coast of Scotland,** is an important fishing port. Scottish waters are especially known for their successful salmon farm.

Northern Ireland

NORTHERN IRELAND
Belfast

Northern Ireland has a population of 1,739,000 in an area of 5,467 square miles (14,160 square kilometers). Belfast, the capital, has more than 277,000 residents.

Northern Ireland consists of the northeastern section of the island of Ireland. It is often called *Ulster.*

Ulster was the name of a large province of British-controlled Ireland until 1920. In 1920, the United Kingdom separated Northern Ireland from the rest of Ireland in order to create separate governments for the predominantly Protestant north and the mostly Roman Catholic south. The majority of the Northern Irish people supported the separation, but many Roman Catholics in both the north and the south refused to accept the division. In 1921, the south became the self-governing Irish Free State, now the independent Republic of Ireland.

Beginning in 1921, militant Irish groups, particularly the Irish Republican Army (IRA), attacked British government installations in Northern Ireland, hoping to force the British to give up control. Protestant groups retaliated. (Although those involved in the conflict are commonly identified according to religion, the dispute is primarily political and economic.) In the 1960's, the Roman Catholic minority held marches demanding an end to discrimination. The police reacted violently, and riots broke out. In 1968, the IRA and Protestant paramilitaries began committing acts of violence. In 1972, the United Kingdom established direct rule over Northern Ireland and sent in troops.

The violence continued until August 1994, when the IRA declared a cease-fire and the Protestant paramilitaries followed suit. However, the IRA ended its cease-fire in 1996.

Another cease-fire was announced in 1997, and formal peace talks, the first to include all parties involved in the conflict, began. In 1998, the talks ended with an agreement that was put to referendums in both Ireland and Northern Ireland. Voters approved the agreement, which included a commitment to using peaceful means to resolve political differences. The agreement also established a power-sharing legislature called the Northern Ireland Assembly.

In 1999, implementation of the agreement was delayed because leaders disagreed about when the IRA should begin to disarm. Then, in 2000, the agreement was suspended when the Protestants withdrew to protest a lack of disarmament by the IRA. The Northern Ireland Assembly was suspended in 2002.

Negotiations on the issues of self-government and disarmament continued.

Land

Northern Ireland is a land of rolling plains and low mountains. The fertile plains cover the central part of the area, and scenic green valleys and low mountains lie along the coast.

The countryside of Northern Ireland is dotted with smooth, clear lakes called *loughs* (pronounced *lahks*). Lough Neagh, the largest lake in the British Isles, covers 150 square miles (388 square kilometers) near the center of Northern Ireland.

Economy

About 15 per cent of the people of Northern Ireland live in Belfast, the capital and largest

A farm woman prepares a picnic lunch for harvest workers near the Mourne Mountains in the district of Down.

The Lusty Man, *right*—a stone carving of an ancient god dating from about A.D. 500—greets visitors coming and going at Lough Erne in the district of Fermanagh. The statue has two faces, carved on opposite sides.

city. Belfast is also Northern Ireland's manufacturing and trading center, and many of its linen mills, shipyards, and aircraft plants are located there. However, the economy depends mainly on service industries.

Since fertile pastureland is Northern Ireland's chief natural resource, agriculture is also an important industry. And Northern Ireland's fishing fleet processes cod, herring, mackerel, shrimp, and whiting, mainly from the Irish Sea.

Life in Northern Ireland is more like British life than like life in the Republic of Ireland. Such sports as soccer, cricket, and golf are popular pastimes, and pubs play an important role in social life.

***Orangemen* are Protestants of Northern Ireland** who belong to an organization called the *Orange Order*. This term dates back to the late 1600's, when King William of Orange, a Protestant, defeated King James II, a Roman Catholic, in a struggle for the English throne. Every July 12, Ulster Orangemen celebrate William's victory in the Battle of the Boyne.

An aircraft plant in Belfast, *left,* manufactures many aerospace products and ranks as a major employer in Northern Ireland.

Channel Islands and Isle of Man

The Channel Islands are a group of islands in the English Channel. Although they lie only about 10 to 30 miles (16 to 48 kilometers) off the coast of France, the islands have been attached to the English Crown since the 1000's.

The Isle of Man lies in the Irish Sea midway between England and Ireland. Great Britain has controlled the Isle of Man since 1765. But British laws do not apply to the island unless it is specifically named in the legislation.

The Channel Islands lie within sight of France, but they have been associated with England since the 1000's. Parliaments in Jersey, Guernsey, Alderney, and Sark administer the internal laws of the islands.

The Channel Islands

The four main islands of the Channel Islands are Jersey, Guernsey, Alderney, and Sark. Along with numerous smaller islets, the islands cover 76 square miles (197 square kilometers). The total population of the Channel Islands is about 150,000. English is the official language, but many islanders speak a French dialect that varies from island to island.

The islands, which have been largely self-governing since the 1200's, are divided into two administrative units called *bailiwicks.* A lieutenant governor assigned to each bailiwick represents the British monarch and handles international affairs. The four main islands—Jersey, Guernsey, Alderney, and Sark—have their own parliaments to regulate internal affairs.

Several of the islands have their own distinguishing characteristics. For example, Jersey, the largest of the islands, is known for its cows and for its sweaters, which are often called "jerseys." In the 1600's, so many men abandoned their farmwork to knit the jerseys that a law was introduced to ban knitting in the summer months.

Alderney and Guernsey are known for their cattle, and Guernsey is also noted for its financial services industry.

Sark, the smallest self-governing unit in the United Kingdom, has a democratic form of

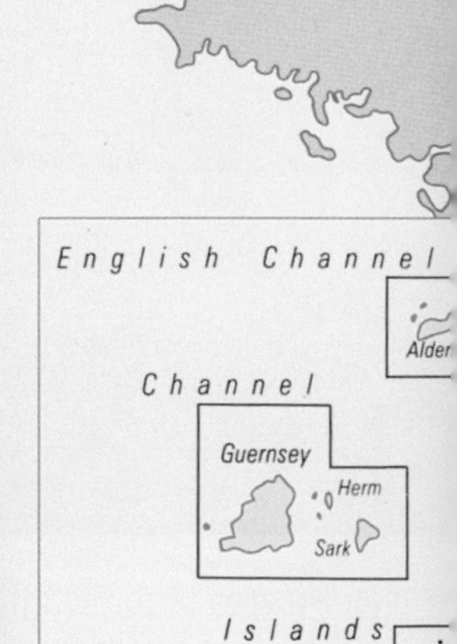

Yachts gather beneath Castle Mont Orguiel in Gorey, *left,* on Jersey's east coast.

Jersey's annual Battle of the Flowers, *far left,* takes place in July or August in St. Helier, the island's chief town.

Isle of Man

A dependency of the British Crown, the Isle of Man is an island in the Irish Sea, about halfway between England and Ireland and about 20 miles (32 kilometers) south of Scotland. The island has an area of 221 square miles (572 square kilometers) and a population of about 78,000. The people speak English and some also speak a Celtic language called *Manx.*

A representative of Great Britain oversees the island's foreign affairs. However a 1,000-year-old parliament called Tynwald Court regulates the island's concerns.

Crowds of tourists visit summer resorts on the Isle of Man, and its international motorcycle race, held each June, draws many enthusiasts. In addition to tourism, important industries include agriculture and fishing. Many new residents and industries have settled on the island since 1961, when the Isle of Man greatly lowered its taxes.

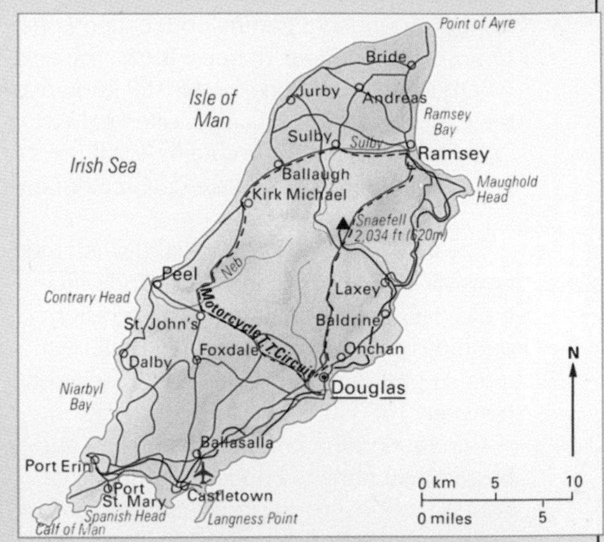

The Tynwald Ceremony, *right,* annually celebrates the Isle of Man's system of self-government. The island's parliament, Tynwald Court, is older than Great Britian's.

government headed by a *seigneur,* or feudal lord. The use of cars is prohibited on Sark, and the people travel by horse-drawn carriage or bicycle. However, the island is only 3 miles (4.8 kilometers) long and 1 1/2 miles (2.4 kilometers) wide.

The leading industry in the Channel Islands is tourism: The pleasant beaches and historic landmarks draw many visitors. The mild climate and fertile soil help make farming important as well. Farmers grow fruits, vegetables, and flowers, and raise cattle. Banking and other financial services are also major economic activities, especially on Jersey.

London

The history of London, the United Kingdom's capital and chief tourist attraction, goes back nearly 2,000 years to A.D. 43, when the Romans founded a trading port, called *Londinium,* on the River Thames. Today, the area where Roman London stood is still known as the City of London. This area, often called "the City," and the 32 *boroughs* (local units of government) around it make up Greater London, or simply London.

The City is London's financial district and consists largely of modern bank and office buildings. But it also contains such historic buildings as St. Paul's Cathedral, built by the great English architect Sir Christopher Wren between 1675 and 1710.

Central London contains the busiest and best-known parts of London. The Tower of London, once a royal prison, is now a museum where the British crown jewels are on display. In Westminster, the center of the United Kingdom's government, are the Houses of Parliament, Westminster Abbey, and Buckingham Palace. The city's main shopping and entertainment districts lie nearby.

London has many professional theaters and four world-renowned symphony orchestras. Numerous art galleries and museums are found in the city, and the South Bank section of central London is the site of a large, modern cultural center.

London is also rich in traditional ceremonies. Every morning, the famous changing-of-the-guard ceremony takes place in Buckingham Palace's courtyard. Trooping the Colour is a traditional part of the queen's official birthday celebration in June. The Lord Mayor's Show, on the second Saturday in November, celebrates the election of a new lord mayor of London.

Like most other large cities today, London has such problems as poverty, crime, and drug addiction, especially in the poorer districts. Yet London is also a thriving city with an economy based on service industries and manufacturing.

The City of London, *right,* is the city's financial center and the oldest part of London, where Roman London once stood. However, towering office blocks now dwarf many of the area's elegant older buildings.

Trafalgar Square, *left,* London's best-known square, honors Lord Horatio Nelson's naval victory over the French at Trafalgar in 1805. The huge column in the square is topped by a giant stone statue of Nelson, who died in the battle. London's National Gallery and the church of St. Martin-in-the-Fields border the square.

The street market on Portobello Road, *right,* sells everything from fresh fruit and vegetables to bargain jewelry and antiques.

The Trooping of the Colour in a parade ground known as the Horse Guards Parade, *above,* is one of London's most spectacular ceremonies. The royal Household Cavalry also changes the guard daily at Horse Guards Parade.

The Lord Mayor of London, *above,* leads the aldermen and the Common Council, who govern the City of London.

Greater London spreads across the River Thames Basin, extending outward from central London (shown on the inset map). The city is divided into the City of London and 32 local boroughs, each with its own government.

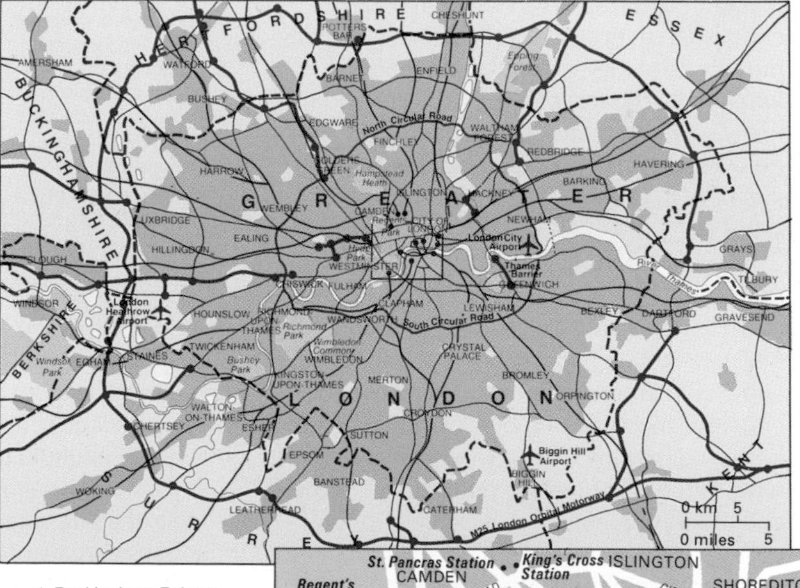

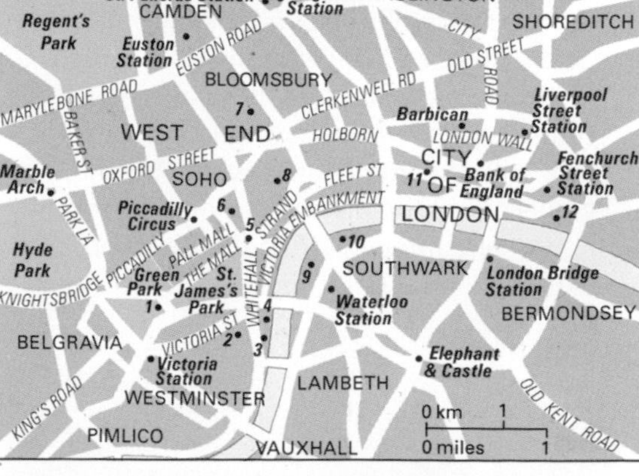

1. Buckingham Palace
2. Westminster Abbey
3. Houses of Parliament
4. Big Ben
5. Trafalgar Square
6. Leicester Square
7. British Museum
8. Covent Garden
9. Royal Festival Hall
10. National Theatre
11. St. Paul's Cathedral
12. Tower of London

Public houses are popular gathering places in London. Pubs serve all kinds of drinks, and usually offer inexpensive meals as well.

A Multiethnic Society

Most of the people of the United Kingdom descend from the many early peoples who invaded the region, including the Celts, Romans, Angles, Saxons, Scandinavians, and Normans. Many other people have come to the United Kingdom seeking religious freedom or better living conditions. During the late 1600's, for example, French Protestants fled to the United Kingdom to escape persecution in their own country. Many Jews from Eastern Europe also sought refuge in the United Kingdom during the late 1800's and early 1900's.

Since the 1950's, however, many immigrants from Commonwealth countries have settled in the United Kingdom. Most of these immigrants came from India, the West Indies, Pakistan, and Africa, with smaller percentages of people from Bangladesh (previously East Pakistan), Hong Kong, and some Arab countries. Today, about a million Commonwealth immigrants live in the United Kingdom.

Many of the immigrants came seeking factory jobs and settled in such industrial cities as London, Birmingham, and Leicester. Some immigrants also found employment in transportation and health services, while others established their own small businesses. Today, most British cities reflect the influence of immigrant cultures. Ethnic goods are sold in many specialty stores, and Indian and Chinese restaurants compete successfully with the more traditional British fish-and-chips shops.

Immigrants flooded into the United Kingdom at the rate of about 3,000 a month in the mid-1950's. In the 1960's, the government began to restrict the number of immigrants allowed to enter the country each year. Nevertheless, the large numbers of immigrants crowding into the United Kingdom's cities have resulted in housing, educational, racial, and other problems.

Racial tension is particularly a problem in inner cities, where the inhabitants must often deal with poor living conditions and high unemployment rates. In these areas, many British people resent having to compete for jobs with the immigrants, while the immigrants resent the discrimination they sometimes encounter. In 1981, this situation exploded in race riots in over 50 British cities and towns. Riots again broke out in urban areas in 1985.

People of many different races and backgrounds shop at an open-air market on a busy London street. They include Muslims from Bangladesh, immigrants from the Caribbean islands, and native Londoners.

A school playground, *lower right,* in inner London reflects the multiracial nature of most large cities of the United Kingdom. The concentration of immigrants in many of the country's inner cities has resulted in crowded living conditions.

Children of Indian ancestry, *right,* light candles during the Divali celebration, the Festival of Lights. Hindu people in many British cities also decorate the streets with lights during this festival, which lasts five days in fall.

Despite these problems, many people in the United Kingdom accept the society's multiracial structure. Citizens of Asian and West Indian descent hold positions of power in both local and central government. In some schools, the languages and histories of ethnic minorities are taught to encourage an appreciation of the country's different cultural heritages.

Nonetheless, many ethnic groups fear that traditional ties of family loyalty and patriotism will be weakened by exposure to British culture. While many immigrants accept British citizenship, they also wish to preserve their cultural and religious identity. Thus, they con-

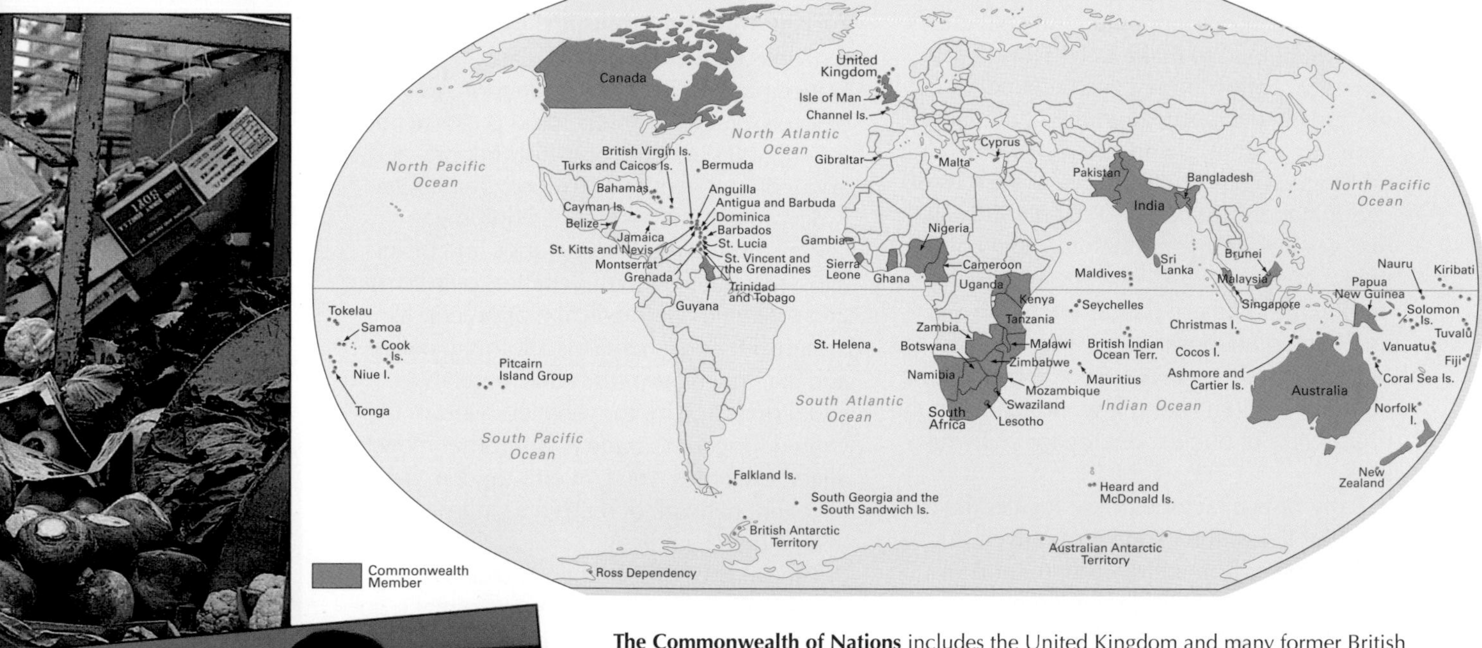

The Commonwealth of Nations includes the United Kingdom and many former British colonies. Many people have immigrated to the United Kingdom from independent Commonwealth countries and also from Commonwealth dependencies.

A reveler dressed in a traditional dragon costume dances through London's Chinatown during a Chinese New Year's Day parade. Most British Chinese have come into the country through Hong Kong.

tinue to worship at Hindu temples and Islamic mosques. In addition, the ethnic peoples in many large British cities publish their own newspapers. They also celebrate their traditional festivals, such as Divali (the Hindu Festival of Lights) and Chinese New Year.

In the past, invaders of the United Kingdom and immigrants have been absorbed by the country and have become part of British society. Some people believe that the United Kingdom's situation today is part of the continuing interaction between new arrivals and established residents. In fact, the children of many immigrant parents already pledge their loyalty to the United Kingdom rather than to the country of their origin.

Economy

The United Kingdom's plentiful coal and iron ore helped to make it the world's first industrial nation. Today, the demand for coal has greatly decreased, and the United Kingdom's iron ore is low in quality, but the United Kingdom has uncovered new sources of raw materials. In the 1960's and 1970's, natural gas fields large enough to supply all of the United Kingdom's needs were discovered in the southern part of the North Sea, and large oil deposits were discovered in the North Sea off the east coast of Scotland. The United Kingdom is also a world leader in the production of nuclear energy for peaceful uses.

Early British factories were located near the coal fields, their source of power. But today, power from petroleum, natural gas, and nuclear energy has enabled many new industries to develop in other areas.

Industry and agriculture

The United Kingdom is a leading producer of manufactured goods. British factories have long been known for their production of automobiles, ships, steel, and textiles. Today, in addition to these goods, the country produces heavy machinery, aerospace equipment, electronics, and chemicals. Other major industries include printing and publishing, the production of clothing, and food and beverage processing. Most major British industries are in central England, the London area, the Scottish Central Lowlands, the Newcastle upon Tyne area and southern Wales.

Service industries, however, account for about 70 per cent of the total value of goods and services produced within the country annually. About three-fourths of the United Kingdom's workers are employed in service industries. These include community, social, and personal services; wholesale and retail trade; and finance, insurance, and real estate.

Only about a third of British land is fertile enough for raising crops, but highly mechanized farming methods help the United Kingdom produce two-thirds of the food it needs. The country's most important crops are barley, potatoes, rapeseed (used to make livestock feed, vegetable oil, and industrial lubricants), sugar beets, and wheat. Sheep are the United Kingdom's chief livestock.

Foreign trade and transportation

Only the United States, Germany, Japan, and France outrank the United Kingdom in value of foreign trade. Manufactured goods account for about 85 per cent of British imports and about 80 per cent of its exports. The United Kingdom exports aerospace equipment, chemicals and medicines, foods and beverages, machinery, motor vehicles, petroleum, and scientific and medical equipment. Its imports include chemicals, clothing, coffee and tea, machinery, fruits and vegetables, metal ores, and motor vehicles. Most of the United Kingdom's trade is with other industrialized countries. The United Kingdom has an excellent network of roads and railroads and a large merchant fleet.

The value of British imports usually exceeds the value of its exports. To make up part of the difference, British banks and insurance companies sell their services to people and firms in other countries. Tourism is another important source of income. The income from all these "invisible" exports exceeds $200 billion a year.

British expertise contributed to the world's aerospace industry. In the 1970's the United Kingdom, together with France, built the Concorde, *right,* a supersonic transport plane that flew faster than the speed of sound. The Concorde stopped flying in 2003 because it proved too expensive.

Party politics

The Labour Party, which campaigned on a socialistic program, came to power in 1945 and converted the United Kingdom to a *welfare state.* The new government, which remained in power until 1951, expanded the social security system and launched a program of free medical care. The Labour government also began to put private industries, such as coal mining, steel, and the nation's railroads, under state control.

These measures were taken in an effort to restore the economy, which had suffered greatly during World War II. By 1955, economic conditions had improved, but by the early 1960's the welfare state had become difficult to finance and the United Kingdom had to borrow an increasing amount of money from other countries.

In 1979, the Conservative Party, under Prime Minister Margaret Thatcher, worked to reduce taxes and government involvement in the

Margaret Thatcher, prime minister from 1979 to 1990, worked to reduce government involvement in the economy and to encourage private enterprise.

The London Stock Exchange in the heart of the city's business district, *right,* is one of the United Kingdom's great financial institutions. The financial services provided by British banks contribute in large measure to the country's "invisible" exports.

In the early 1900's, many British industries were concentrated in the rich coal-mining areas of Wales and Scotland. Today, the use of such materials as petroleum and natural gas has led to the growth of industry in other areas, particularly in southern England.

economy. The Conservative government also began to sell state-owned industries to private buyers in a program known as "privatization." The goal was to replace the welfare state with an "enterprise economy."

The Conservative government's policies and the general decline of manufacturing eventually resulted in high unemployment. The rate peaked in 1986 but dropped significantly by 1988. Today, the government is aided by revenue from North Sea oil. However, Thatcher's policies became increasingly unpopular with the majority of the British people, and she resigned in November 1990. On November 28, John Major—chancellor of the exchequer in Thatcher's government—became prime minister. Major sought a closer union with the Economic Community, a group of countries that formed the European Union in 1993. He was defeated in 1997 elections by Labour Party leader Tony Blair.

Tony Blair became prime minister of the United Kingdom in 1997. He retained the post following elections in 2001 and 2005. The Labour Party, led by Blair, won election by large margins in 1997 and 2001.

Festivals

An extraordinary variety of events is celebrated in the United Kingdom. Some festivals, such as the *eisteddfods* (festivals featuring poetry and music) in Wales and the Highland Games in Scotland, highlight national cultures. Others, such as the Trooping of the Colour, honor the reigning monarch.

Many regional festivals, such as the Up-Helly-A' festival in the Shetland Islands, recall an area's past. In addition, there are music festivals, theater festivals, religious festivals, and flower shows, as well as celebrations that accompany special horse races and sporting events.

Sports events in the United Kingdom draw many spectators. Thousands of fans jam stadiums to watch soccer, the nation's favorite sport. Rugby football, cricket, tennis, and golf are also popular. The United Kingdom participates in international matches in all these sports.

Festive themes

Many British festivals appeal to particular groups, such as the upper class, the working people, or members of other cultures who have immigrated to the country. For example, high society goes on parade at the Ascot Races, particularly on Ladies' Day. Much of the spectacle of this annual event is provided by the exquisitely dressed women and top-hatted men.

On the other side of the social coin, English miners unite to celebrate their profession in the Durham Miners' Gala. And in London, the Notting Hill Carnival celebrates the rich musical tradition of immigrants from the West Indies. The carnival, which began as an effort to soothe the cultural clashes that erupted soon after the immigrants arrived, is probably the largest Afro-Caribbean festival in the world today.

Spring and summer festivities

Many festivals celebrate the return of spring. Whitsuntide, for example, a religious holiday observed throughout the United Kingdom on the seventh Sunday after Easter, celebrates returning hope and faith.

In May, the British, who are traditionally considered to be enthusiastic gardeners, turn out for the Chelsea Flower Show in London. The Glyndebourne opera season also begins in May with pomp. And the theater season in Stratford-upon-Avon starts in the spring.

Summertime brings more flower shows, horse shows, and—particularly in Scotland and Wales—many fascinating sheepdog trials. This is also the time for huge, tourist-attracting festivals, such as the Edinburgh International Festival held from mid-August to early September.

It is impossible to list all the festivals that take place in the United Kingdom because, it seems, there is something to celebrate on almost every day of the year.

Calendar of events

A sampler of British festivals, by location

Ascot:
Royal Ascot Races *(June)*
Badminton:
Horse Trials *(April)*
Braemer:
Braemer Royal Highland Gathering *(September)*
Brighton:
Arts Festival *(May)*
Cambridge:
King's College Carol Festival *(December)*
Epsom:
The Derby *(June)*
Edinburgh:
Edinburgh Festival *(August)*
Farnborough:
Farnborough Air Show *(September)*
Glastonbury:
Glastonbury Fair *(August)*
Glyndebourne:
Opera season *(opens May)*
Henley:
Rowing Regatta *(June)*
Padstow:
Padstow Hobby Horse Festival *(May)*
Shetland Islands:
Up-Helly-A' Celebration *(January)*
London:
Chelsea Flower Show *(May)*, Trooping the Colour *(June)*, Notting Hill Carnival *(August)*, State Opening of Parliament *(October)*, Lord Mayor's Procession *(November)*
Stratford:
Royal Shakespeare Theatre season *(opens March)*
Liverpool:
Grand National Steeplechase *(April)*
York:
Mystery Plays and Festival *(June)*
Wales:
Royal National Eisteddfod *(August)*

The Henley Rowing Regatta is a spectacular social event that takes place in June. It features parties with elegantly dressed women and men in straw hats called boaters. The world's first rowing regatta was held at Henley-on-Thames in 1839.

The Up-Helly-A' festival, *right,* in the Shetland Islands off northern Scotland dates back to the time of the Vikings. Islanders celebrate the "return of the sun" in January with feasting, drinking, and the burning of a *long ship*—a Viking warship.

The Notting Hill Carnival, *left and lower left,* is a high point of the London summer. Each August, thousands of visitors crowd the narrow streets of Notting Hill to enjoy the best in Afro-Caribbean music and dance.

The Royal Ascot Races are run for four days in mid-June, and combine high fashion with fast horses. On Ladies' Day in particular, women's exotic hats and outfits, *above left,* compete with the races for attention.

A morris dancer, *above,* awaits his turn to enter the folk dance at a Whitsuntide festival in Exeter, England. Morris dancers often represent characters from Robin Hood stories.

Early History

In A.D. 43, the Roman emperor Claudius invaded Britannia, as the island was then called, and began the conquest of the Celtic tribes who inhabited it. The Romans ruled until the early 400's, when they withdrew to defend Rome from invaders. The Britons were then attacked and driven into Wales by Germanic tribes, especially the Angles, Saxons, and Jutes, who set up kingdoms throughout southern and eastern England.

The Angles and Saxons soon became the most powerful tribes in England. The tribes developed into seven kingdoms, called the *Heptarchy.* King Egbert of Wessex, the last king to control the Heptarchy, is often considered the first king of England.

During the 800's, the Danish Vikings attacked and conquered all the Anglo-Saxon kingdoms except Wessex. In 886, Alfred the Great, King of Wessex, drove back the Vikings and pushed them into the northeastern third of England.

In 1066, Harold, the last Saxon king, was killed at the Battle of Hastings by Normans led by William, duke of Normandy in France. William was then crowned king of England, and became known as William the Conqueror.

Power struggles

During the late 1000's and early 1100's, the English government was centralized and its courts were reformed by several strong kings, like Henry II. In 1215, a group of barons and church leaders drew up the Magna Carta (Great Charter) and forced King John to agree to it. The document gave the nobles many rights that later became important to all the people.

In the late 1200's, Parliament began to gain greater importance. In 1295, King Edward I called a meeting of town representatives, nobles, and church leaders that became known as the "Model Parliament" because it set the pattern for later Parliaments.

Oxford University, above right, developed during the 1100's. Cambridge University grew up in the 1200's. The two schools helped make England an important center of learning.

After the Romans conquered England, they built walls to protect their settlements from the warlike peoples of Scotland. Hadrian's Wall, right, the most famous of these fortifications, was built by Emperor Hadrian in the A.D. 120's and extended from Carlisle to Newcastle upon Tyne.

A.D. 43 Romans invade Britain.
400's Germanic tribes invade Britain.
886 Alftred the Great, King of Wessex, defeats the Danes.
1066 Normans win the Battle of Hastings and conquer England.
1215 English barons force King John to sign the Magna Carta.
1282 England conquers Wales.
1295 Edward I forms the Model Parliament.
1314 Scotland keeps its independence by defeating Edward II at the Battle of Bannockburn.
1337–1453 England and France fight in the Hundred Years' War.
1455–1485 Two royal families fight for the throne in the Wars of the Roses.
1534 Henry VIII has Parliament make him the supreme head of the church in England.
1588 The English defeat the Spanish Armada.
1588–1603 Elizabeth I reigns during the Golden Age of English history.
1603 England and Scotland are joined under James I.
1649–1659 Charles I is executed and England becomes a Protectorate.
1660 Parliament restores the monarchy.
1688 The Glorious Revolution ends the rule of James II.

Queen Elizabeth I
(1533–1063)

William Shakespeare
(1564–1616)

Oliver Cromwell
(1599–1658)

In 1282, Edward I conquered Wales. He attempted to bring Scotland under English control, too, but the Scots rebelled repeatedly. Their victory over Edward II in 1314 at the Battle of Bannockburn assured Scotland's independence.

The 1300's and 1400's were years of conflict in England. The Hundred Years' War with France (1337–1453) was followed by the Wars of the Roses (1455–1485), a bitter struggle for control of the Crown. King Henry VII emerged as England's new ruler in 1485 and returned the country to stability. He also strengthened England's position among other nations by marrying his children into the families of other European rulers.

Reformation to 1688

Henry VIII, who became king in 1509, initiated the Protestant Reformation in England in 1529 after the pope refused to give Henry permission to obtain a divorce. The king had Parliament pass a law in 1534 naming the king, not the pope, supreme head of the church in England. This move led to the formation of the Church of England, independent of the Roman Catholic Church.

Elizabeth I became queen in 1558, and her reign is often called the Golden Age of English history. Explorers, such as Sir Francis Drake, and writers, such as William Shakespeare, brought glory to the Elizabethan era. In 1588, England defeated the Armada, a huge fleet built by Spain, the most powerful nation in Europe.

After Elizabeth died in 1603, her cousin, James VI of Scotland, also became James I of England, but each country kept its own laws and Parliament. In the 1600's, England was ruled by monarchs who wanted absolute power. Charles I did not allow Parliament to meet from 1629 to 1640. As a result, civil war broke out in 1642, and Charles was beheaded in 1649. England was made a "Protectorate," and Oliver Cromwell became "lord protector."

In 1660, the monarchy was restored under Charles II. However, civil war once again threatened under James II, who became king in 1685. His efforts to restore Catholicism to England were defeated during the Glorious Revolution in 1688, when leading English politicians invited James's daughter Mary and her husband, William of Orange, ruler of the Netherlands, to rule England. William invaded England with Dutch forces, and James fled to France.

Modern History

At the same time that Parliament turned over the throne of England to William and Mary in 1689, it moved to limit the power of the monarchy. The new rulers agreed to accept the Bill of Rights, which granted the people basic civil rights and curbed the king's power in such matters as taxation and keeping a standing army.

In 1707, under Queen Anne, the Act of Union joined England, Wales, and Scotland under one kingdom—the Kingdom of Great Britain. By this time, the English Parliament had won a controlling influence over the monarchy. After Queen Anne died in 1714, her cousin, George, a German prince, became king. However, George I spoke little English, so his chief minister, Sir Robert Walpole, controlled the council of ministers, and thus began the British Cabinet system of government. Walpole is considered Great Britain's first prime minister.

The British Empire

At the end of the Seven Years' War (known in America as the French and Indian War) in 1763, Great Britain acquired many of

The first car rolls off a Channel Tunnel train on Dec. 22, 1994. Work on the tunnel, a railroad between England and France that runs under the English Channel, lasted for seven years.

The Great Exhibition of 1851, *right,* held in London, reflected the national spirit of optimism and energy at a time when the United Kingdom was the most powerful and technologically advanced nation in the world.

1689 William and Mary accept the Bill of Rights and become joint rulers of England.

1707 England, Wales, and Scotland are united, forming the Kingdom of Great Britain.

1756–1763 Great Britain wins many of France's possessions in the Seven Years' War (also known as the French and Indian War).

1775–1783 Great Britain loses its American colonies in the Revolutionary War in America.

1793–1815 Great Britain defeats France in the Napoleonic Wars.

1801 Act of Union unites Great Britain and Ireland to form the United Kingdom of Great Britain and Ireland.

1837–1901 The British Empire reaches its height during the Victorian Age.

1914–1918 The United Kingdom and the Allies defeat Germany and the other Central Powers in World War I.

1921 The southern part of Ireland becomes a separate nation.

1939–1945 The United Kingdom and the Allies defeat Germany and the Axis Powers in World War II.

1947 India and Pakistan gain their independence.

1973 The United Kingdom becomes a member of the European Community, now known as the European Union.

1982 British troops defeat Argentine troops in battles for control of the Falkland Islands.

1987 Work begins on a railroad tunnel under the English Channel linking the United Kingdom and France.

1994 Channel Tunnel opens.

Duke of Wellington (Arthur Wellesley, 1769–1852)

Queen Victoria (1819–1901)

Sir Winston Churchill (1874–1965)

During the London Blitz in 1940, whole blocks of buildings were destroyed, *right.* Many people had to sleep in air-raid shelters and subway stations after their homes were destroyed by bombs.

France's territories in North America and India. The Revolutionary War in America (1775–1783) cost Great Britain the most valuable part of its empire—the colonies that became the United States. However, shortly after the war ended, Great Britain grew richer than ever before through trade with the new nation. In the late 1700's, the Industrial Revolution began in Great Britain and eventually made that country the richest nation in the world.

In 1793, Great Britain once again went to war against France, which from 1799 was led by Napoleon Bonaparte. In 1815, the British victory in the Battle of Waterloo ended the Napoleonic Wars.

From 1837 until 1901, Queen Victoria enjoyed the longest reign in British history. The British Empire reached its height during the Victorian Age.

The 1900's

The United Kingdom was on the winning side both in World War I (1914–1918) and World War II (1939–1945). But the wars devastated the nation's armed forces and its economy.

Between 1940 and 1980, about 40 British colonies became independent nations. Most remained associated with the United Kingdom through the Commonwealth of Nations.

In the early 1960's, the United Kingdom faced mounting economic problems. In 1973, in an effort to recover its economy, the United Kingdom joined the European Community, now known as the European Union.

In 1994, the United Kingdom and France constructed a railway tunnel under the English Channel. The tunnel linked the United Kingdom with mainland Europe.

Economic growth continued during the 1990's. After 18 years in power, the Conservative Party was defeated by the Labour Party, led by Tony Blair, who at age 43 became the youngest prime minister in more than a century. Blair called for referendums in Scotland and Wales to allow these areas to decide whether they wanted their own legislatures, and Scotland and Wales approved the plans for their own parliaments. Negotiations with Northern Ireland on self-government and disarmament continued. The Labour Party was reelected in 2001 and in 2005. Blair continued as prime minister.

The Greater Antilles

The West Indies are an island chain that separates the Caribbean Sea from the rest of the Atlantic Ocean. The larger West Indies, lying on the northern rim of the Caribbean, are called the Greater Antilles.

From west to east, the Greater Antilles consist of Cuba, Jamaica, Hispaniola (divided into the countries Haiti and the Dominican Republic), and Puerto Rico. Two very small island groups, the Cayman Islands and the Turks and Caicos Islands, lie near the Antilles area.

The islands are part of an underwater mountain chain that once linked North America and South America. Most of the islands were formed by volcanic eruptions. Others are coral and limestone formations.

The islands have great natural beauty. Sandy beaches and tall palm trees line the coasts, while lush tropical vegetation covers many of the islands. The many varieties of flowering plants on the islands include bougainvillea, hibiscus, orchid, and poinsettia. Such sport fish as marlin and sailfish, as well as brightly colored tropical fish, swim in the blue-green waters.

The islands have a warm, tropical climate, with temperatures averaging 80° F. (27° C) in the summer and 75° F. (24° C) in the winter. Rainfall averages 60 inches (150 centimeters) a year, with some mountainous areas receiving up to 200 inches (500 centimeters). Hurricanes often strike the islands, chiefly during the late summer and early fall.

The first inhabitants of the Greater Antilles were Carib Indians. In 1492, Christopher Columbus became the first European to see the islands. Various European nations, especially Great Britain, Spain, and France, eventually gained control of different islands. Today, most of the islands are independent nations.

Cayman Islands

The Cayman Islands are a British dependency about 200 miles (320 kilometers) northwest of Jamaica in the Caribbean Sea. Three islands form the group—Grand Cayman, Little Cayman, and Cayman Brac. The capital and largest city, George Town, stands on Grand Cayman, the largest island.

CAYMAN ISLANDS
Capital: George Town.
Islands: Grand Cayman, Little Cayman, Cayman Brac.
Area: 100 sq. mi. (259 km^2).
Form of government: British dependency.
Head of state: British monarch.
Estimated 2005 population: 44,000.
Official language: English.
Religion: Protestant (92%).
Currency: Caymanian dollar.

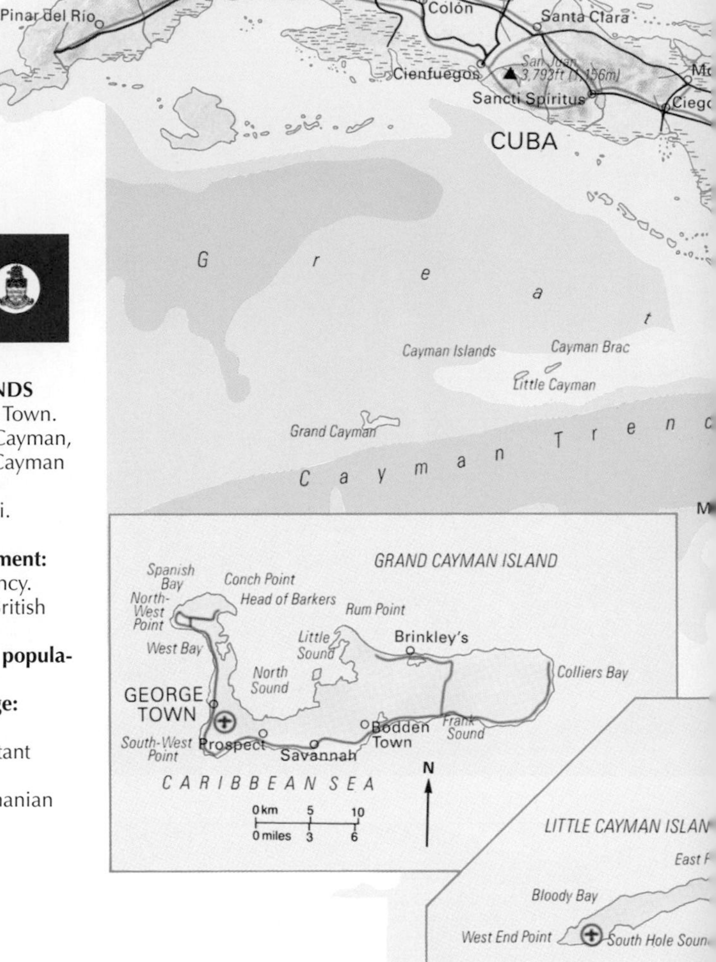

Farm production is low on the islands, and most food must be imported. Taxes are extremely low in the Caymans, so many foreign companies conduct business there. These businesses and the tourist industry are important to the economy. The islands are popular with scuba divers. The Pirate's Week festival in October also draws tourists, who come to see the costumes, parades, treasure hunts, and a staged pirate raid.

The Turks and Caicos Islands

The Turks and Caicos Islands are barren, sandy islands about 90 miles (140 kilometers) north of the Dominican Republic. Like the Caymans, the two island groups are a British dependency in the Commonwealth of

TURKS AND CAICOS ISLANDS
Capital: Grand Turk.
Main islands: Grand Turk, Grand Caicos, South Caicos, East Caicos, North Caicos, West Caicos, Salt Cay, Providenciales.

Area: 166 sq. mi. (430 km^2).
Form of government: British dependency.
Head of state: British monarch, represented by governor general.
Estimated 2005 population: 20,500.
Official language: English.
Religion: Protestant (95%).
Currency: U.S. dollar.

Sandbanks fringe the Turks and Caicos Islands, *above,* and waterways drain the barren landscape.

A tourist in the Cayman Islands, *left,* enjoys the blue sea and sandy beach. Ocean breezes keep the temperature mild, and a little shade offers relief from the tropical sun.

Nations. The capital and largest city is Grand Turk on the island of Grand Turk.

Many of the islanders make their living by fishing. Lobster is the main export.

In 1512, the Spanish explorer Juan Ponce de León sighted the islands. According to legend, the Turks Islands got their name because a red cactus flower on the islands reminded an early settler of a Turkish fez. The Caicos Islands probably got their name from the Spanish word for a *cay,* or small island.

The most important industries on the islands are financial services, fishing, and tourism. Lobster is the chief export.

Greece

Greece is a mountainous country whose landscape ranges from the sunny seacoasts of its many offshore islands to the rocky hills of the mainland. Greece's charming villages, where sea breezes turn windmills and whitewashed cottages glisten in the sunlight, have fascinated artists and poets throughout the ages. Even the rocky soil that covers most of the land sparked the imaginations of its ancient inhabitants.

According to an age-old Greek legend, God made the world by sifting earth through a strainer. He made one country after another with the good soil that sifted through. After He was finished, He threw away the stones that were left, and these stones became Greece.

Greece is a nation with a long and fascinating history, for on this nation's ancient shores, Western civilization began about 2,500 years ago. It was here that the ancient Greeks established the ideals of democracy. In the ancient monuments that dot the landscape, we see the traditions of justice, individual freedom, and representational government carved in stone.

The Greek people have received little benefit from the principles of democracy founded so long ago in their homeland. Shaky governments and political confusion have troubled Greece through most of its modern history. At times, strong military leadership has imposed control over chaos, but often at the expense of individual rights. In the 500 years of Turkish occupation that began in the 1300's, the Greek people suffered terrible blows to their pride and sense of independence.

Since it gained independence from Turkish control in 1829, Greece has rarely been free from political and economic difficulties. In 1897, Greece found itself at war with the Ottoman Empire over Turkish-held Crete. When Greece was declared a republic in 1923, its people were divided between the *republicans,* who supported the republic, and the *royalists,* who wanted a king. Later, the effects of World War II (1939–1945) almost destroyed the country's economy.

The promise of a better life through justice for all has been slow to come to Greece. Farmers still struggle to make a bare living off the land. Many families have lived in extreme poverty for generations.

But high on a hill in Athens, one of the magnificent achievements of the ancient Greeks can still be seen—the Acropolis. It stands as a reminder of the glories of the past and a symbol of hope for the future.

Greece Today

The dress uniform of today's *evzone* (elite guard soldier) is adapted from that of the Klephts, the mountain soldiers who helped fight the Ottomans in the 1800's. It consists of a *fustanella* (kilt), a widesleeved shirt, an embroidered *fermeli* (waistcoat), and a red cap with a long tassel.

The 1950's provided Greece with some relief from the political and financial problems it had suffered since the early days of independence. The Western powers provided massive military and economic aid to the country, and Greece joined the North Atlantic Treaty Organization (NATO) in 1952. During this time, the government improved the country's finances, controlled rising prices, and encouraged agriculture and industry to expand.

Rule of the colonels

In 1963, George Papandreou of the Center Union Party became prime minister of Greece. But Papandreou disagreed with King Constantine II on who should have political power and control over the armed forces. After Constantine dismissed Papandreou in 1965, another period of political confusion rocked the government of Greece. Only a month before new elections were to be held, the government was taken over by Greek army units.

Although King Constantine II remained head of state, he was powerless. A junta of three army officers led by Colonel George Papadopoulos set up a military dictatorship that suspended the rights of the people, prohibited all political activity, and made mass arrests. The junta also imposed harsh controls on newspapers, and dissolved hundreds of private organizations of which it disapproved.

In 1973, Papadopoulos announced the end of the monarchy and declared that Greece would be a republic. But even as the government was preparing to hold parliamentary elections, a group of military officers once again overthrew the government.

In 1974, when Greek officers on the island of Cyprus helped Cypriot troops overthrow their government, Turkey accused Greece of violating the independence of Cyprus and Turkish troops invaded the island. After several days of fighting, a cease-fire was signed.

Shortly after the cease-fire, the government of Greece collapsed under the combined pressure of the Cyprus crisis and economic recessions. Constantine Caramanlis, head of the New Democracy Party, then became prime minister.

In November 1974, the Greek people held their first free elections in 10 years. In December of 1974, they voted to make the country a republic. Political life began to return to normal, as a new Constitution was adopted and civilian control over the military was established.

FACT BOX

COUNTRY

Official name: Elliniki Dimokratia (Hellenic Republic)
Capital: Athens
Terrain: Mostly mountains with ranges extending into the sea as peninsulas or chains of islands
Area: 50,949 sq. mi. (131,957 km^2)
Climate: Temperate; mild, wet winters; hot, dry summers
Main rivers: Vardar, Aliakmon, Pinios, Arakhthos
Highest elevation: Mount Olympus, 9,570 ft. (2,917 m)
Lowest elevation: Mediterranean Sea, sea level

GOVERNMENT

Form of government: Parliamentary republic
Head of state: President
Head of government: Prime minister
Administrative areas: 51 nomoi (prefectures), 1 autonomous region
Legislature: Vouli ton Ellinon (Parliament) with 300 members serving four-year terms
Court system: Supreme Judicial Court, Special Supreme Tribunal
Armed forces: 171,000 troops

PEOPLE

Estimated 2008 population: 11,128,000
Population density: 218 persons per sq. mi. (84 per km^2)
Population distribution: 60% urban, 40% rural
Life expectancy in years:
Male: 76
Female: 81
Doctors per 1,000 people: 4.4
Percentage of age-appropriate population enrolled in the following educational levels:
Primary: 99
Secondary: 96
Further: 68

The mainland of Greece lies on the southern tip of the Balkan Peninsula. The large island of Crete and 437 small islands lie in the Mediterranean Sea. The small islands make up a fifth of Greece's land area.

Languages spoken:
Greek 99% (official)
English
French

Religions:
Greek Orthodox 98%
Muslim 1%

TECHNOLOGY

Radios per 1,000 people: 466

Televisions per 1,000 people: 519

Computers per 1,000 people: 81.7

ECONOMY

Currency: Euro

Gross domestic product (GDP) in 2004: $226.4 billion U.S.

Real annual growth rate (2003–2004): 3.7%

GDP per capita (2004): $21,300 U.S.

Goods exported: Cement and cement products, clothing, metal products, olive oil, petroleum products, prepared fruits, textiles

Goods imported: Chemicals, machinery, manufactured goods, meat, petroleum and petroleum products, transportation equipment

Trading partners: European Union, United States

The Socialist era

In 1981, the Panhellenic Socialist Movement (PASOK) party formed the first Socialist government in Greece. Andreas Papandreou became prime minister.

The Socialists were defeated in elections in 1989 but won control again in 1993. Papandreou served as prime minister until 1996, when he resigned because of ill health. Costas Simitis replaced him. Simitis remained prime minister until 2004, when the New Democracy Party won control of Parliament. Costas Caramanlis became prime minister.

Environment

About 70 per cent of the land in Greece is composed of limestone mountains and hills, which are either bare or covered with patches of thorny, woody shrubs. Yet, however desolate the upland areas may appear, they have a strange, rugged beauty all their own.

Cypress, fir, myrtle, and fig trees grow on the mountainsides. In the spring, blooming red poppies blanket the slopes. The uplands give way to river valleys, narrow coastal plains, and the long fingers of land jutting out into the Mediterranean Sea that give the Greek landscape its unique character.

Ruins cling to the hillsides, *above left,* in what was once the town of Mistras on the Peloponnesian Peninsula. Founded as a Frankish fortress in 1248, the town was taken by Byzantine rulers in 1262. Under their control, Mistras became a center of Byzantine culture.

Time has changed the rugged Greek coastline. The blue waters of the Mediterranean Sea splash against the rocks, but the natural beauty of Greece has attracted throngs of tourists and lined its beaches with hotels and resorts.

Land regions

Greece has often been described as the "Land of the mountains and the sea"—and, except for a few districts in Thessaly, no part of the country is more than 85 miles (137 kilometers) from the ocean. The mountain ranges and the sea divide Greece into several land regions.

The Pindus Mountains, which extend southward down the "backbone" of Greece, are an extension of the Dinaric Alps. The Pindus are composed mainly of limestone, with large areas of *karst* (eroded limestone). Many underground streams run through this striking landscape of steep slopes and deep ravines. Sheep and goats graze on the mountain pastures, while the region's two river valleys—the Ptolemais and the Ioannina—are the main population centers.

The coastal plains and lowlands are the center of the nation's agricultural development. The region of Thessaly, surrounded by tall mountains in the east-central part of the country, is known as the breadbasket of Greece because fields of wheat cover most of its cultivated land.

In northeastern Greece, the river basins and alluvial plains of Macedonia-Thrace produce plentiful harvests of tobacco and other crops. Farmers on the Salonika Plain, located on the southwestern tip of Macedonia-Thrace, grow cotton, fruits, rice, and wheat.

The Peloponnesus is a large peninsula connected to the Greek mainland by an isthmus. The Corinth Canal cuts across the isthmus, and virtually makes the Peloponnesus an island. This region of rugged mountains, small valleys, and rocky coasts is not suitable for farming, but the Peloponnesus is rich in history. The ruins of Corinth, Olympia, and other historic sites still stand on the peninsula.

The most populated region of Greece is the Southeastern Uplands, a region of mountains and hills with many small valleys among them. Athens, Greece's capital and largest city, is located on the southern tip of the Southeastern Uplands.

Climate and vegetation

Greece enjoys a typical Mediterranean climate, with hot, dry summers followed by mild, wet winters. However, the climate varies greatly, depending on altitude and location. Winters can be extremely cold and summers can be very hot in the north and inland regions, with snow and freezing temperatures in the mountains.

The trees, shrubs, and small plants that grow in Greece are well adapted to the long, dry summers. Such typical Mediterranean crops as lemons and olives thrive in the lower altitudes, and the air is scented by the aromatic herbs typical of the

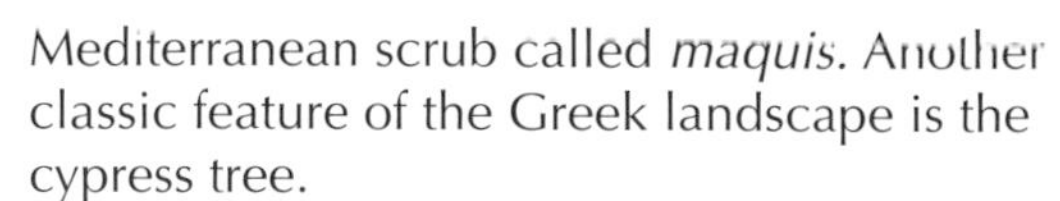

Mediterranean scrub called *maquis*. Another classic feature of the Greek landscape is the cypress tree.

Greece has suffered greatly from the loss of the dense forests of oak, chestnut, beech, and plane trees that once covered the land. Today, only about 20 per cent of the forestland remains uncleared. The destruction of the forests began over 2,000 years ago, when the Greeks cut down trees for fuel and shipbuilding.

In the 1900's, forestland has been cleared for livestock grazing. With financial aid from other countries, the Greek government has now begun to take steps to reforest the land.

Vineyards and olive groves, *above,* form a pattern of cultivation in the valleys and plains that lie among the mountains on the island of Crete. Irrigation must be used on the island's many farms during the dry summer months.

Mouse Island lies off the east coast of Corcyra (Kerkira or Corfu). According to legend, the island, called Pontikonísi in Greek, was once a boat that was turned to stone by Poseidon, the Greek god of the sea, after it carried Odysseus back to Ithaki (Ithaca).

People

About 95 per cent of the people in Greece belong to the same cultural group. Minorities such as Macedonians, Turks, Albanians, and Bulgarians make up the other 5 per cent. Almost all of the people in Greece speak modern Greek, a language that developed from classical Greek, and most of them belong to the Greek Orthodox Church.

Over the centuries, the Greeks have developed a strong cultural unity and a deep sense of national identity. This feeling of community has flourished in spite of the fact that the geography of Greece—a land of mountains divided by valleys and plains—has always made communication difficult.

Waves of emigration

In recent years, many rural people have left their villages, hoping to find jobs and a better life in the cities. Today, only one-third of the people live in rural communities. Almost one-third of the entire population of Greece lives in the Athens metropolitan area.

Many Greeks have left their homeland to live in other countries. Some fled to escape the country's political turmoil, and others left to seek better jobs. After civil war broke out in Greece in 1945, political refugees were forced to flee the country. In the early 1960's, many Greek workers settled in Western Europe, especially in Germany.

Village life

As an increasing number of young people move to the cities in search of higher-paying jobs, the populations of the mountain villages and the islands consist mainly of the older people left behind. Some of those who have remained in the rural areas still live the traditional life of the Greek peasant, but others have turned away from many of the age-old customs.

In the past, Greek villagers maintained strong family ties, and each family member had a specific role and certain responsibilities. Parents made all the major decisions for their children, including selecting or approving the person their child would marry. Daughters had to marry in order of age—with the eldest wed first—and sons were allowed to marry only after all their sisters had married.

These rigid expectations applied to husbands and wives as well. The husband took care of the household's external affairs, while the wife was primarily concerned with taking care of the house and the family.

In rural areas today, people do not follow the old customs as closely as they once did, and in the cities the old ways have all but died out. Modern life styles have gradually loosened the strong family ties of the past.

However, such traditions as the Greek Orthodox festivals remain an important part of Greek life. Almost every city, town, and village has a patron saint, and festivals celebrate the saint's yearly feast day. The people enjoy food and wine after the church service, and there is singing and dancing far into the night.

Some people dress in colorful national costumes during the festivals. The men wear heavily braided jackets and pleated kilts over woolen tights, and the women wear long, brightly colored skirts and full-sleeved white blouses.

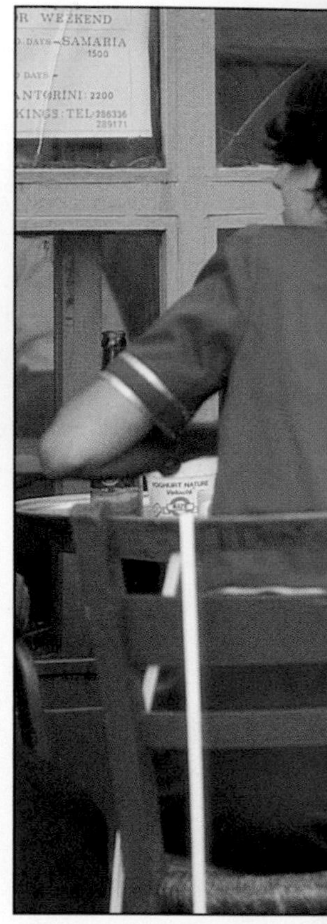

A group of Greek townspeople chat in a *kafenion,* or coffee house, *above.* Conversation with friends—especially about politics—is a highlight of daily life for Greek men. Women, who generally have a less active public role in Greek society, are seen less often in coffee houses.

A shopkeeper on the Ionian island of Corcyra (Kerkira or Corfu), *left,* checks his display of fresh vegetables. Many of the foods on display are ingredients of *horiátiki,* a traditional Greek salad.

Greek food

Eating and drinking has always been an important part of Greek hospitality. Even in the poorest village home, guests are welcomed with something sweet to eat, a cup of strong coffee, and a glass of cold water.

Greeks generally eat their main meal in the middle of the day. It may include such popular Greek dishes as *soupa avgolemono* (lemon-flavored chicken soup), *dolmathes* (vine leaves filled with rice and ground meat), and *souvlaki* (meat cooked on a long skewer, usually with onions and tomatoes).

The Greeks eat dinner very late at night—usually between about 8 and 11 p.m. This late dinner usually is a light meal of salad, cheese, fresh fruit, and a glass of wine.

A member of the congregation kisses the hand of a Greek Orthodox priest at a religious festival, *below left.* The Greek Orthodox Church is a self-governing member of the Eastern Orthodox Churches, a federation of Christian churches in Greece, the former Soviet republics, Eastern Europe, and western Asia.

A woman on the island of Crete greets her neighbor on the way back from the market. The Greeks are a lively people who greatly enjoy conversation and the company of others.

Athens

Modern Athens, one of the world's most historic cities, is a unique combination of past and present. Amid its modern factories, high-rise apartments, shopping centers, and restaurants stand the reminders of the city's ancient beginnings, when Athens was the cradle of Western civilization—the world's first center of culture.

The voice of the great philosopher Socrates once rang through the Agora, the ancient hub of Athens' public life, where he taught philosophy some 2,400 years ago. In the open-air Odeon, where western European comedy and drama were first developed by ancient Greek playwrights, present-day actors play their parts. And high above the city, on a rocky hill called the Acropolis, stands the majestic Parthenon—an ancient temple dedicated to Athena, the city's patron goddess.

The presence of these ancient ruins gives Athens a truly timeless spirit. They are reminders of the Golden Age of Greece—and its magnificent achievements in science, government, philosophy, and the arts—achievements that still influence our lives today.

The ancient city

Historians know little about the history of Athens before about 1900 B.C., when the Greeks first occupied Attica, a peninsula that extends from southeastern Greece into the Aegean Sea. On and around a great flat-topped hill covering slightly more than 10 acres (4 hectares), the Greeks built a city. The hill became known as the Acropolis, from the Greek words *akro* (high) and *polis* (city), and the city became known as Athens, for its patron goddess.

In 480 B.C., most of the buildings on the Acropolis were destroyed by an invading Persian army. But by 447 B.C., the Athenians began to rebuild under the leadership of Pericles. The structures of this period, which is known as the Golden Age of Greece, still dominate the Acropolis. The greatest of these structures—and perhaps the best example of ancient Greek architecture—is the Parthenon.

The Parthenon was originally decorated with brightly painted sculptures that illustrated important events in the life of Athena, but the colors faded centuries ago. Then, like other buildings on the Acropolis, the Parthenon suffered serious damage when Greece was part of the Turkish Empire. The Ottoman Turks used it for storing gunpowder, which exploded and destroyed the central part of the building.

In 1802, Lord Elgin, the British ambassador to Constantinople, began collecting some of the Parthenon's finest sculptures. Between 1803 and 1812, he shipped his collection to England. Today, the collection, known as the Elgin Marbles, remains on display at the British Museum, in spite of Greek demands for its return.

A street vendor sells embroidered cloth at a stall in Monastiraki Square, the heart of Athens' old market district. For centuries, Greek craftworkers have been famous for their fine embroidery.

The city then and now

Although the Acropolis is the most striking reminder of Athens' glorious past, historic sites can be found throughout the city. At the famous old markets of Monastiraki Square, blacksmiths work their trade just as their ancestors did 2,600 years ago.

Around the corner stand the ruins of the Agora and many of the Agora's buildings, including its *stoas* (covered arcades). This

On the Acropolis, high above the city of Athens, stand the ruins of the majestic Parthenon. Built between 447 and 432 B.C., this ancient Greek temple once held a huge gold and ivory statue of the goddess Athena, to whom the temple was dedicated.

Athens, the capital and largest city in Greece, is located about 5 miles (8 kilometers) from the seaport of Piraeus, *below.* Athens was home to many of the world's great writers, philosophers, and artists. Some of the masterpieces they left behind can still be seen, including the Parthenon, the Erechtheum, and the Temple of Olympian Zeus. Many other ancient treasures are displayed in the National Archaeological Museum, the Benaki Museum, the Acropolis Museum, and the Byzantine Museum.

marketplace was once the center for trade, athletic displays, dramatic competitions, and philosophical discussions.

Modern Athens, the leading cultural, economic, and financial center of Greece, has grown up around its historic treasures. The city's National Archaeological Museum houses masterpieces of ancient Greek jewelry, pottery, and sculpture. Its factories manufacture cement, chemicals, clothing, and other products. And its thriving tourist industry welcomes visitors from all over the world.

There is much in modern Athenian life to delight the visitor—from a high-spirited conversation in the traditional *kafenions* (coffee houses) to a sampling of delicious pastries in the *tavernas* (cafes). The city's activities center around its three main squares—Syntagma, Omonoia, and Monastiraki. But no matter where one travels in Athens, the past is always present and alive. Each monument symbolizes the rich intellectual and artistic spirit that gave birth to this city and lives on in the hearts of its people today.

Young Athenians enjoy a social gathering at a *taverna* (cafe), *left.* Taverna patrons may dine on tasty Greek specialties, join in singing *rebetika* (folk songs), and dance to folk music featuring clarinets and *bouzoukis* (stringed instruments similar to mandolins).

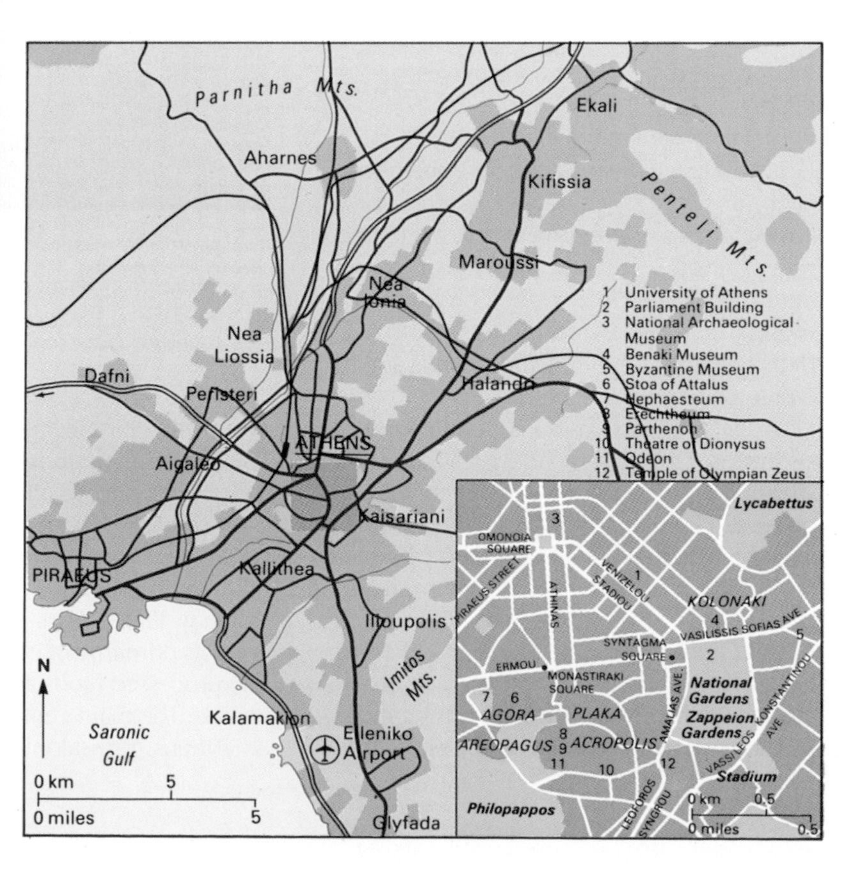

Economy

The economy of Greece was almost destroyed during World War II (1939-1945) and during the Greek civil war (1946-1949). Although still weak by Western European standards, the Greek economy has expanded greatly since the 1950's.

Service industries, taken together, account for more than 60 per cent of Greece's gross domestic product (GDP). Service industries employ more than 50 per cent of the country's workers. Community, social, and personal services produce a larger portion of the GDP than any other industry. This industry includes such economic activities as education and health care. Other important service activities include banking, government services, trade, and transportation. Tourism benefits many of Greece's service industries.

Agriculture

Only about 40 per cent of Greece's entire land area is suitable for farming because of the mountainous landscape and the scarcity of fertile soil. The nation's farmers raise such field crops as cotton, tobacco, vegetables, and grains. Wheat is Greece's main crop. In addition, grapevines, olive trees, and lemon and other fruit trees are grown in such abundance that Greece ranks among the world's leading producers of lemons, olives, and raisins.

For many generations, family lands in Greece were divided among the members of each succeeding generation. As a result, Greek farms are generally small, averaging only 8 acres (3.2 hectares) in size. Not only does the size of the land limit production, but most farmers use old-fashioned methods and tools because the hilly terrain makes it difficult to use modern equipment.

About another 40 per cent of the land consists of pastures and meadows, where cattle, sheep, and goats are grazed, but the quality of livestock in Greece is generally poor. However, higher-quality breeds are now being imported from other countries to improve the stock.

The Greek fishing industry has suffered from a drastic reduction in the catch off coastal waters. At present, the fishing industry has seriously declined, and Greece must now import large quantities of fish to meet domestic demands.

Industry

Industrial development in Greece is limited by the country's lack of raw materials. Mineral deposits are varied but limited. About 90 per cent of the nation's major mining product—a low-quality brown coal called *lignite*—is used to generate electricity. Other important minerals include asbestos, barite, bauxite, chromite, iron ore and pyrite, lead, magnesite, marble, and nickel.

In order to develop its manufacturing industry, Greece needs much more electric power than it now has. The country has no commercially important deposits of natural gas, and its only significant petroleum deposit is located in the northern Aegean Sea, where geological characteristics make it difficult to drill for oil.

A shopkeeper takes stock of his selection of cheeses. Because sheep and goats outnumber cattle on Greek pastures, most Greek cheeses are made from the milk of goats or sheep. Feta is the most popular cheese.

Today, Greece's manufacturing industry consists primarily of cement, cigarettes, clothing, processed foods and beverages, and textiles. The main centers of industrial activity are Athens, Thessaloniki, and Patrai (Patras).

The whitewashed walls of a country church stand out against the green trees in a Greek olive grove. The olive is one of the few crops that thrive in the thin, stony soil of Greece.

Large ships look like toy boats against the towering walls of the Corinth Canal, *below.* This deep channel—just wide enough for a single ship—crosses the land that connects the Peloponnesus with the northern mainland.

Grapes are spread on brown paper to dry in this vineyard on Crete, *right.* In about 10 to 14 days, the grapes will become raisins. The same hot sun that parches the earth provides ideal conditions for cultivating raisins.

History

The first important civilization in the region that is now Greece arose on the Aegean island of Crete about 3000 B.C. Known as the Minoan culture, it flourished until about 1450 B.C. About 2000 B.C., settlers from the north began to develop small farming villages on the mainland of Greece. The culture they developed is called Mycenaean, after the large and powerful town of Mycenae in the Peloponnesus.

The Mycenaeans were in contact with the Minoan culture on Crete and adopted some aspects of that culture, such as their system of writing. Shortly after 1200 B.C., Mycenae and most other settlements in the Peloponnesus were destroyed, although no one knows for sure why this happened.

Soon, a people called the Dorians moved into the region from northern Greece, and many of the Mycenaeans fled to Asia Minor. The Peloponnesus entered a period known as the Dark Age, where people lived in isolated settlements and lost the knowledge of writing.

It was during the Dark Age that independent city-states began to develop. At first, the city-states were ruled by kings, but by about 750 B.C., the nobles in most city-states had overthrown the kings and seized power.

Beginning in 477 B.C., the city-state of Athens reached the height of its power and prosperity as the center of culture in the Greek world. The Golden Age ended when Athens and Sparta, its rival city-state, went to war in 431 B.C. The prolonged struggle, known as the Peloponnesian War, lasted until 404 B.C., ending with the surrender of the Athenians.

The Hellenistic Age

Continuing warfare between the city-states caused them to become so weak that in 338 B.C., they lost their independence to invading armies from Macedonia, a country to the north of Greece. The Macedonian ruler Alexander the Great spread the Greek culture throughout his vast empire, from Greece to India.

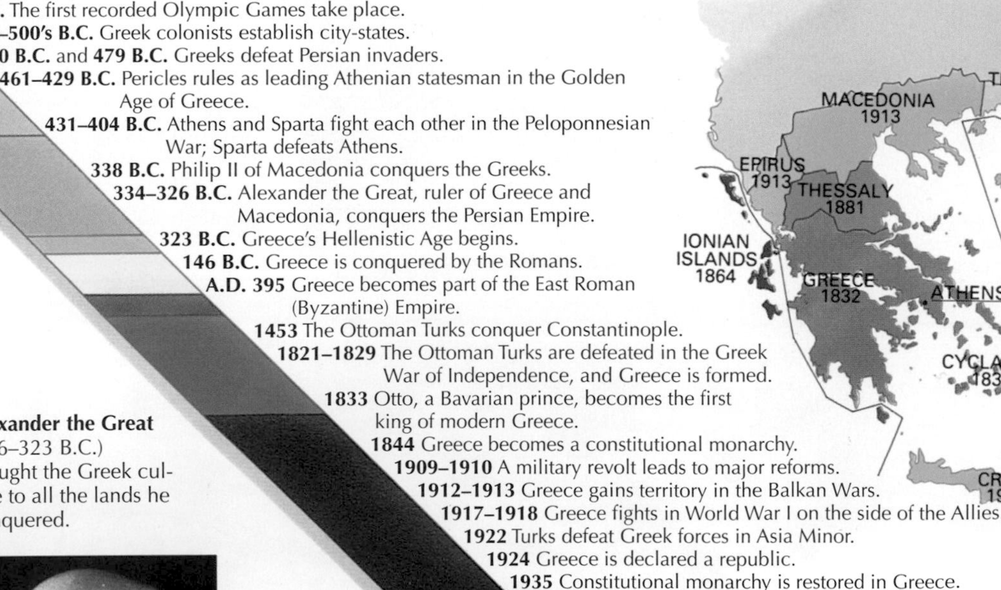

- **c. 3000 B.C.** Minoan culture develops on Crete.
- **1600–1200 B.C.** Mycenaean culture develops on Greek mainland.
- **776 B.C.** The first recorded Olympic Games take place.
- **700's–500's B.C.** Greek colonists establish city-states.
- **490 B.C.** and **479 B.C.** Greeks defeat Persian invaders.
- **461–429 B.C.** Pericles rules as leading Athenian statesman in the Golden Age of Greece.
- **431–404 B.C.** Athens and Sparta fight each other in the Peloponnesian War; Sparta defeats Athens.
- **338 B.C.** Philip II of Macedonia conquers the Greeks.
- **334–326 B.C.** Alexander the Great, ruler of Greece and Macedonia, conquers the Persian Empire.
- **323 B.C.** Greece's Hellenistic Age begins.
- **146 B.C.** Greece is conquered by the Romans.
- **A.D. 395** Greece becomes part of the East Roman (Byzantine) Empire.
- **1453** The Ottoman Turks conquer Constantinople.
- **1821–1829** The Ottoman Turks are defeated in the Greek War of Independence, and Greece is formed.
- **1833** Otto, a Bavarian prince, becomes the first king of modern Greece.
- **1844** Greece becomes a constitutional monarchy.
- **1909–1910** A military revolt leads to major reforms.
- **1912–1913** Greece gains territory in the Balkan Wars.
- **1917–1918** Greece fights in World War I on the side of the Allies.
- **1922** Turks defeat Greek forces in Asia Minor.
- **1924** Greece is declared a republic.
- **1935** Constitutional monarchy is restored in Greece.
- **1936–1941** General Joannes Metaxas rules as dictator.
- **1941–1944** Axis forces occupy Greece during World War II.
- **1952** Greece joins NATO.
- **1960** Cyprus becomes independent.
- **1967** Army officers seize the Greek government and suspend the Constitution.
- **1973** Premier George Papadopoulos abolishes monarchy. Later, military officers overthrow Papadopoulos government.
- **1974** Greece holds free elections, and a civilian government is formed.
- **1993** Greece and other members of the European Community form the European Union.

Alexander the Great (356–323 B.C.) brought the Greek culture to all the lands he conquered.

Plato (427?–347? B.C.), *far left,* was a great philosopher of ancient Greece.

Eleutherios Venizelos (1864–1936) served as prime minister of Greece six times.

The *Tholos* (rotunda) at Delphi, *top,* was built shortly after 400 B.C. Delphi, the oldest and most influential religious site in ancient Greece, was home to the famous *oracle* (prophet). The Greeks believed that the oracle spoke the words of Apollo.

The famous Lion Gate, *above,* leads to the acropolis of Mycenae, the first major civilization to develop on the Greek mainland. A massive beam across the top supports a triangular bas-relief of two rearing lionesses.

After Alexander's death in 323 B.C., Greece entered a period known as the Hellenistic Age, when Greek ideas continued to spread throughout Alexander's empire. When the Romans took control of the city-states in the 140's B.C., they adopted much of the Greek way of life and spread it throughout their empire.

When the Roman Empire was divided in A.D. 395, Greece became part of the East Roman Empire—also known as the Byzantine Empire. Despite centuries of war and invasions from neighboring peoples, the Byzantine Empire continued to control at least part of Greece for over 1,000 years. The capture of the Byzantine capital of Constantinople by the Ottoman Turks in 1453 marked the end of the Byzantine Empire and of Greek independence for nearly four centuries.

The making of modern Greece

Although a Greek independence movement developed in the 1700's, the Greek War of Independence did not begin until 1821. Six years later, Great Britain, France, and Russia sent troops to end the fighting and establish Greece as a self-governing country. In 1833, a Bavarian prince, Otto, became the first king of Greece.

The new kingdom was less than half the size of present-day Greece. About 3 million Greeks lived in Ottoman territory, and 200,000 lived in the British-controlled Ionian islands. The resulting economic confusion and political discontent led to a peaceful revolution in 1862 that forced Otto I to give up his throne. He was replaced by a Danish prince who became George I.

In 1912–1913, Greece and several other Balkan states defeated the Ottoman Turks in the First Balkan War. The Balkan states then fought against each other during the Second Balkan War, in 1913. As a result of these wars, Greece gained the island of Crete, southern Epirus, part of Macedonia, and many Aegean islands. In 1924, Greece declared itself a republic, which lasted until 1935, when the monarchy was temporarily restored.

Ancient Greek Civilization

The modern world owes much to the ancient Greeks. The ancient Greeks developed the first democratic government; drama was born in their huge, open-air theaters; and Greek thinkers developed the reasoning skills necessary for demonstrating important mathematical principles. The splendid Olympic Games—the world's most honored sporting tradition—first took place in the Stadium of Olympia in Greece.

The magnificent achievements of the ancient Greeks are still with us today—in the majestic temples of the Acropolis, in the poetry of Homer's *Iliad* and *Odyssey,* and in our modern Olympic games. But perhaps most important, the ancient Greeks encouraged creative thinking, valued personal freedom, and explored the human potential. They laid the foundation for the continuing search for knowledge, truth, and new forms of expression throughout generations of Western civilization.

The Golden Age of Greece

During the 500's B.C., the Persian Empire conquered the Greek city-states of Asia Minor. From 499 to 494 B.C., the city of Athens aided the captured city-states in rebelling against Persian rule. Athens also played a leading role in preventing the Persians from gaining more Greek territory.

After the wars with Persia, Athens became head of the Delian League, an organization of city-states formed as a continuing defense against the Persians. The league quickly developed into the Athenian empire, and Athens became the literary and artistic center of Greece. Under the leadership of the great statesman Pericles, the Greeks enjoyed a period of outstanding cultural achievement, known as the Golden Age.

Of all the city-states in ancient Greece, Athens had the most successful democracy. Along with its advanced political system, Athens enjoyed great prosperity as an international trading center. Athenian merchant ships carried olive oil, painted pottery, wine, wool, and other goods to ports in Egypt, Sicily, and Scythia—a country on the Black Sea. There, Greek merchants would sell their goods for slaves and for such products as grain, timber, and metals.

Philosophical and artistic triumphs

Philosophy originated in ancient Greece, and Athens nourished its most important teachers and philosophers—Socrates, Plato, and Aristotle. They often gathered with their pupils in the *Agora,* the marketplace in the center of the city, to discuss philosophical issues. The early philosophers wondered about the substance of the universe and how it operated. Later philosophers explored the nature of knowledge and reality.

Socrates was a teacher in Athens during the Golden Age. He believed in the basic goodness of people, and said that evil and wrong actions arise from ignorance. Socrates taught his students by questioning them and exposing the weaknesses of their ideas. But some of the more influential citizens mistrusted his ideas, and he was sentenced to death in 399 B.C.

Plato, a friend of Socrates and one of his most gifted pupils, explored such subjects as beauty, justice, and good government. In 387 B.C., Plato founded a school of philosophy and science called the *Academy.* Some

The ruins of the Temple of Poseidon, *top,* on Cape Sounion overlook the island of Salamis, near the site of a great sea battle between the Greeks and the Persians in 480 B.C. That Greek victory helped save the country from being invaded by the Persians.

Aristotle, *above,* was a philosopher, teacher, and scientist, and once Plato's pupil at the *Academy,* a school of philosophy in Athens.

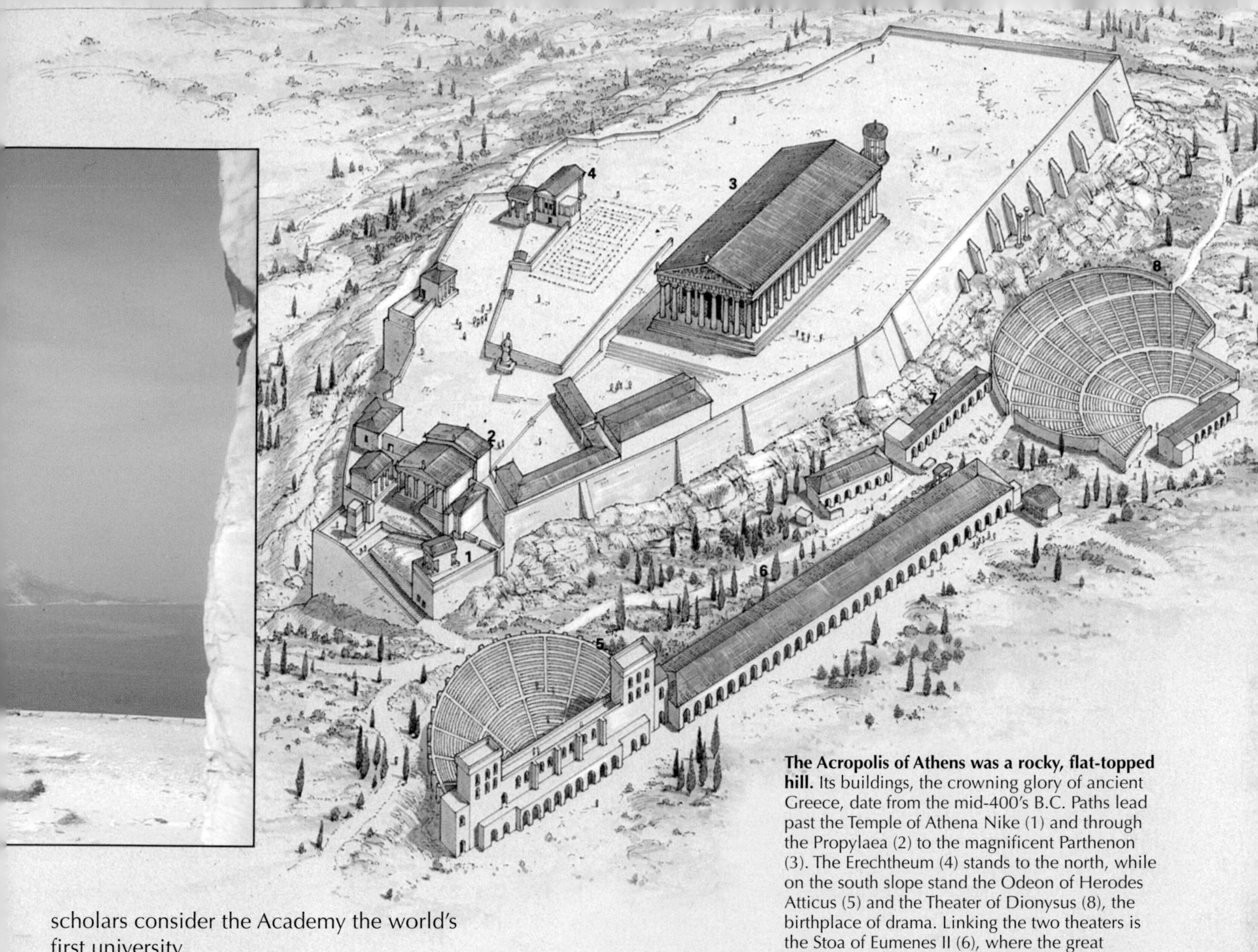

The Acropolis of Athens was a rocky, flat-topped hill. Its buildings, the crowning glory of ancient Greece, date from the mid-400's B.C. Paths lead past the Temple of Athena Nike (1) and through the Propylaea (2) to the magnificent Parthenon (3). The Erechtheum (4) stands to the north, while on the south slope stand the Odeon of Herodes Atticus (5) and the Theater of Dionysus (8), the birthplace of drama. Linking the two theaters is the Stoa of Eumenes II (6), where the great philosophers often walked with their students. Behind it stands another stoa, the Asklepion (7), which was dedicated to the Greek god of healing.

scholars consider the Academy the world's first university.

Aristotle, one of the greatest and most influential thinkers in Western culture, was a student at the Academy. In his writings, Aristotle summed up the rich intellectual tradition he had inherited from his teachers.

The ancient Greeks excelled in the arts as well as in philosophy. Athenian playwrights wrote comedies and tragedies that were performed at religious festivals. Greek writers introduced many new forms to the world of literature, including lyric and epic poetry, philosophical essays and dialogues, and critical and biographical history. Their writings were the model for much of the later literature in the West.

Ancient Greece, *right,* included a peninsula that jutted into the Mediterranean Sea, the Aegean and Ionian islands, and the west coast of Asia Minor.

The Aegean Islands

Scattered in the Aegean and Ionian seas off the Greek mainland lie hundreds of small islands that make up about a fifth of Greece's land area. Many of these islands played an important role in the history and development of ancient civilization, while others are tiny islets virtually untouched by humanity.

Most of the Greek islands, which are known as the Grecian Archipelago, lie in the Aegean Sea, an arm of the Mediterranean Sea between Greece, Turkey, and the island of Crete. Some of these islands are ancient volcanoes and are made of lava, while others are made of pure white marble.

Evvoia, the largest of the central Greek islands, is located just off the coast of mainland Greece. The rest of the Aegean Islands form two main groups, the Cyclades and the Sporades. The Dodecanese Islands are part of the Sporades group.

The Cyclades

The Cyclades lie in the southern Aegean. They are so named because to the ancient Greeks, the islands appeared to lie in a circle (*kyklos* in Greek) around the island of Dhilos, the birthplace of Apollo. The Cyclades include such well-known islands as Ios, Mikonos, Milos, Naxos, Paros, and Thira (Santorin). Geologically, the Cyclades are a continuation of the hills and mountains of the Greek mainland, which linked Greece and Turkey in prehistoric times.

The Cyclades have been populated for more than 4,500 years. On the island of Naxos, anthropologists found evidence of an early Cycladic culture dating from about 3000 B.C. The islands were later influenced by the Minoan and Mycenaean civilizations. Their location on shipping routes between Greece and Asia Minor made the Cyclades important to the ancient Greek world.

The Sporades

The northern Sporades lie in the northeastern Aegean Sea off the coast of Asia Minor (Turkey). The larger islands of the Sporades include Khios (Chios), Lesbos, Limnos, and Samothraki (Samothrace). South of the Sporades, the Dodecanese include the larger islands of Rhodes and Samos, and the

A group of women enjoy a chat, *below,* while embroidering and crocheting on a sunny street on the island of Limnos. Today, the economy of the Aegean Islands depends largely on tourism.

Lions carved from Naxian marble have guarded the Sacred Lake of Apollo on Dhilos (Delos) since the 600's B.C. Dhilos was important in ancient times because it was thought to be the birthplace of Apollo and Artemis.

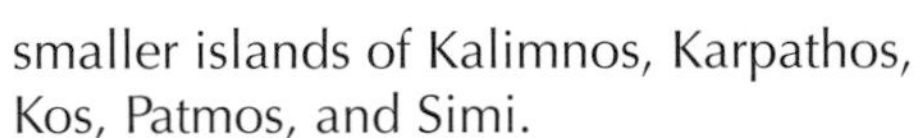

smaller islands of Kalimnos, Karpathos, Kos, Patmos, and Simi.

The Sporades have been Greek in culture since the 400's B.C., when Athens gained control of the islands, and the islanders are proud of their role in the rise of ancient Greece. Many famous Greek thinkers, including the philosopher and mathematician Pythagoras of Samos, came from these islands. The island of Lesbos was a major cultural center from about 600 B.C. to the end of the Golden Age of Greece in 431 B.C.

Lesbos, Khios (Chios), and Samos, which lie only a short distance from the mainland of Turkey, have felt the constant presence of the Turks. The Dodecanese Islands were also influenced by Italian, German, and British rule.

Visiting the islands

The beauty and charm of the Greek islands attract many tourists. Standing out against the deep blue Aegean Sea, the white houses of the islands reflect the sun's dazzling rays. The narrow, winding streets are lined

The cloud of ash produced when the volcanic island of Thira (Santorin) erupted about 3,500 years ago spread to surrounding islands. The volcanic ash was so destructive that it ruined crops and caused a famine in central and eastern Crete.

On Karpathos, where many traditions survive, women and children display their finest garments, *above.* In the island's villages, houses are decorated inside and out with elaborate designs.

Gleaming white houses line the coast of the Cyclades island of Mikonos, a major tourist center. The whitewash on the houses reflects the sun's rays and keeps the interiors cool.

with colorful village houses—most with blue doors and many decorated with tiles. The sunny, Mediterranean climate makes walking a pleasure, while the beaches provide a refuge from the hectic routines of city life. The many ancient temples and historic monuments of the Aegean Islands also attract visitors from around the world.

Tourism has become a major part of the economy of the Aegean Islands. The island of Rhodes in particular has developed into a major tourist center. During the tourist season, the islands are crowded with tourists and with temporary workers employed in the hotel and resort industry. But at the end of the season, the pace of island life slows considerably. Jobs are scarce, and many young people must leave the islands to seek higher-paying jobs in the mainland cities.

Crete and Rhodes

Crete, the largest of the Greek islands, lies about 60 miles (97 kilometers) south of the Peloponnesus. Crete is famous for its scenic landscape and colorful traditions, as well as for the Minoan ruins that still dot its landscape.

Early history

Crete has an important place in Greek history. The first major European civilization, the Minoan culture, developed on this island. The Minoans are named for Minos, the king of Crete in Greek mythology. According to legend, Minos kept a Minotaur—a monster with the head of a bull and the body of a man—in the Labyrinth, a building designed as a maze from which no one could escape.

After Minos conquered much of Greece, including Athens, he sacrificed seven Athenian youths and seven Athenian maidens to the Minotaur every year. Finally, Theseus—one of the intended victims—killed the Minotaur and eloped with Minos' daughter Ariadne.

The Minoan culture flourished on Crete from about 3000 B.C. until about 1450 B.C. Some scholars believe that the effects of a volcanic eruption on the nearby island of Thira (Santorin) may have weakened the Minoan culture. Some towns on the island were never reoccupied. The culture began to decline in the early 1300's B.C., and by the mid-1100's B.C. it had completely disappeared.

In 68 B.C., the Romans invaded Crete, and in 66 B.C. it became a Roman province. After the Roman Empire was divided in AD. 395, Crete came under the rule of the East Roman (Byzantine) Empire. Between 1204 and 1669, Venice ruled Crete.

The Ottoman Turks occupied Crete from 1669 to 1898 and outlawed the Christian Orthodox religion of the islanders. Many of the people fled to the mountains to take up the struggle against Turkish occupation. After the Turks were forced to leave the island, Crete was independent until it became a part of Greece in 1913.

Relics of ancient times

The Palace of Minos at Knossos has the most important remains of Crete's Minoan culture. The palace stands about 3 miles (5 kilometers) southeast of Iraklion, Crete's largest city. Sir Arthur Evans, a British archaeologist, began unearthing the enormous palace in 1900 and had it partially rebuilt.

Evans' discoveries showed that the king's residence was surrounded by impressive villas decorated with wallpaintings and plaster reliefs. These artworks include scenes of *bull-leaping*—young men and women leaping over the backs of bulls. Seafaring themes are also depicted.

In addition to the Minoans, the Romans, Venetians, and Muslim Turks also left their mark on the island. Rethimnon, Crete's third-largest city, is an attractive mixture of

The setting sun casts a warm glow over the harbor at Khania (Canea), *center.* The colorful old-town waterfront features Venetian buildings with red-tiled roofs.

Playful dolphins decorate the walls of the Queen's Apartments in the Palace of Minos at Knossos. These frescoes were painted on wet plaster, a technique that gives the colors a vibrant quality.

Rhodes

Rhodes, the largest of the Dodecanese Islands, lies 12 miles (19 kilometers) off the southwestern coast of Turkey. Rhodes features beautiful scenery and historic buildings.

The fertile soil on Rhodes yields abundant crops of olives, tobacco, grapes, and other fruit. The coastal waters provide large quantities of sponges, the main export.

Rhodes was once a wealthy state of Greece, home to poets, artists, and philosophers. The Colossus of Rhodes, a huge bronze statue of the god Helios, stood in its harbor. One of

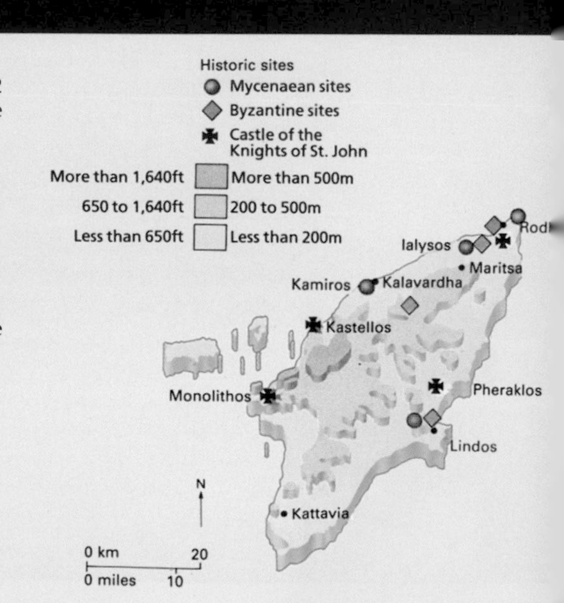

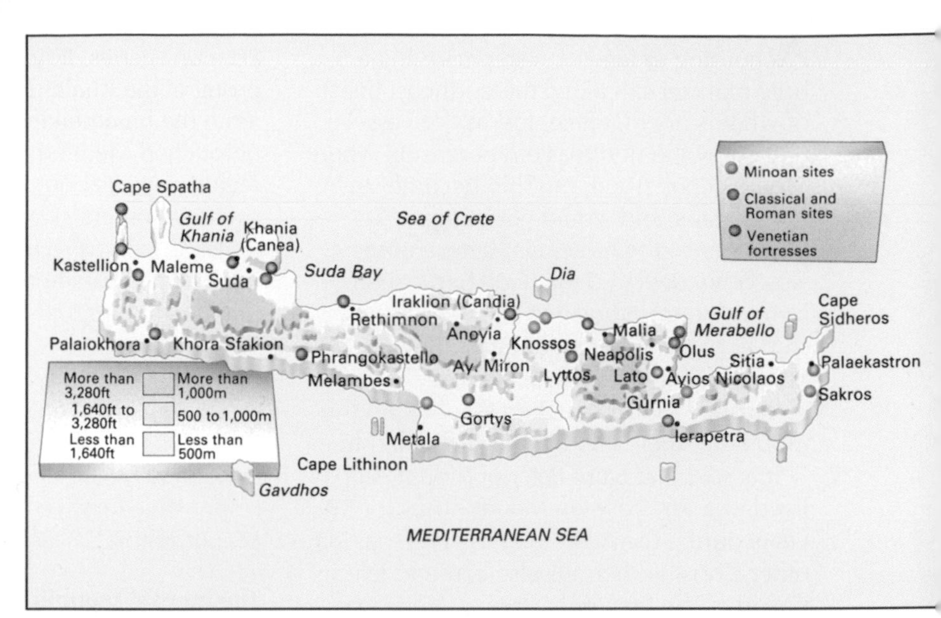

The island of Crete covers 3,217 square miles (8,332 square kilometers). Khania (Canea) is the capital of Crete, and Iraklion (Candia), the island's chief commercial center and port, is the largest city.

these different cultures. The old part of the town contains a Venetian castle as well as Turkish houses with their latticed wooden balconies. And on the site of Gortys, the ancient Roman capital of Crete, stand the foundations of a Roman palace.

A simple island life

The island's largely mountainous landscape makes it difficult to use modern equipment, so much of the work is done by hand. Most crops are grown in the fertile river valleys in the upland areas. Farmers in Crete grow grapes, olives, oranges, vegetables, and nuts. Some villagers make a living selling hand-crafted items, such as baskets, metalwork, and pottery.

The northern cities have a growing food-processing industry, and factories in Iraklion manufacture soap and leather goods. The tourist trade provides employment for a large number of islanders.

The people of Crete are very proud of their ancient heritage. Most of them speak Greek, belong to the Eastern Orthodox Church of Crete, and follow many age-old customs.

the Seven Wonders of the Ancient World, it was destroyed by an earthquake in 224 B.C.

From 1310 till 1522, the island was occupied by the Knights Hospitallers of St. John, crusaders who sought to remove the Muslims from Jerusalem. Turkish forces held Rhodes from 1522 till 1911, when Italy took it. Rhodes became part of Greece in 1947.

The massive walls of the Residence of the Grand Master rise high above the town of Rhodes. The palace was built by Helion de Villeneuve in the 1300's and was later used as a prison by the Turks.

Mount Athos

According to legend, a ship carrying a holy man set sail along the northeast coast of what is now Greece. Just as the vessel passed by the northeastern peninsula where Mount Athos stood, the ship became stuck in the water and would not budge.

Suddenly, the holy man, whose name was Peter, declared that God had called him to Mount Athos. He jumped off the ship at present-day Karavostasi (Bay of Standing Ships) and swam ashore. As the astonished sailors watched Peter climb the mountain, their ship once again sailed free.

It is said that Saint Peter of Mount Athos lived in a tiny cave on Mount Athos for 50 years during the A.D. 700's. As time passed, other Christian hermits also came to live in the area. The first monastery of Athos was probably founded in 963.

Over the next 200 years, 40 monasteries were established, and Athos became a center for Christian Orthodox learning. Monks from Russia, Romania, and Bulgaria came to study and live in Athos.

Today, Athos is a self-governing monastic republic, where about 2,000 monks live in 20 monasteries. It lies on the easternmost prong of the Khalkidhiki Peninsula. Set amid the breathtaking beauty of the untouched Mediterranean coast, the region covers about 130 square miles (335 square kilometers). Mount Athos, known to the Greeks as *Ayion Oros* (Holy Mountain) dominates the rocky peninsula, rising to a height of 6,670 feet (2,033 meters).

Athos is governed by the Holy Community, located in Karyas, a group of 20 representatives who are elected annually by the monastic communities. A governor appointed by the Greek government is responsible for civil administration on Mount Athos.

The monks' republic

Most of the monasteries on Athos were founded in the 900's and the 1000's. Clinging to the edges of steep, inaccessible cliffs, the monasteries were easy to defend against pirates. Today, their well-preserved Byzantine and medieval architecture, framed by lush vegetation and dense woodlands, attracts scholars and tourists alike.

The libraries of Athos hold many elaborate Byzantine manuscripts and sacred objects. Services take place in cross-shaped chapels decorated by frescoes and mosaics and lined with religious statues set with silver and precious stones.

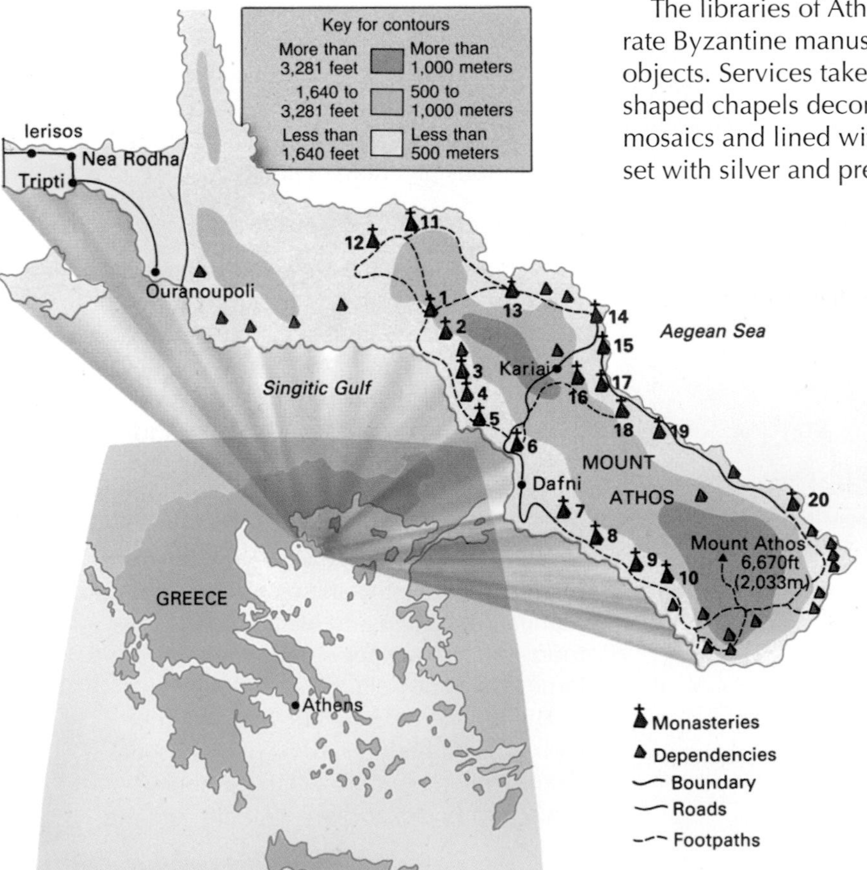

Monasteries

1 Zografou
2 Konstamonitou
3 Dohiariou
4 Xenofontos
5 Ag. Panteleimonos
6 Xiropotamou
7 Simonos Petras
8 Ossiou Grigoriou
9 Dionyssiou
10 Ag. Pavlou
11 Esfigmenou
12 Hiliandariou
13 Vatopediou
14 Pantokratoros
15 Stavronikita
16 Koutloumoussiou
17 Iviron
18 Filotheou
19 Karakalou
20 Meg. Lavras

Icon painting, *above,* is one of the tasks carried out by the monks on Mount Athos.

Mount Athos is a self-governing religious community that is the site of 20 monasteries.

The Monastery of St. Gregorius, *left,* clings to a cliff on the slopes of Mount Athos. Many Athos monks live outside the monasteries in *sketes* (semi-independent communities) scattered throughout the Khalkidhiki Peninsula.

Christian hermits probably have lived on Athos since the 700's when legend says that Saint Peter of Mount Athos arrived. Some monks still live as hermits in caves scattered throughout the forest.

A serious problem confronting Athos today is the decreasing number of new monks. The monastic population, once as high as 40,000, has declined to 2,000, close to the minimum required to keep the monasteries functioning. The upkeep and repair of the old stone buildings require a great deal of time. In addition, the lack of manpower has been blamed for fires that have destroyed entire wings of monasteries.

With the occasional help of Greek farmworkers, the monasteries support themselves by producing olives, vegetables, and wine. The monks also keep chickens and cats. Some of the monks also produce handicrafts and woodcarvings, which are sold to visitors. The monasteries also receive income from lands in Greece and from private donations.

The domestic tasks of the monks are as much a part of their routine as their hourly prayers. The monks hope to free themselves from worldly concerns and become closer to God through worship and labor.

Visiting Athos

Women and children are not permitted to set foot on Mount Athos beyond the town of Ouranoupoli. Male visitors who have obtained permission from the authorities in Athens or Thessaloniki (Salonika) usually take a boat to Dhani. Once on Athos, most travel is done on foot or by mule. Upon arrival, visitors are reminded to follow the rules and behave in a respectful manner—smoking, singing, and whistling are frowned upon, for example.

To outsiders, life on Athos is a world away from their own experience. Most of the monasteries follow the Julian calendar, which is 13 days behind the calendar used by the rest of the world. And most still use an old Byzantine method of reckoning time, in which their days begin at sunset, when clocks read 12 o'clock. At that time, the gates close and everyone retires to their quarters. The drum-beat that calls the monks to prayer can be heard throughout the night.

Greenland

The largest island in the world, Greenland is situated in the North Atlantic Ocean, only about 10 miles (16 kilometers) from Canada. Although geographically it is part of North America, Greenland is a province of Denmark, with the constitutional right of *home rule* (local self-government).

Landscape of ice

Viking explorers gave Greenland its name, hoping to attract Norwegian settlers to the island, but most of Greenland lies north of the Arctic Circle, and only the coastal areas are green during the island's short summers. The landscape consists of a low inland plateau surrounded by coastal mountains. The plateau is covered by a thick, permanent icecap averaging 1 mile (1.6 kilometers) in thickness.

Hundreds of long, narrow sea inlets called *fiords* penetrate the coastal mountains that surround the plateau, providing a transportation link between the interior and the coastal cities and towns. Glaciers, formed on the icecap by compressed snow, flow slowly down coastal valleys and into the fiords, often breaking up into enormous icebergs.

Early settlement

The first Greenlanders were Arctic tribes of the Sarqaq culture, who came to the island from North America. Norwegian Vikings are believed to have sighted what is now Greenland about A.D. 875. The Viking explorer Eric the Red brought the first settlers to the island about 985.

In 1261, the Greenland colonists voted to join Norway, and when Norway united with Denmark in 1380, the island came under Danish rule. The colony died out during the 1400's, and Greenland was lost to the outside world for many years. It was rediscovered in 1578 by the English navigator Martin Frobisher, whose expedition had set out to find the Northwest Passage. The colonization of the island began again in 1721, when Hans Egede, a Norwegian missionary, established a mission and trading center near what is now the capital city of Nuuk.

Snow blankets wooden houses in Upernavik, an island off the southwestern coast of Greenland. Almost all Greenlanders live along the country's southwestern coast.

The busy seaport of Umanak, *right,* is one of Greenland's largest towns. Like most of the other settlements, Umanak lies on the western coast, the warmest region of Greenland.

When the union between Norway and Denmark ended in 1814, Greenland remained with Denmark. Norway disputed the move, but in 1933, a world court upheld Denmark's claim, and in 1953, a new Danish Constitution changed Greenland from a colony to a province.

A changing environment

The climate of Greenland is very cold, but has been gradually getting warmer since the early 1900's. The warming of the sea has brought great numbers of fish to Greenland's coastal waters, while causing seal herds to migrate northward. This shift in natural patterns led the Danish government to promote a change in Greenland's economic base from seal hunting to fishing. More than 30 per cent of Greenland's people now work in the fishing industry, though seals are still hunted in the north.

About 80 per cent of the people of Greenland were born there, and most of the others are Danish immigrants. Relatively

The tundra near Thule, *above,* is typical of the partly thawed Greenland landscape in summer. Temperatures are higher on the country's west coast because warm currents pass between Greenland and North America.

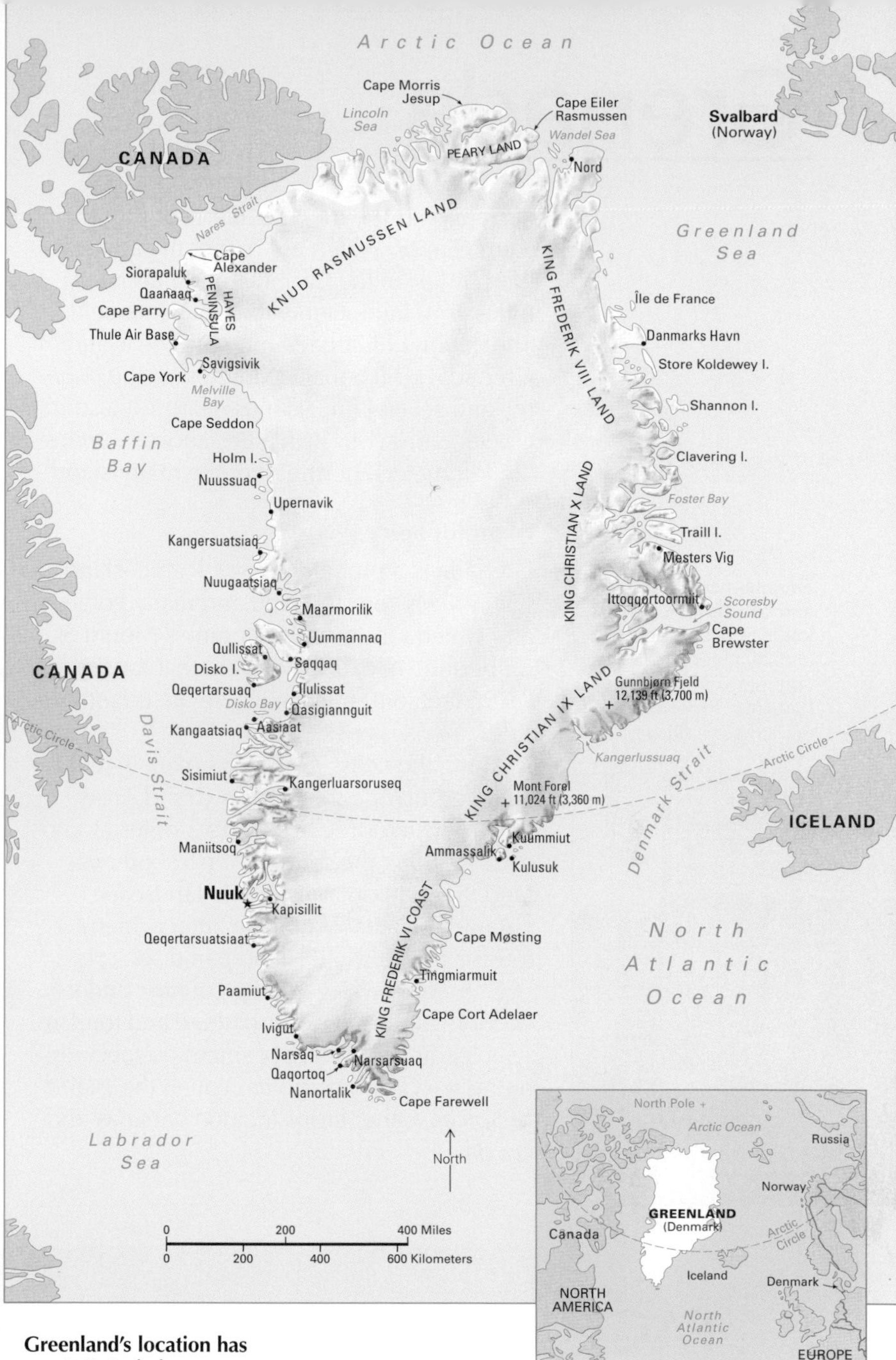

Greenland's location has great strategic importance. Scientists on the island can forecast storms on the North Atlantic Ocean, and U.S. military bases there form a major part of the North American defense system.

Fishing crews off the coast of Greenland reel in cod, halibut, salmon, shrimp, and wolf fish in the shadow of two giant icebergs. A large part of the catch from the icy waters is canned, frozen, or salted for export.

few Greenlanders have entirely Inuit (Eskimo) ancestry. Most have Danish ancestors as well and no longer follow the old Inuit ways. They now live mainly in towns, where they work in the fishing industry, live in wooden houses, and wear European clothing.

Most of the Greenlanders who have mainly Inuit ancestry live in the far northwest regions and hunt seals. Most of these people follow the traditional Inuit way of life. They use the meat of the seals for food, their blubber for oil, and their skins to make clothing and kayaks.

Grenada

Grenada is the most southerly of the Windward Islands, which are part of the Lesser Antilles. It lies in the Caribbean Sea about 90 miles (140 kilometers) north of Venezuela. The nation of Grenada includes the island of Grenada, which makes up most of the country, and several tiny islands nearby. It also includes Carriacou, which lies about 17 miles (27 kilometers) northeast of the main island.

Island of Spices

Grenada is a mountainous volcanic island, with thickly forested land and many gorges and waterfalls. Its highest point is Mount St. Catherine, an extinct volcano that towers 2,756 feet (840 meters) above the countryside.

Along the coast, Grenada's magnificent beaches include the famous Grand Anse Beach, which stretches about 2 miles (3 kilometers) along the southwest part of the island. In addition to its scenic landscape, Grenada's pleasant climate attracts many tourists to this small island paradise.

Grenada is a developing country, and its economy is based on agriculture and tourism. As a leading producer of spices—especially nutmeg—Grenada is often called the *Island of Spices.* Other crops include bananas and cocoa.

Shoppers crowd a lively and colorful street market in St. George's, the capital, chief port, and commercial center of Grenada. Tourism is the city's main industry.

About 95 per cent of Grenada's people have African or mixed African and European ancestry. The official language is English, but many people speak an English or French dialect.

History

In 1498, when Christopher Columbus landed in what is now Grenada, he found Carib Indians living there. Columbus named the island *Concepcion,* but later explorers called it Grenada. In 1650, the French claimed Grenada and later slaughtered many Indians.

France and Great Britain fought for control of Grenada until the island became a British colony in 1783. In the mid-1900's, the British gave Grenada some control over its own affairs. In the early 1970's, Prime Minister Eric M. Gairy led a movement for independence from the United Kingdom.

Grenada gained independence in 1974, and the new country became a constitutional monarchy. Gairy served as prime minister until 1979, when rebels led by Maurice Bishop overthrew his government. Bishop was a Marxist who had close ties to Cuba, but some rebels felt that Bishop did not go far enough in adopting a complete Marxist system in Grenada. In October 1983, they took over the government and killed him.

FACT BOX

COUNTRY

Official name: Grenada
Capital: St. George's
Terrain: Volcanic in origin with central mountains
Area: 133 sq. mi. (344 km²)
Climate: Tropical; tempered by northeast trade winds
Main river: Great
Highest elevation: Mount Saint Catherine, 2,756 ft. (840 m)
Lowest elevation: Caribbean Sea, sea level

GOVERNMENT

Form of government: Constitutional monarchy
Head of state: British monarch, represented by the governor general
Head of government: Prime minister
Administrative areas: 6 parishes, 1 dependency
Legislature: Parliament consisting of the Senate with 13 members and the House of Representatives with 15 members serving five-year terms
Court system: West Indies Associate States Supreme Court
Armed forces: N/A

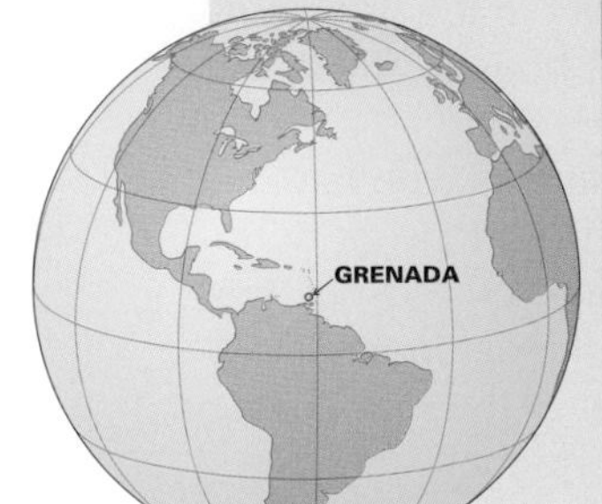

PEOPLE

Estimated 2008 population: 106,000
Population density: 797 persons per sq. mi. (308 per km²)
Population distribution: 61% rural, 39% urban
Life expectancy in years:
Male: 63
Female: 66
Doctors per 1,000 people: N/A
Percentage of age-appropriate population enrolled in the following educational levels:
Primary: N/A
Secondary: N/A
Further: N/A

Workers gather ripe nutmegs, a harvest that provides two of Grenada's most valuable exports—nutmeg and mace. Nutmeg comes from the inner part of the nutmeg tree's brown seeds, while the bright-red membrane that partly covers the seed produces mace.

A banana plantation lies in the shelter of the volcanic mountains that cover much of central Grenada, *below.* The island's beautiful scenery—lush vegetation, picturesque villages with pastel-colored houses, and yacht-filled harbors—draws many tourists.

Soon after Bishop's death, a number of other Caribbean nations—Antigua and Barbuda, Barbados, Dominica, Jamaica, St. Lucia, and St. Vincent and the Grenadines—called upon the United States to help restore order in Grenada. They feared that Grenada would be used as a base by Cuba and the Soviet Union to support terrorism and revolution throughout Latin America.

On Oct. 25, 1983—two days after Bishop was killed—U.S. troops invaded Grenada. Small numbers of troops from six Caribbean nations also took part in the invasion. After several days, the multinational force took complete control of the country. By December 15, all U.S. troops had been pulled out of the country. In elections held in December 1984, a centrist coalition called the New National Party won a clear majority. Its leader, Herbert A. Blaize, became prime minister, and set about restoring stability to his country.

In September 2004, Hurricane Ivan struck Grenada, causing wide-spread destruction. Nearly 40 people were killed and valuable nutmeg crops were ruined.

The island nation of Grenada, *below,* consists of the main island of Grenada and the southern islands of the Grenadines. Since becoming independent in 1974, Grenada has suffered many political problems.

Languages spoken:
English (official)
French patois

Religions:
Roman Catholic 53%
Anglican 14%
Other Protestant 33%

TECHNOLOGY

Radios per 1,000 people: N/A

Televisions per 1,000 people: N/A

Computers per 1,000 people: N/A

ECONOMY

Currency: East Caribbean dollar

Gross domestic product (GDP) in 2002: $440 million U.S.

GDP per capita (2002): $5,000 U.S.

Goods exported: Bananas, cocoa, nutmeg, fruit and vegetables, mace

Goods imported: Food, manufactured goods, machinery, chemicals, fuel

Trading partners: Caricom, United States, United Kingdom, Canada

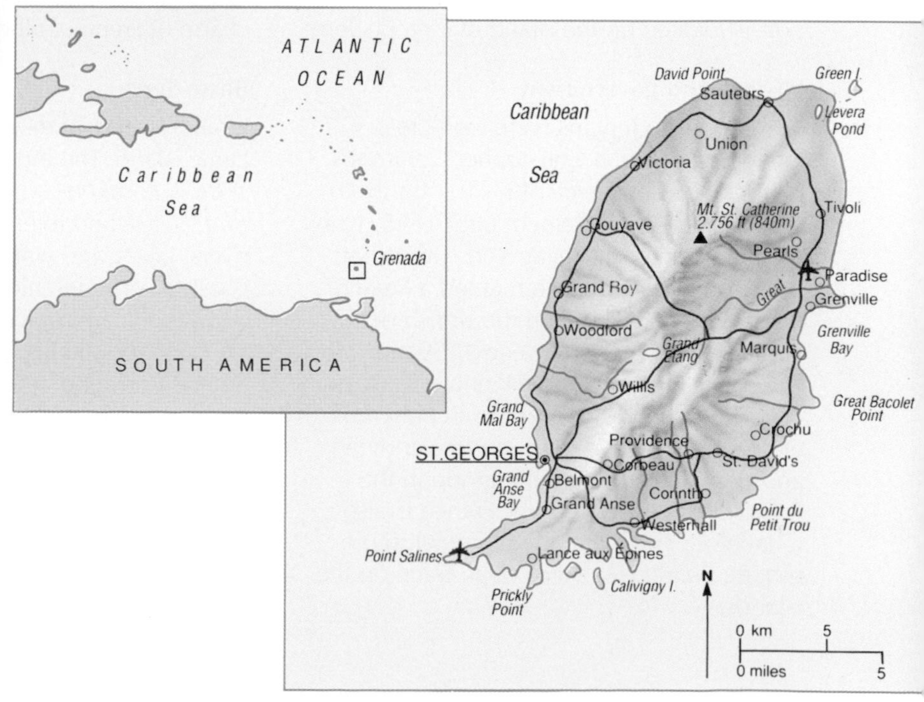

Guadeloupe

Guadeloupe, a group of islands in the Lesser Antilles, consists of two main islands, a smaller island group called Îles des Saintes, and five small islands. Together, the islands make up an overseas *department* (administrative district) of France within the French Community. Guadeloupe covers 687 square miles (1,780 square kilometers).

The larger of the two main islands—Guadeloupe, or Basse-Terre—is separated from the other main island—Grande-Terre—by Rivière Salée, a narrow, bridged strait. From the air, Basse-Terre and Grande-Terre resemble a butterfly.

The five small islands are Marie-Galante, La Désirade, St.-Barthélemy, the northern part of St. Martin, and Petite-Terre. The town of Basse-Terre, on Basse-Terre Island, is the capital of Guadeloupe.

Guadeloupe has a population of about 442,000. Most of the people are of mixed black and white ancestry. The largest all-white community lives in the Îles des Saintes group. These people are descendants of the original Norman and Breton settlers.

Agriculture provides the chief source of income in Guadeloupe. Leading farm products include bananas, cocoa, coffee, and sugar cane. Farmers also grow vegetables and tobacco for local markets. Several distilleries export rum.

Tourism provides another source of income for Guadeloupe. An international airport is located on the island of Grande-Terre.

History and government

Warlike Carib Indians were living in Guadeloupe when Christopher Columbus reached the islands in 1493. The Carib resisted European settlement until 1635, when the first French settlers arrived. Since that time, Guadeloupe has remained a French possession, except when the British occupied the territory between 1759 and 1813.

Guadeloupe became a French overseas department in 1946. A general council of elected members governs the department, and deputies represent the group in the French National Assembly. France is also Guadeloupe's chief trading partner, representing nearly two-thirds of all trade on the islands.

Grande-Terre

Part of the outer arc of the Lesser Antilles, Grande-Terre consists of a low plateau with *karst* (limestone) formations. Dense rain forests once covered the island, but now the landscape is dominated by large plantations and small villages where the plantation workers live.

A sandy beach about 25 miles (40 kilometers) long stretches along the Caribbean coast of Grand-Terre between Point-à-Pitre and the Pointe-des-Châteaux, the easternmost tip of the island. Along this coast, a number of outstanding resorts welcome sunbathers and water-sports enthusiasts. The main center of activity on the southern coast is Gosier, where vacationers enjoy fine restaurants and exciting nightlife.

In Point-à-Pitre, Guadeloupe's chief port and largest city, the streets are lined with peddlers selling exotic fruits, spices, and seafood of all kinds, along with mysterious ingredients for folk medicines. The harbor is another center of activity, as oceangoing ships arrive with goods from France and other countries. Bananas, rum, and sugar cane are then loaded back on board for the return trip.

Although Point-à-Pitre has suffered fires, earthquakes, enemy attacks, and hurricanes, many buildings dating from 1900 still survive. Their carved wooden facades and elaborate wrought-iron balconies display the charm of French colonial architecture.

Basse-Terre

Unlike Grand-Terre, Basse-Terre is a volcanic island. The Indian name for Basse-Terre is *Karukera*, which means *Isle of Beautiful Waters* and refers to the island's many rivers, lakes, and waterfalls. Near the west coast, divers can enjoy beautiful coral reefs and schools of tropical fish in the Caribbean. Volcanic mountains, including the active Soufrière, extend across Basse-Terre from north to south. In the central part lies a nature park, where a road leads partway up to the volcano. Tropical rain forests, exotic flowering plants, hot springs, cinder cones, and sulfur fields cover the land.

Local children, like the increasing numbers of tourists who visit the islands, enjoy Guadeloupe's magnificent beaches, *below.* These young swimmers are of mixed black and white ancestry, as are most of Guadeloupe's people.

The islands of Guadeloupe, *right,* lie in the Caribbean Sea about 370 miles (595 kilometers) north of Venezuela. Guadeloupe has a hot, damp climate from June to December, but the trade winds tend to moderate the heat. From January to May, the islands enjoy cool, dry weather. Dense tropical vegetation, *left,* blankets much of central Basse-Terre, the mountainous western island of Guadeloupe. Along the island's west coast, divers enjoy some of the finest coral reefs in the southern Caribbean.

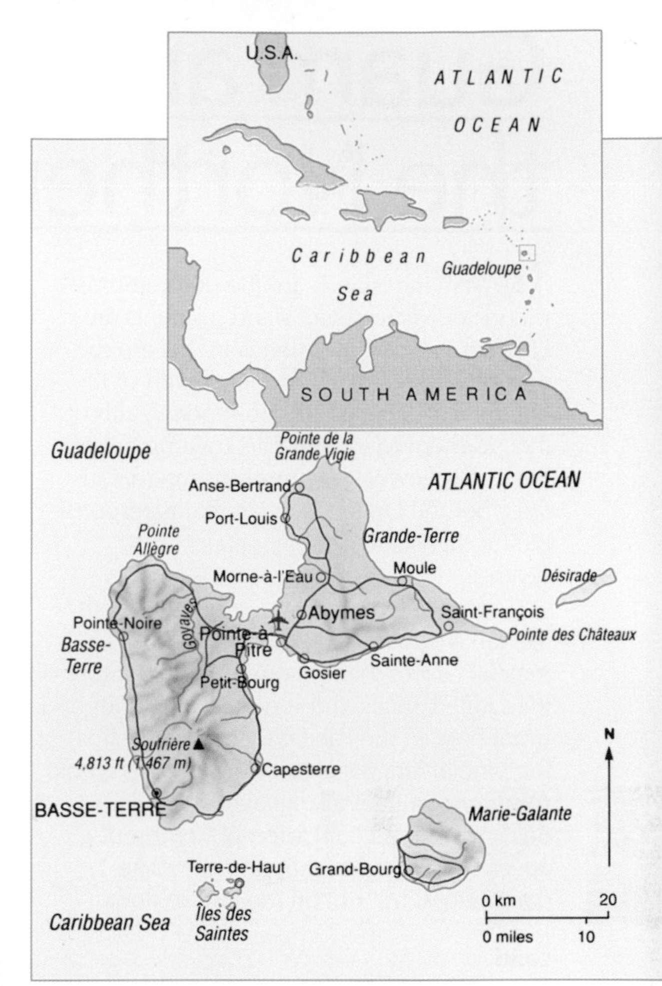

Islanders run for shelter from the fierce waves caused by Hurricane Hugo, which swept across Guadeloupe in 1989. Hugo left a long trail of destruction across the islands. About 10,000 people lost their homes in the devastating storm.

Guadeloupe's hot, damp climate is ideal for growing sugar cane and tropical fruit such as bananas. The sugar cane is processed on Martinique. After harvesting (1), the cane is transported (2) to the processing plant. Here, it is washed (3), crushed (4), shredded (5), and pressed (6) to extract the juice. The juice is heated and clarified (7) before being filtered (8) and evaporated (9). Treatment in a vacuum pan (10) removes excess water, and cooling (11) leaves sugar crystals and *syrup* (molasses). These are separated in a *centrifuge* (12), which separates solids from liquids, before export (13).

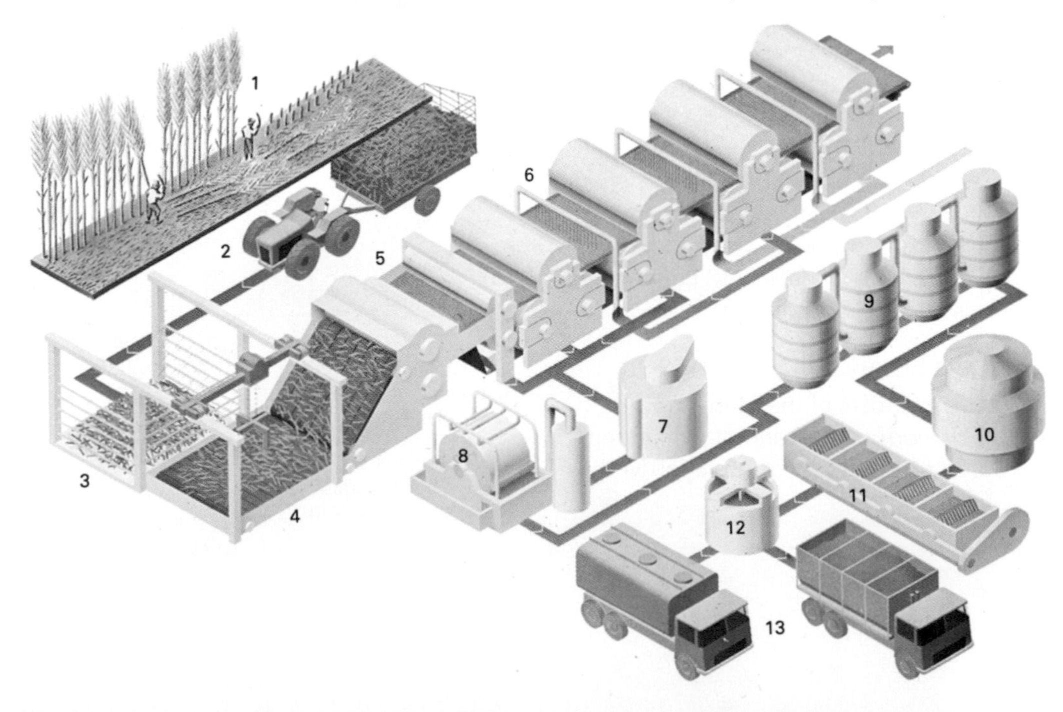

Guam and the Northern Marianas

The Mariana Islands are the northernmost islands of Micronesia. Island residents are U.S. citizens. All the islands in this group, except Guam, are a commonwealth of the United States called the Commonwealth of the Northern Marianas. The government of the commonwealth controls its internal affairs, but the United States remains responsible for the island's foreign affairs and defense.

Guam, which lies at the south end of the Marianas about 1,300 miles (2,100 kilometers) east of the Philippines, is a territory of the United States and serves as a vital air and naval base in the Pacific Ocean. English is the official language, but most people speak *Chamorro,* a native language. They elect a delegate to the U.S. House of Representatives. The delegate can vote in House committees, but not on the House floor.

Land

The Marianas are the southern part of a submerged mountain range that extends 1,565 miles (2,519 kilometers) from Guam almost to Japan. The summits of 15 volcanic mountains in this range form the Marianas. The 10 northern islands have a rugged landscape, and some have volcanoes that erupt periodically. The five southern islands have limestone or reef rock terraces on volcanic slopes that show they are older than the northern group. Guam is the largest of the southern islands.

Coral reefs lie off the coast of Guam. Many forests in northern Guam have been cleared for farms and airfields. The south has a range of volcanic mountains where rivers originate. Earthquakes occasionally strike the island. Guam is warm most of the year, and annual rainfall averages 90 inches (230 centimeters). Typhoons frequently hit the island.

People and history

The native islanders of the Marianas are called *Chamorros*. Their ancestors, among the earliest settlers of Micronesia, arrived from Asia thousands of years ago. The Chamorros, who have intermarried with Europeans, Filipinos, and other peoples, now practice many Western customs.

The Marianas have a population of about 250,000. About 165,000 of the people live on Guam, and about 77,000 live on Saipan, the capital of the Northern Marianas. Many of the people on Guam are U.S. military personnel and their dependents. The island's major sources of income are the U.S. military service and tourism, an important economic activity in Saipan as well.

The Portuguese explorer and navigator Ferdinand Magellan led the first European expedition to the area in 1521. The islands received their name, however, from Spanish Jesuits who arrived in 1668. Spain governed the islands from 1668 to 1898.

After the Spanish-American War, the United States kept Guam as a naval base. Spain sold the rest of the islands to Germany. A League of Nations mandate gave Japan control of the Marianas after World War I (1914–1918), and it was the scene of heavy fighting during World War II (1939–1945).

Japan captured Guam in 1941, and U.S. forces recaptured it in 1944. In 1954, the Strategic Air Command of the U.S. Air Force

Surfers wait for the perfect wave in Guam's coastal waters. Each year, thousands of tourists, mostly Japanese, visit the island. Tourism is a major source of income for Guam.

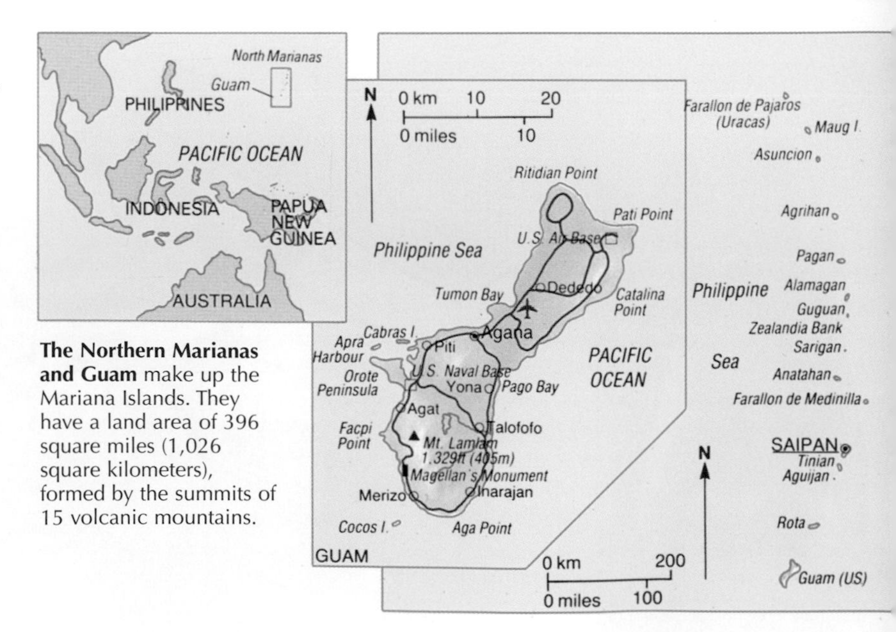

The Northern Marianas and Guam make up the Mariana Islands. They have a land area of 396 square miles (1,026 square kilometers), formed by the summits of 15 volcanic mountains.

At an outdoor festival on Guam, *above,* islanders display sea turtles, crabs, and an array of shells. Tuna is the most important commercial fish, but fishing is now only a minor economic activity on the island.

A group of teen-age girls, *left,* on Guam reflect the influence of Western styles on the culture of their island. Many Guamanians are *Chamorros,* islanders of mainly Micronesian, Filipino, and Spanish descent.

Talofofo Waterfall, *center,* one of the island's main tourist attractions, is the highest waterfall on Guam. The southern half of the island has mountains of volcanic origin. Several rivers have their source in the mountains and run to the coast.

established Andersen Air Force Base and made Guam its Pacific headquarters. In 1950, the United States declared Guam a territory. In 1986, the agreement to form the Commonwealth of the Northern Mariana Islands went into effect.

The Mariana Trench is the deepest point in the world's oceans. It lies 35,840 feet (10,924 meters) below sea level, 200 miles (320 kilometers) southwest of Guam.

Guatemala

Little is known about the earliest people in what is now the Central American country of Guatemala. Farmers lived in the Highlands about 1000 B.C., but it was the Mayas who built the highly developed Indian civilization whose ruins bring tourists to Guatemala.

The Maya people flourished between A.D. 300 and 900. They built beautiful limestone religious centers, mainly on the Northern Plain. These centers included palaces, temples, and pyramids. The Mayas also carved important dates on tall stone blocks and used a kind of picture writing.

For unknown reasons, the Mayas abandoned their centers. When the Spaniards arrived in Guatemala in 1523, most of the Mayas were living in the Highlands.

The Spanish invader Pedro de Alvarado came to Guatemala from the Spanish colony in Mexico. Alvarado conquered all the major Indian groups in Guatemala and established Spanish rule.

On Sept. 15, 1821, Guatemala and other Central American states declared their independence. The states later became part of Mexico, but they broke away in 1823 and formed the United Provinces of Central America. The union, which had liberal political and economic policies, established civil rights and tried to curb the power of rich landowners and the Roman Catholic Church.

However, conservative wealthy people and church officials worked against the union to regain their powers, and Guatemala withdrew from the union in 1839. The country then came under the rule of the first of its many dictators, Rafael Carrera.

Carrera was a conservative army general who restored privileges to the wealthy. After his death in 1865, a number of liberal presidents ruled Guatemala, but they were dictators too. They promoted economic

The Temple of the Giant Jaguar rises 150 feet (45 meters) at Tikal. These ruins are all that remains of a great Mayan city and ceremonial center.

Colorful garlands decorate the church at Chichicastenango, *below,* an example of the country's Spanish Catholic heritage.

The presence of soldiers is a constant reminder of the influence of the military on Guatemalan society. Army officers have supported or led many conservative governments, while leftist rebels have fought them.

Stone figures of Mayan gods, such as the one shown above, mark ancient ceremonial sites. The Mayas worshipped gods and goddesses that influenced daily life, such as the corn god and the moon goddess. During their ceremonies, the Mayas sacrificed animals and sometimes humans.

Major centers of Mayan civilization, *below,* were located in what is now Guatemala, especially in the tropical rain forests of the Northern Plain. The Mayas developed picture writing and a calendar.

development, especially coffee production, and supported immigration and investment by foreigners.

The Guatemalan people, still seeking political freedom, started a 10-year political and economic revolution in 1944. In 1945, a new Constitution gave the people liberties they had never known. A free press developed, and political parties formed. Under President Juan José Arévalo, the government promoted education, health care, and labor unions.

In 1952, President Jacobo Arbenz Guzmán began taking privately owned land and giving it to poor landless peasants. When the government began to take land from the United Fruit Company, however, the U.S. government feared that Guatemala's government was being influenced by Communists. The United States supported a successful revolt against Arbenz in 1954, and a temporary military government was set up.

Political confusion marked Guatemala from the 1960's through the mid-1980's. Between 1970 and 1982, four presidential elections were held. Many people claimed they were dishonest. Sometimes military dictators were elected; other times they seized power. In the late 1970's, violence became widespread when various rebel groups of *leftists*—people who hold socialist or Communist beliefs—fought the government. These rebel groups included poor rural people with little economic or political power.

In 1986, a civilian government was finally elected after one general had overthrown another. A new Constitution was written, and the Congress was reestablished.

A new civilian president, Jorge Serrano Elías, was elected in 1991. In May 1993, Serrano suspended the Constitution, the Supreme Court, and the Constitutional Court. He also placed news organizations under censorship. The military promptly forced Serrano out of office. Days later, democratic rule was restored when Ramiro de Leon Carpio, one of Guatemala's leading human rights advocates, was sworn in as president. The violence that had begun in the late 1970's when various groups of leftists fought government forces for economic and political power ended in 1996 when the opposition groups and the government signed a peace agreement.

Guatemala Today

Guatemala is the most populated of the Central American nations. Most of its people live in Guatemala's rugged central highlands, the site of Guatemala City, the nation's capital and largest city.

The land and farming

The land of Guatemala is divided into three main regions. The Northern Plain, also called El Petén, is an area of thick tropical rain forests and some grasslands. The region gets from 80 to 150 inches (200 to 381 centimeters) of rain a year, and the temperature averages 80° F. (27° C) year-round. Few people live in this area now, but many ancient Mayan ruins still stand in the forests. The country's largest lake, the 228-square-mile (591-square-kilometer) Lake Izabal, lies near the eastern Caribbean coast. Bananas are grown in this area.

The Highlands are a chain of mountains that stretch across Guatemala from east to west. This region has many volcanoes as well as the highest mountain in Central America—Volcán Tajumulco, which rises 13,845 feet (4,220 meters) above sea level. Earthquakes sometimes shake the region.

Most Guatemalans live in the Highlands because the soil is rich and the climate is mild. Mountain valleys have yearly average temperatures of 60° to 70° F. (16° to 21° C).

FACT BOX

COUNTRY

Official name: Republica de Guatemala (Republic of Guatemala)
Capital: Guatemala City
Terrain: Mostly mountains with narrow coastal plains and rolling limestone plateau
Area: 42,042 sq. mi. (108,889 km^2)
Climate: Tropical; hot, humid in lowlands; cooler in highlands
Main rivers: Río Motagua, Río Negro, Río de la Pasión, Río San Pedro
Highest elevation: Volcan Tajumulco, 13,845 ft. (4,220 m)
Lowest elevation: Pacific Ocean, sea level

GOVERNMENT

Form of government: Constitutional democratic republic
Head of state: President
Head of government: President
Administrative areas: 22 departamentos (departments)
Legislature: Congreso de la Republica (Congress of the Republic) with 113 members serving four-year terms
Court system: Corte Suprema de Justicia (Supreme Court), Court of Constitutionality
Armed forces: 31,400 troops

PEOPLE

Estimated 2008 population: 13,532,000
Population density: 322 persons per sq. mi. (124 per km^2)
Population distribution: 53% rural, 47% urban
Life expectancy in years: Male: 63, Female: 69
Doctors per 1,000 people: 0.9
Percentage of age-appropriate population enrolled in the following educational levels:
Primary: 106*
Secondary: 43
Further: 9

The perfectly cone-shaped Tolimán, *left,* is one of the inactive volcanoes that rise near beautiful Lago de Atitlán. The lake bed is believed to be an ancient valley that was dammed by volcanic ash.

Most of the coffee-growing and corn-growing farmland is in the Highlands too. A great deal of coffee is exported, and corn is Guatemala's basic food crop. Some workers live on the coffee plantations that lie on the southern edge of the Highlands.

The Pacific Lowland runs along the coast between the ocean and the Highlands. Many forest-lined streams flow from the mountains through the lowland. The region is hot and humid and thinly populated, consisting mainly of farmland. The production of cotton for export is expanding rapidly. Farmers there also grow corn, rubber-bearing trees, and sugar cane, and raise beef cattle.

The cities and manufacturing

Many rural people have moved to Guatemala's cities, and more keep coming. Guatemala City is now the largest city in Central America. The nation's manufacturing industry is growing, but it cannot keep up with urban population growth. As a result, the cities have much unemployment. Guatemala's manufacturers produce mainly consumer goods, such as processed foods, beverages, and clothes.

Guatemala is a developing Central American country. Thick forests cover almost half this tropical land. Two blue stripes on its flag represent the Atlantic and Pacific oceans, which lie to the east and southwest.

Languages spoken:
Spanish (official)
Amerindian languages (more than 20 Amerindian languages, including Quiche, Cakchiquel, Kekchi, Mam, Garifuna, and Xinca)

Religions:
Roman Catholic
Protestant
Indigenous Mayan beliefs

TECHNOLOGY

Radios per 1,000 people: 79

Televisions per 1,000 people: 145

Computers per 1,000 people: 14.4

ECONOMY

Currency: Quetzal

Gross domestic product (GDP) in 2004: $59.47 billion U.S.

Real annual growth rate (2003–2004): 2.6%

GNI per capita (2004): $4,200 U.S.

Goods exported: Coffee, sugar, bananas, fruits and vegetables, meat, apparel, petroleum, electricity

Goods imported: Fuels, machinery and transport equipment, construction materials, grain, fertilizers, electricity

Trading partners: United States, El Salvador, Mexico

*Enrollment ratios compare the number of students enrolled to the population which, by age, should be enrolled. A ratio higher than 100 indicates that students older or younger than the typical age range are also enrolled.

Economy

Agriculture is the leading goods-producing industry. Guatemala's economy depends heavily on the export of farm products.

In the early 1990's, Guatemala faced serious economic problems. Rising inflation dramatically reduced earnings. In 1993, a 55 per cent rise in non-industrial electricity rates sparked protests by students and workers.

The protests were part of the reason President Jorge Serrano Elías gave for seizing control of the government in 1993. But Serrano's action drew international criticism and a halt to U.S. aid. The move further threatened the economy, which is heavily dependent on foreign aid. The crisis ended when Ramiro de Leon Carpio was sworn in as president after the military ousted Serrano.

People

Like many other Latin–American countries, Guatemala was colonized by Spain. Today, most of Guatemala's people belong to one of two groups—Indians or people of mixed Indian and European descent. In Guatemala, people of mixed descent are called *Ladinos.*

The Indians

Almost half the people of Guatemala—about 45 per cent—are Indians. Their ancestors, the Maya, built a highly developed civilization hundreds of years before Europeans came to the Americas. Today, the Indians live in small peasant communities apart from the mainstream of Guatemalan life. Their ways differ greatly from those of other Guatemalans.

In Guatemala, as in Mexico, being called an Indian depends more on an individual's way of life and self-image than on the person's race. A Guatemalan is considered an Indian if that individual speaks an Indian language, wears Indian clothing, and lives in a community that follows the Indian way of life.

The Indians think of themselves more as part of their own small community than as part of the country of Guatemala. And there is little sense of political or social unity among the various Indian communities. Almost every Indian community has its own colorful style of clothing. Indians often travel far from home to trade in local markets or to find work. Most of them are extremely poor, and about 80 per cent cannot read and write. Almost all of Guatemala's Indians speak one of the many Mayan languages. In addition, many Indian men and some women speak Spanish.

Although most Indians are considered Roman Catholic, they also follow many of the religious practices of their ancestors. They worship local gods and spirits along with the Christian God, the Virgin Mary, and the saints. The Indians believe that their gods are present in nature, and they pray to them, especially during planting and harvesting time.

Religious feast days are the main recreation among Indian peasants. On these holidays the people hold processions, set off fireworks, and play marimba music. They also perform colorful dances that tell stories from history or legends.

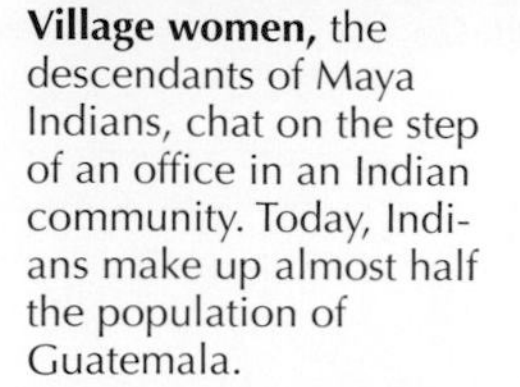

Village women, the descendants of Maya Indians, chat on the step of an office in an Indian community. Today, Indians make up almost half the population of Guatemala.

Mam Indians, *right,* dress in brightly colored clothing of handwoven cloth. Almost every community has its own traditional style of dress, lending color and variety to Indian life.

A colorful flower market, *left,* is held in the central square in the village of Chichicastenango. On market days, peasants bring their crops and handcrafted wares such as pottery to sell or trade. Sunday is the main market day.

An Indian mother and child, *far left,* dress in their finest clothes for a religious festival in the village of Zunil. Such holidays provide relief from the hardships of an everyday life limited by poverty and lack of education.

The Ladinos

About 55 per cent of Guatemala's people are of both Indian and Spanish ancestry and follow Spanish-American customs and traditions. These Ladinos speak Spanish, the official language of the country. Again, whether a person is called a Ladino or an Indian does not depend entirely on that person's racial background. An Indian who drops the Indian way of life, speaks Spanish, and joins a Ladino community is considered a Ladino.

Some Ladinos are peasant farmers and laborers, particularly in the east and south where few Indians live. The homes of Ladino peasants are simple, much like those of the Indians. Most live in one- or two-room houses made of adobe or poles lashed together, with palm, straw, or tile roofs. Their farm tools may include an ax, a digging stick, a hoe, and a machete. Farmers with flat, fertile land may have oxen and plows.

Ladinos make up the middle and upper classes in Guatemala, as they do in other Central American countries. Most Ladinos live in cities and towns. That is especially true in Indian areas, such as the western Highlands.

Ladinos control much of the economy and government of Guatemala. Most Ladinos feel superior to the Indians. There are even a few very wealthy Ladinos who have a high standard of living. The customs and clothing of Ladinos vary little by region, but differ according to their income, occupation, and social class.

Guatemala's population is growing rapidly. The number of Ladinos is increasing much faster than the Indian population, largely because the Ladinos receive better health care in the cities. There has, however, been a slow shift among Indians toward the Ladino way of life.

In the Ladino cities, the people enjoy such sports as basketball, bicycling, and soccer. On religious feast days, Ladinos in rural areas celebrate as the Indians do.

Guinea

Guinea is a West African country with a tropical climate and a variety of landscapes. A swampy coastal strip rises to the Fouta Djallon, a central plateau of hard, crusty soil. In the north lie grassy plains called *savannas,* while in the southeast the forested hills of the Guinea Highlands rise more than 5,000 feet (1,000 meters).

Mangrove trees line the mouths of Guinea's many rivers. Antelope, buffalo, crocodiles, elephants, hippopotamuses, leopards, lions, and monkeys are among the wildlife that make their home in the country.

The discovery of ancient stone tools has led scientists to believe that people have lived in the area of Guinea since prehistoric times. Hunters and then farmers inhabited the area.

Several powerful empires ruled parts of Guinea between 1000 and the mid-1400's. One of these, the Mali Empire founded by the Malinke people, ruled the region during the 1200's. Later, the Songhai Empire conquered the area. During the 1600's, Muslim Fulani people from the north moved to Guinea and fought a *jihad* (holy war) against the Malinke for control of the Fouta Djallon.

The Portuguese were the first Europeans to reach Guinea, beginning in the mid-1400's. They captured and enslaved many Guineans.

FACT BOX

COUNTRY

Official name: Republique de Guinee (Republic of Guinea)
Capital: Conakry
Terrain: Generally flat coastal plain, hilly to mountainous interior
Area: 94,926 sq. mi. (245,857 km^2)
Climate: Generally hot and humid; monsoonal-type rainy season (June to November) with southwesterly winds; dry season (December to May) with northeasterly harmattan winds
Main rivers: Niger, Sénégal, Gambie
Highest elevation: Mont Nimba, 5,748 ft. (1,752 m)
Lowest elevation: Atlantic Ocean, sea level

GOVERNMENT

Form of government: Republic
Head of state: President
Head of government: President
Administrative areas: 7 regions administrative (administrative regions), 1 zone speciale (special zone)
Legislature: Assemblee Nationale Populaire (People's National Assembly) with 114 members serving five-year terms
Court system: Cour d' Appel (Court of Appeal)
Armed forces: 9,700 troops

PEOPLE

Estimated 2008 population: 10,044,000
Population density: 106 persons per sq. mi. (41 per km^2)
Population distribution: 67% rural, 33% urban
Life expectancy in years: Male: 48 Female: 50
Doctors per 1,000 people: 0.1
Percentage of age-appropriate population enrolled in the following educational levels: Primary: 81 Secondary: 24 Further: N/A

A typical Guinean village is made up of round huts with mud walls and thatched roofs, the traditional homes of rural Guineans. Urban dwellers live in one-story rectangular houses made of mud bricks or wood. Few houses have electricity or indoor plumbing.

Guinea, a small country on the western bulge of Africa, is a land of coastal swamps, plateaus, grassy plains, and forested hills. The capital, Conakry, is also the largest city in this underdeveloped nation.

By the mid-1800's, France had gained some territory in the region through treaties and some through conquest. In 1891, Guinea became a French colony called French Guinea.

In 1947, a political party called the *Parti Démocratique de Guinée* (Democratic Party of Guinea), or PDG, was formed. Sékou Touré became head of the PDG in 1952, and by 1957, the party won control of the Guinean legislature. Guineans gained full independence from France in 1958, and Touré became the country's first president.

Under Touré and the PDG, the only legal political party, the government took nearly complete control of the economy and tried to create a socialist state. Throughout the 1960's and early 1970's, Touré also crushed all opposition to his policies and threw many of his opponents into prison. By the late 1970's, Touré began to relax some of the government's political restrictions and release political prisoners, but economic problems remained.

Touré died on March 26, 1984, and army officers took control of the government less than a month later. Colonel Lansana Conté then became president. His government abandoned the socialist policies of Touré and adopted free enterprise policies for the Guinean economy.

In 1990, voters approved a new Constitution that provided for a return to civilian rule. In a multiparty election held in 1993, Lansana Conté was elected president. He was reelected in 1998 and 2003.

Languages spoken:
French (official)
Each ethnic group has its own language

Religions:
Muslim 85%
Christian 1%
Indigenous beliefs

TECHNOLOGY

Radios per 1,000 people: 52

Televisions per 1,000 people: 47

Computers per 1,000 people: 5.5

ECONOMY

Currency: Guinean franc

Gross domestic product (GDP) in 2004: $19.5 billion U.S.

Real annual growth rate (2003–2004): 1.0%

GDP per capita (2004): $2,100 U.S.

Goods exported: Bauxite, alumina, diamonds, coffee, bananas, pineapples

Goods imported: Petroleum products, building materials, machinery, transport equipment, textiles, grain and other foodstuffs

Trading partners: Canada, Japan, United States, various European countries

People and Economy

Almost all Guineans are black Africans, and about 75 per cent of the people belong to one of three main ethnic groups. Most of the Fulani, or Peul—the largest group—live in the central plateau region called the Fouta Djallon. Members of the second largest group, the Malinke, live mainly in the northeastern section of the country, especially in the towns of Kankan, Kouroussa and Siguiri. Members of the third largest ethnic group, the Sopso, live along the coast.

About two-thirds of all Guineans live in rural areas. They wear traditional clothing and, like their ancestors, live in round huts. In the cities and towns, however, some Guineans wear Western-style clothes. Most urban dwellers live in one-story rectangular houses made of mud bricks or wood. A serious housing shortage exists in the capital city of Conakry, however.

While French is the nation's official language, most Guineans speak one of eight African languages. The people have a rich popular culture that they express through oral folk tales, dramas, and music. History is recited by storytellers called *griots.*

Only about 35 per cent of Guinean adults can read and write. Public schools are free, and all children between the ages of 7 and 19 are required to attend. But few actually go to school, partly due to a shortage of teachers and classrooms.

Most of Guinea's people are farmers—about 80 per cent of the nation's workers are employed in agriculture. Many raise barely enough food for their families.

Rice is the main food of Guineans who live near the coast, while corn and millet are the basic foods of people on the northern savannas. The grain is usually pounded into meal, mixed with water, and boiled into a porridge that is served with a hot, spicy sauce. Fruits, such as bananas and pineapples, and vegetables, such as cassava and plantains, are sometimes included in meals. Meat and fish are served occasionally, but the diet of most Guineans lacks protein and vitamins.

Thatched huts and a single tin-roofed building tucked amid tropical vegetation make up a rural settlement in the forested hills of the Guinea Highlands. Most Guineans live in such small villages.

A herdswoman drives her cattle along a rural road. Some farmers in the plains and highlands of Guinea raise livestock for a living. Guineans occasionally eat meat, and some drink milk mixed with water.

Balancing loaves of bread on her head, a young woman threads her way through automobile traffic in Conakry. Poor transportation limits Guinea's economic development.

Guinean children sit on a boxcar carrying bauxite, the country's major source of wealth. Used in making aluminum, bauxite and its processed form—alumina—together with diamonds, make up 95 per cent of Guinea's export income.

Although Guinea is one of the least developed nations in the world, it has many natural resources that could make it prosperous. For example, Guinea has about one-third of the world's reserves of bauxite, a mineral used to make aluminum, as well as deposits of iron ore, diamonds, gold, and uranium

Mining, along with manufacturing and construction, employs about 10 per cent of Guinean workers. Factory workers manufacture food products and textiles. Craftworkers create woven baskets, metal jewelry, and leather goods.

Economic development in Guinea is also hampered by the country's poor transportation systems. The roads, usually unpaved, are in poor condition. Most of the country's railroad tracks also need repair. However, an international airport operates in Conakry, and Conakry and Kamsar are international shipping ports.

Communications are also limited in Guinea. The government controls the country's newspapers, radio, and television. The largest general-interest newspaper is published several times a week.

Wearing similar cotton dresses, Guinean women took part in a political rally during the rule of Sékou Touré. Touré tried to create a feeling of unity among the nation's various ethnic groups. Most Guineans today wear traditional clothing consisting of a loose robe called a *boubou* for men, and a blouse with a skirt made of brightly colored cloth tied around the waist for women.

Guinea-Bissau

The republic of Guinea-Bissau is a tiny nation wedged between Senegal to the north and Guinea to the south and east. Many rivers flow through this very warm, rainy land, which slopes upward from a forested and swampy coast to inland grasslands called *savannas.*

Most of the people of Guinea-Bissau are farmers. They live in straw huts with thatched roofs, and grow beans, coconuts, corn, palm kernels, peanuts, and rice. Only a small percentage of the people work in the country's few industries—mainly in building construction and food processing.

About 85 per cent of Guinea-Bissauans are black Africans who belong to about 20 different ethnic groups, including the Balanta, Fulani, Manjako, and Mandinka—the largest groups. The rest of the population are *mulattoes,* people of mixed black African and Portuguese ancestry.

Although education has improved since independence, only about a third of all adults in Guinea-Bissau can read and write. The law requires all children from the ages of 7 to 13 to attend school, but only about half actually do so.

Many ethnic groups of black Africans were already living in what is now Guinea-Bissau when the Portuguese explorers arrived in the region in 1446. From the 1600's to the 1800's, the Portuguese used the area as a base for their slave trade. It became a Portuguese colony called Portuguese Guinea in 1879, and an overseas province in 1951.

In 1956, African nationalist leaders founded the African Party for the Independence of Guinea and Cape Verde, or PAIGC (its initials in Portuguese). The party, headed by Amilcar Cabral, sought independence for both Portuguese Guinea and the island group called Cape Verde. During the early 1960's, the PAIGC trained many farmers in the hit-and-run tactics of guerrilla warfare. It also established many schools and adult education programs.

A war for independence began in 1963, and by 1968, the PAIGC controlled about two-thirds of the province. The people in these areas elected the first National Popular Assembly in 1972, and in 1973, the Assembly declared the province to be an independent nation called Guinea-Bissau. Amilcar Cabral was assassinated, but his brother, Luis Cabral, became the new nation's first president. The war ended in 1974.

The government then tried to increase farm production—during the war farming had

A Mandinka girl, a member of one of the larger ethnic groups in Guinea-Bissau, balances a bottle on her headdress. The nation's official language is Portuguese, but most of the nation's people speak *crioulo,* a local dialect that combines Portuguese with African languages.

FACT BOX

COUNTRY

Official name: Republica da Guine-Bissau (Republic of Guinea-Bissau)
Capital: Bissau
Terrain: Mostly low coastal plain rising to savanna in east
Area: 13,948 sq. mi. (36,125 km^2)
Climate: Tropical; generally hot and humid; monsoonal-type rainy season (June to November) with southwesterly winds; dry season (December to May) with northeasterly harmattan winds
Main rivers: Cacheu, Corubal, Geba
Highest elevation: Unnamed location in the northeast corner of the country, 984 ft. (300 m)
Lowest elevation: Atlantic Ocean, sea level

GOVERNMENT

Form of government: Republic
Head of state: President
Head of government: President
Administrative areas: 9 regioes (regions)
Legislature: Assembleia Nacional Popular (National People's Assembly) with 102 members serving a maximum of four years
Court system: Supremo Tribunal da Justica (Supreme Court), Regional Courts, Sectoral Courts
Armed forces: 9,250 troops

PEOPLE

Estimated 2008 population: 1,454,000
Population growth: 2.4%
Population density: 104 persons per sq. mi. (40 per km^2)
Population distribution: 64% rural, 36% urban
Life expectancy in years:
Male: 47
Female: 51
Doctors per 1,000 people: 0.2
Percentage of age-appropriate population enrolled in the following educational levels:
Primary: 50
Secondary: N/A

been disrupted and many crops destroyed. Government leaders planned to farm unused land and modernize farming methods. They also wanted to develop the nation's mineral resources, but because of political instability and a shortage of skilled workers, these plans have had little success.

The PAIGC worked to unite Guinea-Bissau and Cape Verde under one government, but in 1980 army officers who opposed such a union overthrew Guinea-Bissau's government and abolished the National Assembly. In 1984, however, a new Constitution was adopted. A new National Assembly was established and elected Brigadier General João Bernardo Vieira as president. The country held its first multiparty elections in 1994, and Vieira remained as president. In 1999, rebels removed Vieira from office and named a temporary acting president. In 2000, voters elected a president who was forced out of office by military leaders in 2003. A transitional government as established.

A village chief on the Bijagós Islands off the coast of Guinea-Bissau displays his symbols of office. Most of the people of Guinea-Bissau live in rural areas and make a bare living by farming.

Further: N/A
Languages spoken:
- Portuguese (official)
- Crioulo
- African languages

Religions:
- Indigenous beliefs 50%
- Muslim 45%
- Christian 5%

TECHNOLOGY

Radios per 1,000 people: 178
Televisions per 1,000 people: 36
Computers per 1,000 people: N/A

ECONOMY

Currency: Communaute Financiere Africaine franc
Gross domestic product (GDP) in 2004: $1.008 billion U.S.
Real annual growth rate (2003–2004): 2.6%
GDP per capita (2004): $700 U.S.
Goods exported:
Mostly: cashew nuts
Also: shrimp, peanuts, palm kernels, sawn lumber
Goods imported: Foodstuffs, machinery and transport equipment, petroleum products
Trading partners: Portugal

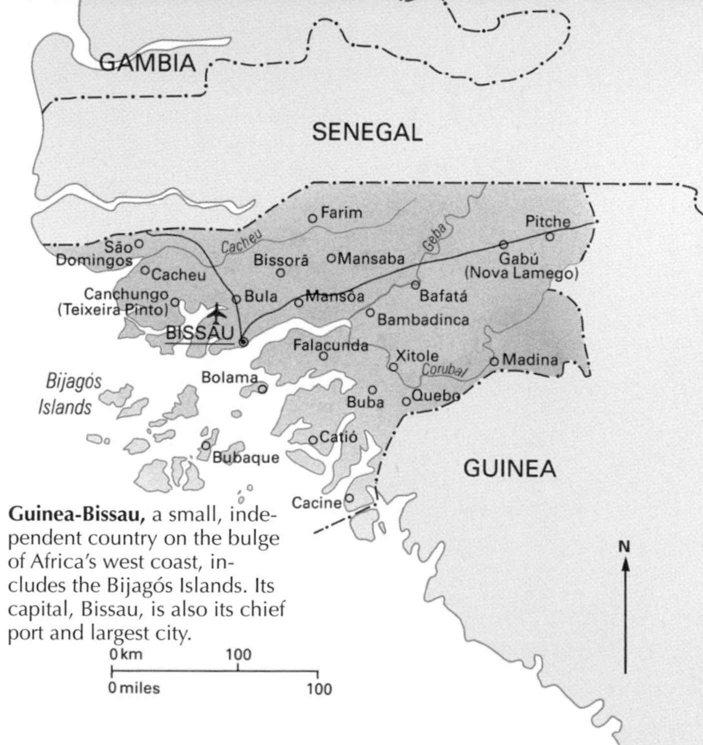

Guinea-Bissau, a small, independent country on the bulge of Africa's west coast, includes the Bijagós Islands. Its capital, Bissau, is also its chief port and largest city.

Guyana

The republic of Guyana, on the northeast coast of South America, is a tropical land of sugar cane plantations and rice farms. Rain forests and mountains cover much of the interior of the country, making most of Guyana difficult to reach. Isolated from the rest of South America by its terrain, Guyana more closely resembles the islands of the Caribbean in its culture, history, and economy.

Guyana is an *Amerindian* (American Indian) word meaning *Land of Waters.* The country's official name is the Cooperative Republic of Guyana. Guyana is made up of people from several national and racial groups. East Indians and blacks form the largest groups.

Early days

European explorers first arrived in Guyana during the late 1500's and early 1600's. In 1581, the Dutch founded a settlement in what is now Guyana and claimed the area. Later, Great Britain and France also claimed it. The United Kingdom gained control of the land in 1814, and formed the colony of British Guiana in 1831.

The early settlers found Arawak, Carib, and Warrau Indians living in the area, but there were not enough Indians to work the sugarcane plantations, so the plantation owners brought in black African slaves. When slavery was abolished in the 1830's, many of the newly freed blacks went to live in the towns, and the plantation owners then began to import laborers from India.

During the 1940's and 1950's, the British became more active in preparing the colony for self-government. More of the people were allowed to vote, and more members of the legislature were elected by the people.

Independence and after

In 1953, a new Constitution was adopted, and British Guiana held its first election based on *universal suffrage* (the right of all adults to vote). However, when the People's Progressive Party (PPP), led by Cheddi B. Jagan, won most of the seats in the legislature, the British government suspended the Constitution, believing that Jagan's administration would turn British Guiana into a Communist state.

In 1955, the PPP split apart, and in 1957, Jagan's deputy, Forbes Burnham, founded the People's National Congress (PNC). The PNC attracted the support of the black population, while the East Indians favored the PPP.

British Guiana became the independent nation of Guyana in 1966, and Forbes Burnham became the country's first prime minister. From 1968 to the early 1990's, the PNC

FACT BOX

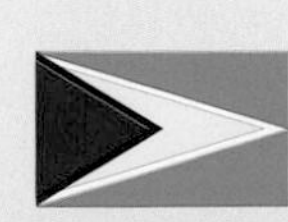

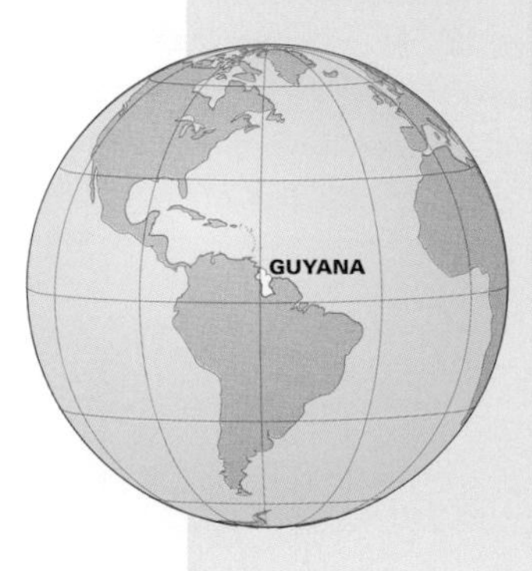

COUNTRY

Official name: Cooperative Republic of Guyana
Capital: Georgetown
Terrain: Mostly rolling highlands; low coastal plain; savanna in south
Area: 83,000 sq. mi. (214,969 km²)
Climate: Tropical; hot, humid, moderated by northeast trade winds; two rainy seasons (May to mid-August, mid-November to mid-January)
Main rivers: Demerara, Essequibo, Courantyne, Berbice
Highest elevation: Mount Roraima, 9,094 ft. (2,772 m)
Lowest elevation: Atlantic Ocean, sea level

GOVERNMENT

Form of government: Republic
Head of state: President
Head of government: Prime minister
Administrative areas: 10 regions
Legislature: National Assembly with 68 members serving five-year terms
Court system: Supreme Court of Judicature, Judicial Court of Appeal, High Court
Armed forces: 1,600 troops

PEOPLE

Estimated 2008 population: 753,000
Population density: 9 persons per sq. mi. (4 per km²)
Population distribution: 64% rural, 36% urban
Life expectancy in years: Male: 60 Female: 67
Doctors per 1,000 people: N/A
Percentage of age-appropriate population enrolled in the following educational levels: Primary: N/A Secondary: N/A Further: N/A

held the majority of seats in the nation's legislature and, as a result, blacks held most of the leading positions in the government and armed forces. However, East Indians continued to dominate in rice production and commerce.

Burnham ruled as prime minister, and later as president, until he died in 1985. His successor, Hugh Desmond Hoyte, continued Burnham's policy of fostering close relations with industrial nations.

In 1989, the government announced it would sell some state-owned companies. In 1992 elections, the PPP won a majority and Cheddi B. Jagan became president. Upon his death in 1997, his widow, Janet Jagan, was elected president. She was succeeded in 1999 by Bharrat Jagdeo, also of the PPP. Jagdeo was reelected in 2001.

A butcher in the capital city of Georgetown displays meats that comply with the dietary requirements of the nation's Muslims. East Indians—both Muslim and Hindu—make up about 50 per cent of the population. They are descendants of people brought from India in the 1830's to work on plantations.

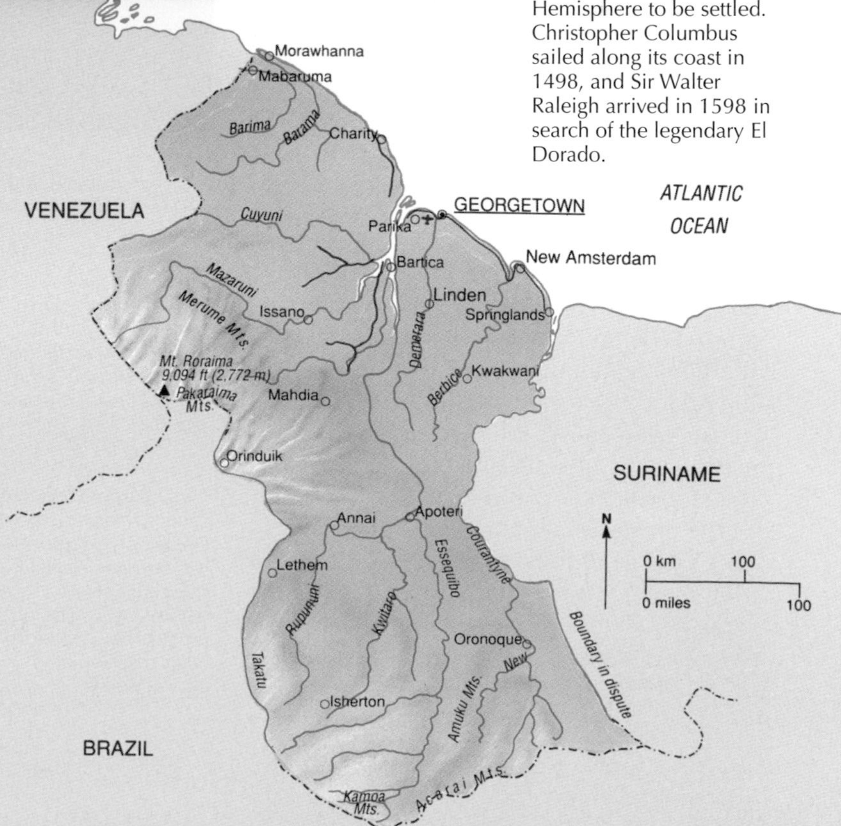

The republic of Guyana lies nestled between Venezuela, Brazil, and Suriname. *Guyana* is an American Indian word meaning *Land of Waters.* Although it is a new nation, the region was one of the first in the Western Hemisphere to be settled. Christopher Columbus sailed along its coast in 1498, and Sir Walter Raleigh arrived in 1598 in search of the legendary El Dorado.

Languages spoken:
- English (official)
- Amerindian dialects
- Creole
- Hindi
- Urdu

Religions:
- Christian 50%
- Hindu 35%
- Muslim 10%

TECHNOLOGY

Radios per 1,000 people: N/A

Televisions per 1,000 people: N/A

Computers per 1,000 people: N/A

ECONOMY

Currency: Guyanese dollar

Gross domestic product (GDP) in 2004: $2.899 billion U.S.

Real annual growth rate (2003–2004): 1.9%

GDP per capita (2004): $3,800 U.S.

Goods exported: Sugar, gold, bauxite/alumina, rice, shrimp, molasses, rum, timber

Goods imported: Manufactured goods, machinery, petroleum, food

Trading partners: United States, Netherlands Antilles, Canada, Trinidad and Tobago, United Kingdom

Land and People

Much of Guyana's coastal region, which lies along the North Atlantic Ocean, was swampland until the early European settlers drained it. Most of this area lies as much as 4 feet (1.2 meters) below sea level at high tide. This coastal strip, which is only 2 miles (3.2 kilometers) wide in some places and only about 30 miles (48 kilometers) at its widest point, is home to 90 per cent of the country's population.

Rice fields and sugar-cane plantations cover the coastal plains. *Dikes* (dams built to prevent flooding), sea walls, and a system of drainage canals prevent the sea from flooding the land and protect people and their crops.

South of the coastal plain lies the inland forest region, which covers about 85 per cent of Guyana and contains about a thousand different types of timber, as well as many plant and animal species. This vast region is almost uninhabited.

Beyond the forest lie the highland areas, which consist of mountains and *savannas* (high treeless plains). The Rupununi savanna stretches along the Brazilian border in the southwest, while a smaller savanna lies in the northeast. Like the forests, the savannas are largely uninhabited, except for a few surviving tribes of *Amerindians* (American Indians) who lead a primitive life almost completely isolated from the rest of the population.

Rivers and waterfalls

Guyana's four main rivers—the Essequibo, the Demerara, the Berbice, and the Courantyne—flow northward from the savannas through the rain forests before emptying into the Atlantic Ocean. Heavy rainfall in the forests often causes the rivers to swell, and spectacular waterfalls and rapids can be found along their courses.

The greatest waterfall in Guyana is the King George VI Falls, which drops 1,600 feet (488 meters) on the Utashi River. Great Falls drops 840 feet (256 meters) on the Mazaruni River, and Kaieteur Fall on the Potaro River drops 741 feet (226 meters).

A timber-framed building in British colonial style is typical of the architecture of Georgetown, *right,* the capital and chief city of Guyana. Georgetown, on the east bank of the Demerara River facing the Atlantic Ocean, is the main outlet for Guyana's exports.

Canal barges carry sugar cane, a major cash crop, to a factory for processing, *far right.* An extensive canal network, along with sea walls and dikes, protects Guyana's low-lying coastal plains against floods.

An aerial view of Guyana's landscape shows the country's major land regions and geographical features. Here, a river flows in a winding course through mountains, grassy savannas, and rain forests.

Guyana's wildlife includes the giant otter (1) and the hoatzin (2), a brightly colored bird that lives in the marshy areas. Hoatzins are born with claws on their wings (3), which they use to climb on tree branches until they learn to fly. The claws fall off as the birds mature.

Guyana's rivers—often the only route through the dense forests of the interior—provide an important means of transportation and communication. Farther north, the rivers serve as shipping routes for the mining industry. Although the town of Linden lies about 65 miles (105 kilometers) inland, it is Guyana's most important mining center because the Demerara River is wide enough at that point for oceangoing ships.

People

Guyana's East Indians—descendants of people who were brought to work on the sugar plantations—make up the nation's largest ethnic group. They account for slightly more than 50 per cent of the population. Guyana's second largest ethnic group are blacks, whose ancestors were brought from Africa as slaves. Tension between the East Indians and blacks caused much political turmoil during the mid-1900's.

Most East Indians live in the rural areas and work on sugar plantations or on small farms. However, an increasing number of East Indians have moved to the cities and become merchants, doctors, and lawyers. The blacks, who live in cities and towns, usually work as teachers, police officers, government employees, and skilled workers in the sugar-grinding mills and bauxite mines.

English is Guyana's official language and is spoken by most of the people. Many of the people speak a broken form of English called Creole. Hindi and Urdu are widely used by Guyana's East Indians.

Haiti

Haiti is a small country on the western third of the island of Hispaniola, which lies in the Caribbean Sea between Cuba and Puerto Rico. It is one of the most densely populated countries in the Western Hemisphere, and one of the least developed.

Founded in 1804, Haiti is the oldest black republic in the world, and the second oldest independent nation in the West. Only the United States is older.

The land

Two rugged mountain chains run across Haiti, one in the north and one in the south. Each forms a peninsula at the western end of the country. The northern peninsula extends about 100 miles (160 kilometers) into the Atlantic Ocean, and the southern peninsula juts about 200 miles (320 kilometers) into the Caribbean Sea. Tropical pine and mahogany forests cover some of the mountains, while tropical fruit trees grow on others.

A gulf, the Golfe de la Gonâve, separates the western peninsulas, and an island, Île de la Gonâve, lies within the gulf. The broad valley of the Artibonite River runs between the mountains in the eastern part of the country.

Haiti has a mild tropical climate. Destructive hurricanes sometimes strike the country between June and October.

A portrait of former President Jean-Claude Duvalier can be seen in Haiti's capital; Port-au-Prince. Duvalier was overthrown in 1986. Haiti has often been ruled by dictators.

History

Christopher Columbus landed on the island of Hispaniola in 1492 and claimed it for Spain. Spanish settlers soon arrived and eventually were responsible for the deaths of almost the entire Arawak Indian population. Many Spaniards later left Hispaniola for more prosperous settlements, and pirates took over the western coast, where Haiti lies today. In 1697, Spain recognized French control of that part of the island.

France named its new colony Saint Domingue. Slaves were brought from Africa to work on plantations. In 1791, the slaves rebelled, and, in 1803, they defeated the French army. On Jan. 1, 1804, General Jean Jacques Dessalines—the leader of the rebels—proclaimed the land an independent country named Haiti.

After Dessalines's death in 1806, a decades-long power struggle began. From 1844 to 1914, 32 different men ruled the nation. U.S. troops occupied Haiti from 1915 to 1934.

François Duvalier was elected president in 1957. In 1964, he declared himself president for life and ruled as dictator. Upon his death in 1971, his son, Jean-Claude Duvalier, succeeded him. Both Duvaliers ruled with the aid of violent secret police called the *Tontons Macoutes* (bogeymen).

In 1986, Haitians overthrew Duvalier. The

FACT BOX

COUNTRY

Official name: Republique d'Haiti (Republic of Haiti)
Capital: Port-au-Prince
Terrain: Mostly rough and mountainous
Area: 10,714 sq. mi. (27,750 km^2)
Climate: Tropical; semiarid where mountains in east cut off trade winds
Main rivers: Artibonite, Les Trois Rivieres
Highest elevation: Chaine de la Selle, 8,783 ft. (2,677 m)
Lowest elevation: Caribbean Sea, sea level

GOVERNMENT

Form of government: Elected government
Head of state: President
Head of government: Prime minister
Administrative areas: 9 departements (departments)
Legislature: Assemblee Nationale (National Assembly) consisting of the Senate with 27 members serving six-year terms and the Chamber of Deputies with 83 members serving four-year terms
Court system: Cour de Cassation (Supreme Court)
Armed forces: None

PEOPLE

Estimated 2008 population: 9,037,000
Population density: 843 persons per sq. mi. (326 per km^2)
Population distribution: 61% rural, 39% urban
Life expectancy in years:
Male: 50
Female: 53
Doctors per 1,000 people: 0.3
Percentage of age-appropriate population enrolled in the following educational levels:
Primary: N/A
Secondary: N/A
Further: N/A
Languages spoken:
French (official)
Creole (official)

A stony hillside south of Port-au-Prince in the Massif de la Selle shows the ruggedness of Haiti's land. Its name means *high ground* in an Indian language.

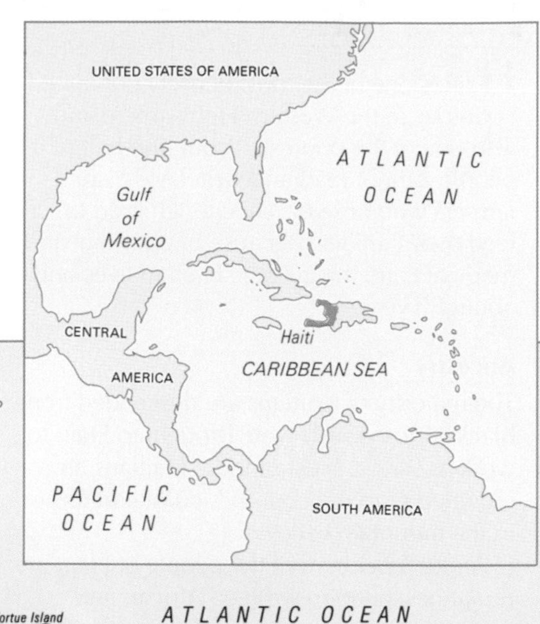

Haiti is on the western end of the island of Hispaniola in the Caribbean Sea. It is the Western Hemisphere's poorest nation.

Religions:
Roman Catholic 80%
Protestant 16% (Baptist 10%, Pentecostal 4%, Adventist 1%, other 1%)
(Note: Roughly 50% of the population also practices voodoo)

TECHNOLOGY

Radios per 1,000 people: 55
Televisions per 1,000 people: 5
Computers per 1,000 people: N/A

ECONOMY

Currency: Gourde
Gross domestic product (GDP) in 2004: $12.05 billion U.S.
Real annual growth rate (2003–2004): -3.5%
GDP per capita (2004): $1,500 U.S.
Goods exported: Manufactures, coffee, oils, mangoes
Goods imported: Food, machinery and transport equipment, fuels
Trading partners: United States, European Union

next three leaders of Haiti, including Jean-Bertrand Aristide, who was elected president in 1990, were also overthrown.

In 1994, the UN Security Council, citing human rights abuses, the expulsion of UN rights monitors from the country, and the mass exodus of Haitian refugees, authorized a U.S.-led invasion of Haiti to oust the military regime and reinstate Aristide. A negotiated settlement allowed U.S. troops to occupy the country without resistance and return power to Aristide. In 1995, the U.S. troops were relieved by UN peacekeepers. The UN peacekeeping forces left Haiti in December 1998.

Aristide was again elected president in 2000. In 2004, an armed rebellion overthrew Aristide. A U.S. peacekeeping force maintained order in the country until the arrival of UN peacekeepers.

People and Economy

Haiti is one of the most densely populated countries in the Western Hemisphere and also one of the poorest. About one-half of its people cannot read and write. Most are farmers who raise barely enough food to feed their families. Because of poor diet and medical care, the average Haitian lives only about 50 years.

Ancestry

The majority of Haitians are descended from black Africans who were brought to Haiti to work as slaves. Most of these Haitians are crowded onto the country's coastal plains or in the mountain valleys.

About 5 per cent of the people of Haiti are *mulattoes,* people of mixed African and European ancestry. Most of the mulattoes belong to the middle or upper class, and many have been educated in France. Most live in comfortable, modern homes. Some are prosperous merchants, doctors, or lawyers, and a few own large plantations. Haiti is also home to small populations of Americans, Europeans, and Syrians.

Middle- and upper-class Haitians speak French, the official language of the country. However, most Haitians speak a language called Haitian Creole, which is partly based on French.

Way of life

About 16 per cent of the Haitian people are Protestants. But most Haitians are Roman Catholics, and many of them practice *voodoo.* Voodoo blends African and Christian beliefs. Voodoo followers believe in many gods, such as gods of rain, love, war, and farming. They also believe that they can be possessed by gods if they perform certain ceremonies. In one such ceremony, for example, a voodoo priest called a *houngan* draws a design on the ground with flour and the people dance until they believe a god has possessed one or more of them.

About two-thirds of Haiti's people live in rural areas, and most are farmers. A typical Haitian family farm is a small plot of land less than 2 acres (0.8 hectare) in size that was once part of the plantation where the family's slave ancestors worked.

In the 1970's, many Haitians began to leave their country because of the poor eco-

The Iron Market in Port-au-Prince, *top,* is a large market place where people from many parts of Haiti sell vegetables, handicrafts, and other goods. The city also has large, overcrowded slum areas.

Farmworkers harvest rice, *above,* in the Artibonite River Valley in central Haiti. Rice is a major food crop, but most farm families have barely enough land to raise rice for their own use.

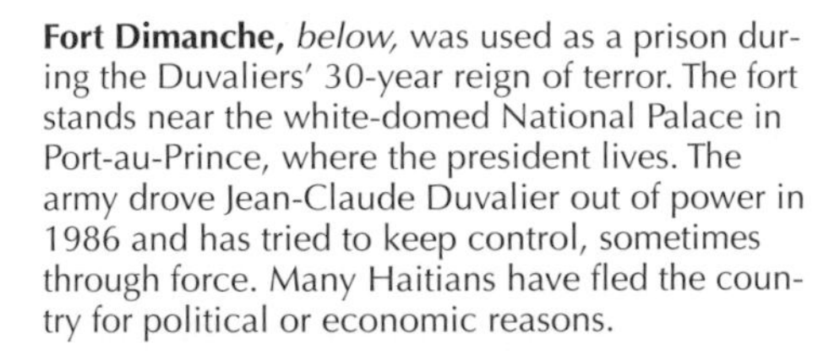

Fort Dimanche, *below,* was used as a prison during the Duvaliers' 30-year reign of terror. The fort stands near the white-domed National Palace in Port-au-Prince, where the president lives. The army drove Jean-Claude Duvalier out of power in 1986 and has tried to keep control, sometimes through force. Many Haitians have fled the country for political or economic reasons.

A brightly colored vehicle—part bus, part truck—loads passengers and goods, *left.* Only about 375 miles (600 kilometers) of Haiti's roads can be used in all kinds of weather.

A local fisherman, carrying a fish trap made of split bamboo, *above,* paddles his dugout canoe through the shallow, clear-blue waters off Haiti's coast.

nomic conditions as well as political oppression. Following the 1991 coup, in which Aristide was overthrown, many Haitians attempted to flee to the United States in small boats, but the U.S. government forced them to return to Haiti or sent them to the U.S. military base at Guantanamo, Cuba.

Agriculture

Haitian farmers raise barely enough food for themselves—mainly beans, corn, rice, and yams. If they are fortunate, they also have some chickens, a pig, or a goat. The family lives in a small, one-room hut with a thatched roof and walls made of sticks covered with dried mud.

The people farm as much of their land as they can. In some areas, they raise crops on slopes so steep that the farmers must anchor themselves with rope to keep from sliding down the hillside. Coffee, fruits, and *cacao*—a seed used to make chocolate—are grown in the mountains. The coffee beans and fruits are sold in the markets.

Most Haitians still follow some of their ancestors' African customs. Much of the work on the small farms of Haiti is done by groups of neighbors. They move from field to field, planting or harvesting crops to music and song in a combination of work and play called a *combite.*

On large plantations, laborers help raise coffee, sugar cane, or *sisal,* a plant used to make twine. Sugar cane is the main crop in the black, fertile soil of the Artibonite Valley. Many Haitians also work on plantations in the Dominican Republic or Cuba.

Haiti has few industries. Some factories process coffee and sugar cane for sale to the United States, France, and other countries, but the international trade embargo against Haiti caused serious hardship before it was lifted in 1994. Haiti has a few cotton mills. Craft workers weave objects from sisal or carve figures out of mahogany.

Honduras

Little is known about what is now the small Central American country of Honduras before the 1500's. Centuries earlier, the Mayas lived in western Honduras and built the magnificent ceremonial center at Copán with its beautiful palaces, pyramids, and temples. The Mayas studied astronomy and invented a calendar. They also developed a number system and picture writing. The Mayas played ball games too, and a Mayan ball court can still be seen at Copán.

The center thrived until the 800's. By the time Europeans arrived, Copán lay in ruins and the Indians of the region had forgotten the city.

Christopher Columbus was the first European to see Honduras. On his fourth voyage, he sailed to what are now the Bay Islands off the northern coast. Columbus landed at Cabo de Honduras (Cape Honduras) on the coast on July 30, 1502, and claimed the land for Spain.

A number of Spanish explorers soon visited the region and founded settlements. Many Indians were killed by the Spaniards or died of diseases brought by the Spanish colonists. Some Indians were shipped as slaves to plantations on the Caribbean islands; others worked in the gold and silver mines started by the Spaniards.

The Spaniards also brought people from Africa to work as slaves in the mines. But the mines were never profitable enough to attract many colonists.

Honduras was a Spanish colony for about 300 years. On Sept. 15, 1821, Honduras and four other Central American states claimed independence. The states became part of Mexico for a short time, but in 1823 they broke away and formed the United Provinces of Central America.

This union had liberal political and economic policies. It established civil rights and tried to curb the power of rich landowners and the Roman Catholic Church.

Honduras withdrew from the union in 1838 after the union began to fall apart under various pressures, including efforts by conservative wealthy people and church officials to regain their powers. Because Honduras was the weakest Central American country, it soon came under the influence of its more powerful neighbors. Guatemala, particularly, interfered in Honduran affairs and started or supported several revolts in Honduras during the 1800's.

In the late 1800's, U.S. fruit companies began to arrive in Honduras, establishing banana plantations on the Northern Coast. Because of the income they brought to the country, the banana companies had a strong influence over the Honduran government. Honduras became known as a "banana republic" because of this relationship.

Until 1933, most Honduran presidents served short terms because they were overthrown in the nation's frequent revolutions. In that year, General Tiburcio Carías Andino became president and ruled as a dictator until 1948, despite several revolts.

During the 1950's, more political violence occurred. Then in 1957, Ramón Villeda Morales, a doctor, was elected president. Villeda started a land reform program and built hospitals, roads, and schools.

In 1963, however, the government was overthrown by Colonel Osvaldo López Arellano and other army officers. A new

The ruins of Copán include a ball court, where the Mayas played a game similar to soccer.

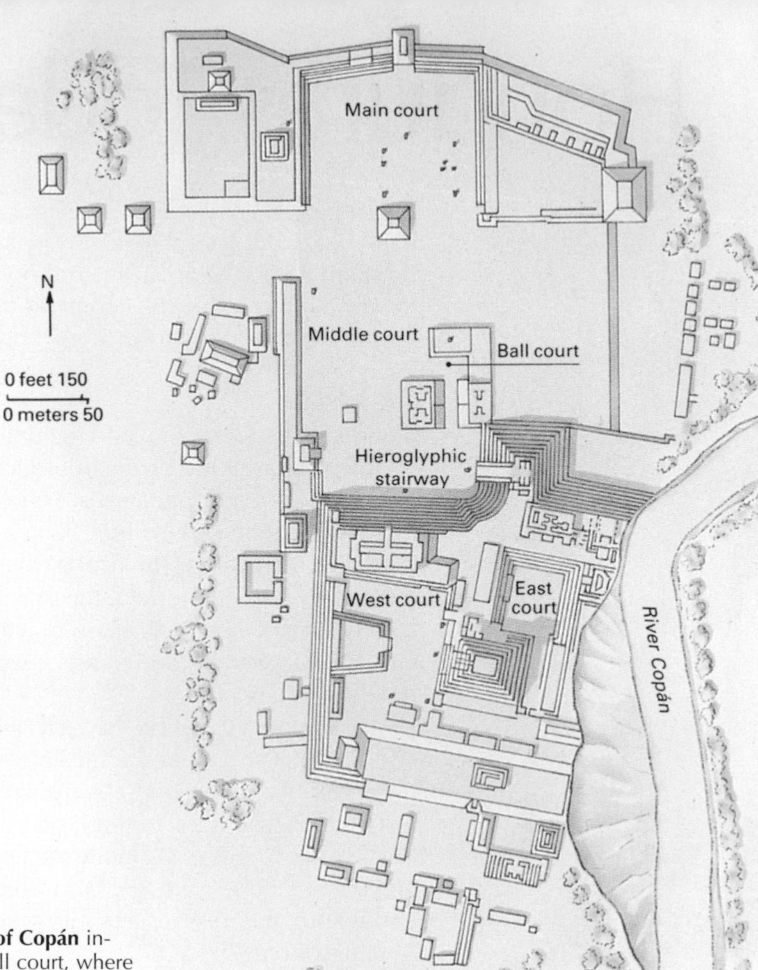

Copán, *above,* was a major Mayan ceremonial center from the A.D. 400's to the late 700's. The center included huge pyramid-shaped temples and spacious courts where the Mayas gathered for special events.

Land titling ceremonies attract farmers in rural areas. With about half of its people living in rural regions, Honduras continues its agricultural heritage. The Mayas farmed, Spain established cattle ranches, and the U.S. companies started banana plantations.

A Miskito child's facial features reveal the group's mixed Indian and black African ancestry, *below*. About 10,000 Miskito live in northeastern Honduras.

constitution was written, allowing López to become president. In 1971, voters again elected a civilian—a lawyer—to the presidency, but still another military revolt followed, and López returned to power.

López himself was overthrown in 1975, after a government scandal involving a bribe from a U.S. company. Two more military-led governments followed. In the 1980's, however, the people elected a civilian president and also chose a new legislature. Since then, voters have elected a new president and legislature every four years.

During the 1980's, Honduras experienced decade-long problems with its neighbor, Nicaragua. Rebels fighting the Nicaraguan government crossed into Honduras and established bases from which they raided Nicaragua. Nicaraguan government troops sometimes entered Honduras to attack these rebels. In 1988, the rebels and the government signed a cease-fire agreement. In 1990, the Nicaraguan government was voted out of office, and the rebel bases in Honduras were shut down.

Honduras Today

Honduras was named by an explorer—perhaps Columbus—for the deep waters off its northern coast. The Spanish word *honduras* means *depths.* Today, the country is known for the bananas it produces.

The land

Honduras has four main land regions. The Northern Coast is the banana-producing region of the country. Bananas are especially important in the fertile Ulua-Chamelecón River Basin and along the coastal plain near the port of Tela. East of Tela, the region is largely undeveloped and sparsely populated. Grasslands, swamps, and forests cover the hot, humid land.

The Southern Coast on the Gulf of Fonseca, a small arm of the Pacific Ocean, is lined with mangrove trees. Narrow plains lie just inland. The largest of these plains, along the Choluteca River, is a fertile area of farms and cattle ranches. The Southern Coast is hot and humid, but receives less rain than the Northern Coast.

The Mountainous Interior covers more than 60 per cent of the nation. One peak in the Cerros de Celaque mountain range rises 9,347 feet (2,849 meters) above sea level, but most of the mountains are much lower. Forests cover many slopes. Honduras has no live volcanoes, so its soils are not enriched by

Slum areas, like this one in Tegucigalpa, reveal the poverty of Honduras.

Columbus landed on Isla Guanaja, *right,* one of the Bay Islands off the coast of Honduras, in 1502.

FACT BOX

COUNTRY

Official name: Republica de Honduras (Republic of Honduras)
Capital: Tegucigalpa
Terrain: Mostly mountains in interior, narrow coastal plains
Area: 43,277 sq. mi. (112,088 km^2)
Climate: Subtropical in lowlands, temperate in mountains
Main river(s): Ulua, Chamelecón, Coco, Choluteca, Aguán
Highest elevation: Cerro Las Minas, 9,347 ft. (2,849 m)
Lowest elevation: Caribbean Sea, sea level

GOVERNMENT

Form of government: Democratic constitutional republic
Head of state: President
Head of government: President
Administrative areas: 18 departamentos (departments)
Legislature: Congreso Nacional (National Congress) with 128 members serving four-year terms
Court system: Corte Suprema de Justicia (Supreme Court of Justice)
Armed forces: 12,000 troops

PEOPLE

Estimated 2008 population: 7,691,000
Population density: 178 persons per sq. mi. (69 per km^2)
Population distribution: 53% rural, 47% urban
Life expectancy in years: Male: 67 Female: 74
Doctors per 1,000 people: 0.8
Percentage of age-appropriate population enrolled in the following educational levels: Primary: 106* Secondary: N/A Further: 15

Honduras, a small country on the land bridge of Central America, has a very short coastline on a gulf of the Pacific and a much longer coastline on the Caribbean.

volcanic ash. However, some of the smaller highland valleys are fertile enough to support farms.

The highlands have a milder climate than the coasts. The capital city of Tegucigalpa in the central mountains has an average temperature of 74°F. (23° C).

The Northeastern Plain, sometimes called the Mosquito Coast, or Mosquitia, is Honduras's least developed and most thinly populated region. Tropical rain forests cover much of this hot, wet area. The plain has some grasslands and forests of palms and pines, and a few little towns.

Languages spoken:
Spanish (official)
Amerindian dialects

Religions:
Roman Catholic 97%
Protestant

TECHNOLOGY

Radios per 1,000 people: 411

Televisions per 1,000 people: 119

Computers per 1,000 people: 13.6

ECONOMY

Currency: Lempira

Gross domestic product (GDP) in 2004: $18.79 billion U.S.

Real annual growth rate (2003–2004): 4.2%

GDP per capita (2004): $2,800 U.S.

Goods exported: Coffee, bananas, shrimp, lobster, meat; zinc, lumber

Goods imported: Machinery and transport equipment, industrial raw materials, chemical products, fuels, foodstuffs

Trading partners: United States, Guatemala, Japan, Germany, Netherlands Antilles

The people and their government

More than 95 per cent of the Honduran people are *mestizos,* people with both Indian and European ancestors. Almost all mestizos speak Spanish, and many people in the northern areas and ports also speak English.

Also living on the Northern Plain, on the northwestern coast, are people called Garifuna, or Black Caribs. These people are the descendants of African slaves and Arawak Indians who lived on the Caribbean island of St. Vincent. In 1797, the British rulers of the island forced some Africans and Indians to live in Honduras because they were considered rebellious. The Garifuna, who number about 80,000, speak an Arawak language.

More than 70,000 Miskito Indians live in small communities on the Northeastern Plain. They are a mixture of Indians, freed slaves, and other groups, and they speak the Miskito language.

Today, Honduras is an independent nation with an elected president and national legislature. Since 1981, the voters of Honduras have elected a new president and legislature every four years. Ricardo Maduro Joest was elected president in 2001.

Economy

Honduras is a poor country, and its people have a low average income. More than half live in rural areas, and most own or rent small farms. They live in small houses made of adobe or wood, or erected on a wooden frame packed with mud and stones.

Transportation in Honduras is limited, and communication between rural and urban areas is poor. Some modernization is taking place in cities as industry and education expand, but such changes are slow to reach the rural areas.

Construction of an airport extension on the Bay Islands north of the mainland occupies a tractor operator, *below.* Honduras has two international airports, at Tegucigalpa and San Pedro Sula.

Agriculture

Honduras has few resources, and the nation's economy is one of the most underdeveloped in Latin America. Agriculture is by far the most important economic activity, with more than half of all Honduran workers involved in farming.

Large numbers of Hondurans on the Northern Coast grow bananas, the nation's leading source of income. The banana industry was developed by U.S. companies about 1900. At that time, fruit companies cleared forests and drained swamps for banana plantations. They built railroads and ports to ship the fruit, and established towns, hospitals, and schools for the workers.

Export taxes paid by the fruit companies to the Honduran government took care of most of the country's expenses. In return, the government gave the fruit companies special privileges. Because of this close relationship, Honduras became known as a "banana republic"—a term still used today to refer to Honduras and some other Latin American countries whose economies depend on a single product.

Today, most of the banana plantations are owned by Honduran companies, but the firms that buy the bananas and ship them to foreign markets are generally owned by foreign companies. Bananas now account for about a third of the nation's exports.

Coffee, which is grown in the inland mountains, makes up about a fourth of all Honduran exports. Other mountain crops include corn and beans. Corn, the country's basic food crop, covers more land than any other crop in Honduras.

Some cattle are raised in the mountains as well as on the plains of the Southern Coast, and beef is an important export. The fertile

Banana cultivation

Bananas thrive in the warm, humid parts of the world. Banana farmers start a crop by cutting growths called *suckers* from the underground stems of mature banana plants. The suckers are planted in the ground. In three to four weeks, tightly rolled leaves sprout up and unroll as they grow. The "trunk" of a banana plant is actually the rolled stalks of its leaves. About 10 months later, a large bud at the end of an underground stem grows from the leaf bundle. When the stem reaches the top of the plant, small flowers appear on the bud. These flowers develop into tiny green bananas. As the bananas grow, they begin to curve upward. The fruit is harvested four or five months later. Cut when still green, the bananas are packed and shipped to markets.

Coconut palms in Honduras not only yield delicious fruit, but also provide building material for houses and leaves for thatching roofs and weaving baskets.

southern soils support cotton farms too.

Honduras exports shrimp caught near the Bay Islands off its northern coast. Timber from pines and tropical hardwood trees is also exported.

Service industries and manufacturing

About a third of Honduran workers work in service industries. Wholesale and retail trade is the most important service industry. Many people are involved in the distribution of agricultural products, and others work in education and health care. These services are mainly centered in the largest cities, especially Tegucigalpa.

Processing bananas for export employs many Hondurans, *below.* Bananas are picked when green so that they will ripen during shipping to distant markets.

Manufacturing employs only about a seventh of all the workers in Honduras. Tegucigalpa and San Pedro Sula are the major manufacturing centers. The main products are consumer goods, such as processed foods and beverages, clothing, and textiles. Pottery-making is a craft industry in rural homes. Sawmills provide lumber for furniture, paper, and wood products industries.

Hungary

A small, landlocked nation in the heart of central Europe, Hungary is bordered to the north by the Czech Republic and Slovakia, to the northeast by Ukraine, to the east by Romania, to the south by Croatia and Serbia, and to the west by Austria. Before World War II (1939–1945), Hungary was an agricultural nation. When the Communists took over the government in the late 1940's, they introduced an economic plan that encouraged industrialization. Today, manufacturing and other industries contribute more to the national income than does farming.

Magyars, whose ancestors settled in what is now Hungary during the late 800's, make up about 95 per cent of Hungary's present-day population. The early Magyars, known as *On-Ogurs* (people of the ten arrows), were nomadic warriors trained from infancy as riders, archers, and javelin throwers. They raided the Danube and Elbe valleys and many towns throughout Europe.

The transformation of the warlike Magyars into a peace-loving people began with their conversion to Christianity. In 1000, Stephen, a Roman Catholic leader of the Magyar tribe, was crowned the first king of Hungary by Pope Sylvester II, and he made Catholicism the nation's official religion. The skill of fighting on horseback lives on today in the Hungarian culture: the ancient Magyar art of fencing with a cavalry saber is still enjoyed as a national sport.

Although the country in which the Magyars settled was surrounded by Slavic and Germanic peoples, the Magyars retained their own language, which has survived to the present day as the official language of Hungary. The Magyar language—also called Hungarian—belongs to the Uralic-Altaic family of languages, which includes Estonian and Finnish. Magyar is totally unrelated to other major East European languages, such as Romanian, Polish, Czech, or Serbo-Croatian. While Magyar is spoken throughout the country, many Hungarians also speak German.

Way of life

As the country became more industrialized, the traditional life of the Hungarian peasant was replaced by more modern, urban ways. Today, more Hungarians work in industry

than on farms, and nearly two-thirds of the Hungarian people live in cities and towns. Budapest, the nation's capital and largest city, is home to about 20 per cent of Hungary's people.

Most Hungarians, especially city dwellers, dress in Western-style clothing, but rural people still wear colorfully embroidered costumes on special occasions. Villagers who once carved their own wooden utensils and embroidered their linens now buy manufactured household items.

But present-day Hungarians still display a traditional love of fine wines and of good, spicy food in large quantities. Their most famous dish is *goulash,* a thick soup, or stew, consisting of cubes of meat, gravy, onions, and potatoes and flavored with a seasoning called *paprika.*

A divided nation

Because the Hungarians share few linguistic, cultural, or racial links with their neighbors, they often consider themselves the most isolated people in Europe. They also regard themselves as a divided nation, since more than three million Hungarians—almost one-third of Hungary's population—live in neighboring countries.

Under the 1920 Treaty of Trianon, which was imposed on the defeated Hungarians by the victorious Allies after World War I (1914–1918), two-thirds of Hungary's territory was given to Austria, Czechoslovakia, Romania, and Yugoslavia. The borders drawn at that time remain much the same today.

The majority of the Hungarians living in neighboring lands are located in Romania, where, under the now-deposed Ceausescu regime, they were denied fundamental human rights. The situation was a constant source of political conflict between the two countries. Hundreds of thousands of ethnic Hungarians also live in the Czech Republic, Slovakia, Serbia, and western Ukraine.

Hungary Today

October 23, 1989, saw the birth of the new Republic of Hungary—a parliamentary democracy based on a free market economy. The nation's gradual movement toward democracy began with the economic reforms of János Kádár in the 1960's. Kádár was head of Hungary's Communist Party and served as premier from 1956 to 1958 and from 1961 to 1965. He tried to win the support of the people by easing some of the restrictions on cultural, economic, and social life. In 1968, the government introduced a new economic program that combined features of a free market system with the country's socialized economy.

These policies came to be known as *goulash Communism* because of the way elements from different systems were combined—just like the Hungarian national dish. Kádár's reorganization soon began to improve the economy, and living standards rose. However, as a result of the new economic freedoms, some people became considerably more wealthy than others, and the prices of many popular goods rose beyond the reach of the poorest people.

In addition to the economic reforms, Hungarians also enjoyed more personal freedom. They were allowed to travel to Western

Hungary, a landlocked nation in central Europe, *far right,* fell under Communist control after World War II. A people's revolt in 1956 was quickly crushed by Soviet troops. On the 33rd anniversary of the 1956 uprising, Hungary declared itself a parliamentary democracy.

A ferryboat on the Danube River passes by the domed Parliament Building, one of Budapest's historic landmarks, *right.* Even under the Communist regime, Hungary enjoyed a thriving tourist trade, attracting more than 10 million visitors a year.

FACT BOX

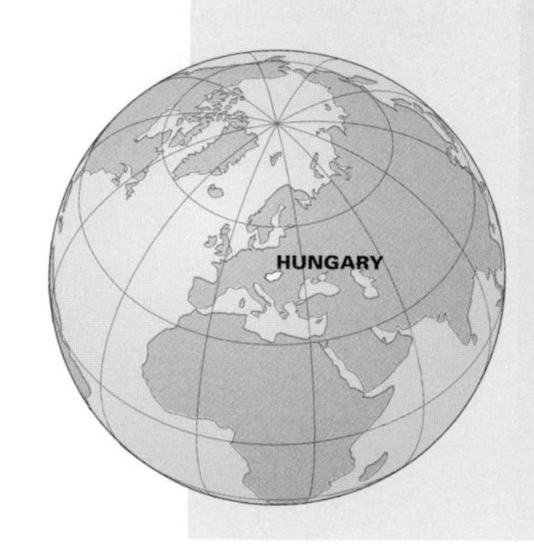

COUNTRY

Official name: Magyar Koztarsasag (Republic of Hungary)
Capital: Budapest
Terrain: Mostly flat to rolling plains; hills and low mountains on the Slovakian border
Area: 35,920 sq. mi. (93,032 km^2)
Climate: Temperate; cold, cloudy, humid winters; warm summers
Main rivers: Danube, Tisza
Highest elevation: Kekes, 3,330 ft. (1,015 m)
Lowest elevation: Tisza River, 259 ft. (79 m)

GOVERNMENT

Form of government: Parliamentary democracy
Head of state: President
Head of government: Prime minister
Administrative areas: 19 megyek (counties), 6 cities that rank as counties
Legislature: Orszaggyules (National Assembly) with 386 members serving four-year terms
Court system: Constitutional Court
Armed forces: 45,000 troops

PEOPLE

Estimated 2008 population: 10,020,000
Population growth: -0.33%
Population density: 279 persons per sq. mi. (108 per km^2)
Population distribution: 66% urban, 34% rural
Life expectancy in years: Male: 68 Female: 77
Doctors per 1,000 people: 3.2
Percentage of age-appropriate population enrolled in the following educational levels: Primary: 101* Secondary: 104 Further: 44
Language spoken: Hungarian (official)

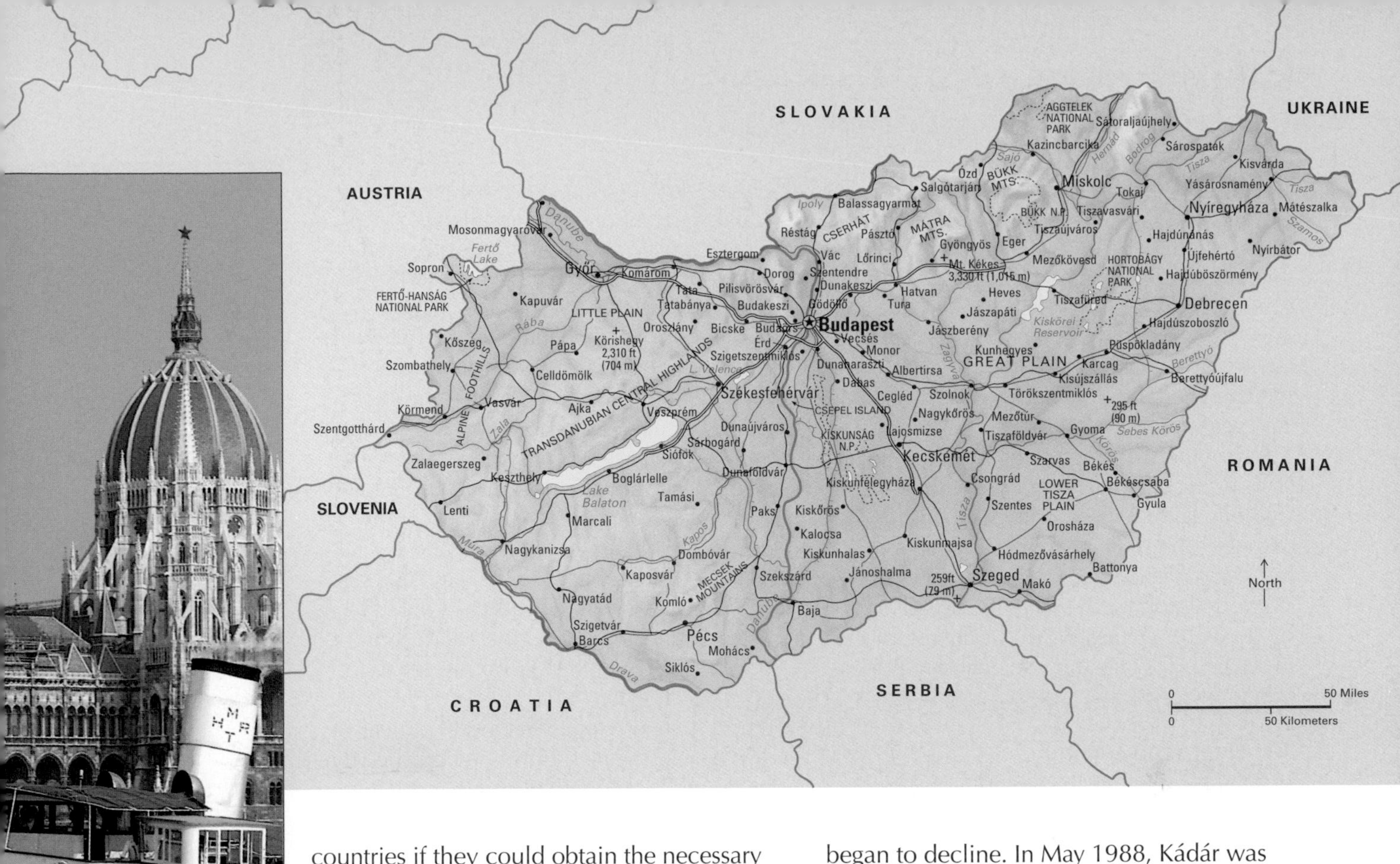

countries if they could obtain the necessary foreign currencies—a major drawback for most Hungarians. Cultural life was also relatively free from government restrictions during this period, and Hungarians were allowed to express their views without fear of censorship or punishment.

During the 1970's, however, the Hungarian economy worsened, and living standards began to decline. In May 1988, Kádár was forced to step down, and Károly Grósz replaced him as head of the Communist Party.

In January 1989, the National Assembly, Hungary's legislature, passed a law guaranteeing Hungarians the right to demonstrate freely. Another new law allowed Hungarians to form associations and political parties independent of the Hungarian Socialist Worker's Party (HSWP), the Communist Party in Hungary.

Hungary's first multiparty elections were held in early 1990 and resulted in a bitter defeat for the Communists and a victory for the Democratic Forum. Arpád Göncz became president, and Jozsef Antall became prime minister.

Over the next four years, Hungarians became dissatisfied with the hardships of economic reforms as the country moved from a state-run system to privatization. In the 1994 parliamentary elections, the Hungarian Socialist Party—formerly the Communist Hungarian Socialist Party—won a clear majority and chose its leader, Gyula Horn, to be premier. Göncz was reelected president in 1995. He was succeeded by Ferenc Madl in 2000 and by Laszlo Solyom in 2005. Hungary became a member of the North Atlantic Treaty Organization (NATO) in 1999 and of the European Union in 2004.

Religions:
Roman Catholic 68%
Calvinist 20%
Lutheran 5%

*Enrollment ratios compare the number of students enrolled to the population which, by age, should be enrolled. A ratio higher than 100 indicates that students older or younger than the typical age range are also enrolled.

TECHNOLOGY

Radios per 1,000 people: 690

Televisions per 1,000 people: 475

Computers per 1,000 people: 108.4

ECONOMY

Currency: Forint

Gross domestic product (GDP) in 2004: $149.3 billion U.S.

Real annual growth rate (2003–2004): 3.9%

GDP per capita (2004): $14,900 U.S.

Goods exported: Alumina, electronic equipment, agriculture and food products, pharmaceuticals, steel, transportation equipment, and wine

Goods imported: Machinery, automobiles, chemicals, electric power, iron ore, natural gas, petroleum

Trading partners: Germany, Austria, Italy, France

History

The history of the Hungarian state began in the late 800's with the arrival of the Magyars, who settled in the middle Danube Basin—the great lowland region bordering the Danube River that makes up most of present-day Hungary. The Magyar tribes were led by Árpád, whose great-grandson, Géza, organized the tribes into a united nation about 100 years later.

When Géza's son Stephen became the first king of Hungary, he brought his Roman Catholic faith to the country, and during Stephen's reign, Hungary became closely identified with the culture and politics of Western Europe. When the last Árpád king died in 1301, Hungary was firmly established as a Christian state and remained an independent kingdom for another 225 years.

The Hungarian Empire

Hungary reached the height of its political power and cultural influence under Matthias Corvinus Hunyadi, who ruled from 1458 to 1490. During that time, the country became a center of the Italian Renaissance, the great artistic and cultural movement that spread across Europe during the 1400's and 1500's.

A period of conflict and disorder followed the death of Matthias Corvinus Hunyadi in 1490, and the Hungarian Empire became seriously weakened. In 1526, the Turks seized control of the eastern third of the country, called Transylvania, and made it a principality dependent on them. The Austrian Habsburgs took the country's western and northern sections. Then, in the late 1600's, the Habsburgs drove the Turks out of Hungary and seized control of the entire country.

The harsh rule of the Habsburgs led to a nationwide revolt in 1703, led by Francis Rákóczi II, whose family included princes of Transylvania. The Habsburgs had crushed the rebellion by 1711, but they were persuaded to relax their rule and improve conditions in Hungary.

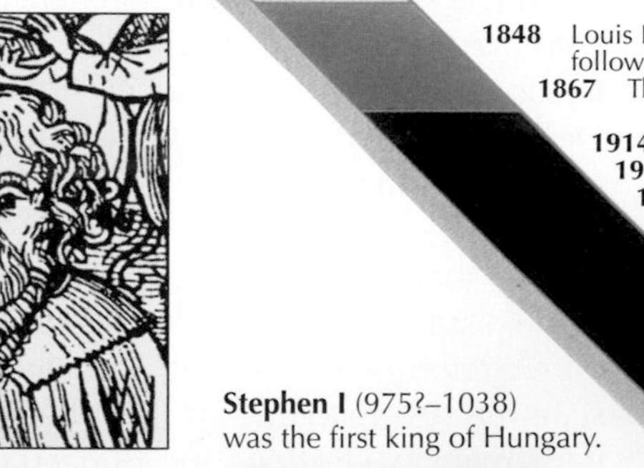

Late 800's The Magyars conquer Hungary.
1000 Stephen I becomes Hungary's first king and converts the country to Roman Catholicism.
1241 The Mongols invade Hungary.
1458-1490 Matthias Corvinus Hunyadi rules Hungary, which becomes a center of Renaissance culture.
1514 Hungarian nobles crush a peasant revolt.
1526 The Ottoman Turks defeat Hungary in the Battle of Mohács and soon occupy central and eastern Hungary.
1600's Austrian Habsburg forces drive the Turks out and take control of Hungary.
1703-1711 Francis Rákóczi II leads an unsuccessful uprising for independence.
1848 Louis Kossuth leads an anti-Habsburg revolution, which is defeated the following year.
1867 The Dual Monarchy of Austria-Hungary is established.
1914-1918 Austria-Hungary is defeated in World War I.
1918 Hungary becomes a republic.
1919 Béla Kun establishes the first Hungarian Communist government, which lasts only a few months.
1919-1944 Admiral Nicholas Horthy rules Hungary as regent.
1920 Under the Treaty of Trianon, Hungary loses two-thirds of its territory.
1941 Hungary enters World War II on Germany's side.
1944 Germany occupies Hungary.
1945 Hungary and the Allies sign an armistice.
1946 Hungary becomes a republic, and the new government introduces political, economic, and social reforms.
1946-1949 Hungarian Communists gradually gain control of the government.
1947 The Allies sign a peace treaty with Hungary that confirms the terms of the 1945 armistice.
1955 Hungary becomes a member of the United Nations.
1956 Soviet forces crush an anti-Communist revolution in Hungary.
1988 Hungary's Communist Party agrees to allow other political parties to operate in the country.
1989 Hungarians gain the right to demonstrate freely.
1990 Democratic elections are held.
1999 Hungary becomes a member of the North Atlantic Treaty Organization.

A magnificent crown now preserved in the National Museum of Budapest is said to be that of King Stephen I, who was made a saint in 1083 for bringing Roman Catholicism to Hungary.

Stephen I (975?–1038) was the first king of Hungary.

Franz Liszt (1811–1886), *far left,* was a Hungarian pianist, composer, and teacher.

Imre Nagy (1896–1958) served as premier of Hungary from 1953 to 1955.

Armed citizens patrol Budapest during the anti-Communist uprising of October 1956. A month later, Soviet forces invaded Hungary and brutally crushed the revolt. Many Hungarians were killed or imprisoned, and about 200,000 people fled the country.

A grotesque figure of a Turkish warrior, *above,* stands at the site of the Battle of Mohács, where the Ottoman Turks defeated the Hungarians in 1526.

During the 1840's, revolutions broke out across Europe, and Hungary tried to break away from Austria. Led by Louis Kossuth, the Hungarians proclaimed their independence. The revolution was eventually crushed by the Habsburgs, but only with the help of Russian troops. Once again, Hungary found itself under Austrian rule.

As Austrian power declined during the 1800's, the Hungarians, led by Francis Deák, were able to force the emperor of Austria to establish the *Dual Monarchy.* Under this arrangement, both Austria and Hungary had the same monarch and conducted foreign, military, and certain financial affairs jointly. But each country had its own government to handle all other matters.

The creation of the Dual Monarchy of Austria-Hungary in 1867 led to great prosperity in Hungary. Over the next 50 years, the nation's economy, educational system, and cultural life developed rapidly.

The end of Austria-Hungary

In 1918, Austria-Hungary's defeat in World War I brought an end to the Dual Monarchy. Thirteen days after Austria-Hungary signed an armistice with the Allied powers, the Hungarian people revolted, declaring Hungary a republic. After a brief period under a Communist dictatorship, Hungary again became a monarchy under Admiral Nicholas Horthy, who ruled as *regent* (in place of a monarch) from 1919 until 1944, when Hitler's troops seized the country and set up a Nazi government.

The Soviet Union invaded Hungary late in 1944, and Hungary and the Allies signed an armistice in January 1945. Early in 1946, Hungary was declared a republic, but the Communists gradually gained control of the government. By 1947, Matthias Rákosi, head of the Communist Party, ruled as dictator. Rákosi's policies resulted in a period of severe persecution for the Hungarian people that did not end until the 1960's.

Hungary once stood at the center of the empire of Austria-Hungary—the Dual Monarchy that reached its greatest height before World War I. After the heir to the Austro-Hungarian throne was assassinated by a Bosnian student in 1914, Austria-Hungary declared war on Serbia, touching off World War I.

Environment

Most of Hungary's land is low. Eastern Hungary is almost entirely flat, while western Hungary consists mainly of rolling hills and low mountains. The Tisza—the country's longest river and a tributary of Hungary's most important river, the Danube—flows from north to south through eastern Hungary.

Land regions

Hungary's four main land regions are the Great Plain, Transdanubia, the Little Plain, and the Northern Highlands. The Great Plain stretches across all of Hungary east of the Danube, except for the mountains in the north. This flat plain, broken only by river valleys, sand dunes, and small hills, covers about half of Hungary's total land area.

In the 1500's and 1600's, when Hungary fell under Turkish rule, the fertile farmland of the Great Plain was a *puszta* (steppeland), where cattle grazed on the tall prairie grasses and wild horses roamed. Today, parts of the puszta are preserved at Hortobágy, a national park about 109 miles (175 kilometers) east of Budapest.

Crops of sugar beets, melons, sunflowers, and wheat now flourish on the Great Plain. With modern irrigation, orchards and vineyards now flourish in the sandy areas. And swamps that once provided a habitat for waterfowl have been drained to provide farmland for corn production.

Transdanubia, a region that consists mostly of hills and mountains, covers all of Hungary west of the Danube, except for the northwest corner of the country. The Transdanubian Central Highlands—a chain of low, rounded mountains that includes the Bakony and Vertese ranges—extend along the entire northern side of Lake Balaton.

A region of great scenic beauty, the Transdanubian Central Highlands feature mountain streams, great oak forests, and picturesque ravines plunging between chalk and dolomite cliffs. The foothills of the Austrian Alps rise in the western region of Transdanubia, while the southeastern part is a major farming region.

The smallest region, the Little Plain, occupies the northwest corner of Hungary. The land in this area is flat, except for the foothills of the Austrian Alps along the western boundary. The steep slopes of the

Horses, *right,* gallop across Hungary's puszta conservation area at Hortobágy.

A country market, *above,* provides villagers with a gathering place as well as a place to sell their poultry.

Farmworkers use long-handled hoes to clear the ground of weeds. When the Communists controlled Hungary, most farming was done on state farms owned by the government and on collective farms where many families worked together.

Often called the Hungarian Sea, Lake Balaton, *right,* is central Europe's largest lake, covering about 230 square miles (596 square kilometers). Averaging 10 feet (3 meters) deep, the lake is easily warmed by the sun, making it a popular recreation spot.

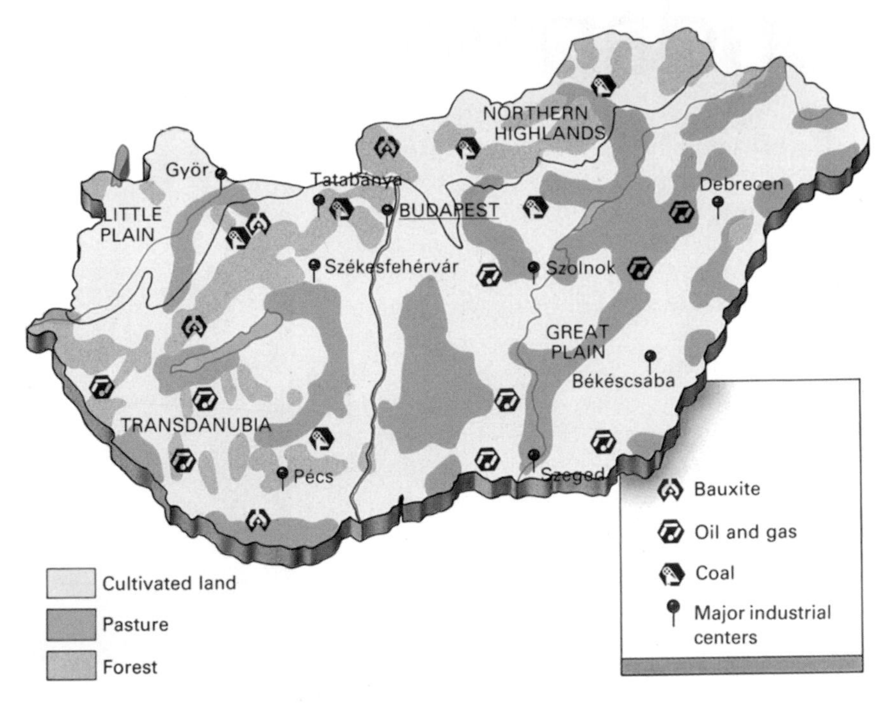

Farmland covers about three-fourths of Hungary, and agriculture is an important industry. Only about 15 per cent of Hungary is forested. Large amounts of timber, as well as coal and iron, must be imported.

Green paint distinguishes the gates of the steelworks in Ózd, a large industrial town in northern Hungary, close to the Czechoslovakian border. The factory uses locally mined iron ore to produce iron and steel.

Northern Highlands rise northeast of the Danube and north of the Great Plain. This region is densely forested, with small streams and dramatic rock formations.

Agriculture and industrialization

Hungary's most important natural resources are its fertile soil and a climate that is generally favorable for agriculture. Although the nation has been heavily industrialized since World War II, farming remains an important industry. In addition to growing crops such as corn, potatoes, grapes, sugar beets, wheat, and wine grapes, Hungary's farmers raise chickens, hogs, and other livestock.

Factories in Hungary produce iron and steel, buses and railroad equipment, electrical and electronic goods, food products, pharmaceuticals, medical and scientific equipment, and textiles. Hungary is also one of the world's leading producers of bauxite.

Since 1988, Hungary has been in the process of changing from a socialized economy to a free market economy. Private industry is now encouraged, and the government hopes foreign investors will bring businesses and jobs to the country.

Some Hungarians hope that their country will eventually join the European Union (EU). However, the government faces many economic challenges, including large foreign debts, out-of-date and inefficient factories, declining productivity, unemployment, and inflation.

Budapest

Perhaps no other city in Hungary can match the splendor of Budapest. The capital and largest city, it is also the center of Hungary's culture and industry, and represents the spiritual and cultural home of the 6 million Hungarians living outside the country.

A union of cities

Situated on both banks of the Danube River in northern Hungary, modern Budapest consists of the once adjoining cities of Buda, Pest, and Óbuda. The city also includes Margaret Island, in the Danube River.

Budapest covers 203 square miles (525 square kilometers). Eight bridges connect the eastern and western banks of the Danube in Budapest.

The part of the city that used to be the city of Buda rests high on the west bank of the river. Buda was the permanent residence of the Hungarian kings beginning in the 1300's. Standing on steep, wooded hills and crowned by Castle Hill, Buda reflects Hungary's ancient military past. The Royal Palace, which includes the remains of an ancient fort, stands on top of Castle Hill.

In contrast, Pest, on the east side of the river, represents Hungary's coming of age as a modern European state. Most of Pest was developed in the 1800's on a series of plateaus. Its public buildings and boulevards rival those of Paris and Vienna in their style and elegance. The Hungarian Academy of Sciences and the House of Parliament are located in the old Inner City, called Belváros.

Budapest was united in 1873, when Hungary was part of the Dual Monarchy (1867–1914). At that time, the city was the scene of rapid developments in mechanical engineering, as well as in the milling and iron industries. Shipyards, breweries, and tobacco-processing factories also pros-

The ornate splendor of a Budapest restaurant dates from the late 1800's, when the city was an important center of Austria-Hungary.

The relaxing thermal waters of the Gellért Baths, *right,* built between 1912 and 1918 in the Art Nouveau style, are one of Budapest's famous tourist attractions. The city's 123 hot springs feed 12 baths, where people swim in the mineral waters.

Budapest is situated on both banks of the Danube River. A total of eight bridges connect the city's eastern and western sections. With the return of democracy to Hungary, Budapest has an opportunity to regain its traditional place as an important European center of commerce.

In A.D. 100, the Romans founded a town called Aquincum on the site of present-day Budapest. Today, Budapest, *below,* boasts a number of historic buildings reconstructed in their original style after being damaged during World War II. Budapest's most famous landmarks include the House of Parliament, the Academy of Sciences, St. Stephen's Basilica, the Royal Palace, the National Museum, and the Museum of Fine Arts.

Budapest also boasts several houses and hotels in the Art Nouveau style of the late 1800's and early 1900's, which have been greatly admired for their beauty and brilliant colors.

In 1887, the first electric trams appeared in Budapest, followed only nine years later by a 2-1/2-mile (4-kilometer) underground railway beneath the city's main boulevard. Residents of Budapest were the first Europeans to travel underground. Today, yellow carriages in their old livery still clatter along the line. However, the modern Metro, which runs under the Danube between Buda and Pest, is quite different. Modeled after the impressive Moscow Metro, it serves as one of the main arteries in the city's transport system.

Today, Budapest has an old-world, turn-of-the-century charm. Budapest's distinguished past lingers on in the city's coffee houses, with their ornate ceilings, gold columns, and marble tables. The citizens of Budapest enjoy meeting in these old coffee houses, where they read or chat with friends.

pered. As agricultural and foreign workers flocked to find jobs in the city, Budapest's population increased from about 178,000 in 1850 to about 733,000 in 1900.

The golden age

The era of the Dual Monarchy is often regarded as Budapest's architectural golden age. The Hungarians wanted Budapest to be as splendid as Vienna, the capital of the Habsburg Empire, and Budapest took on its present-day character during this period.

The old heart of the city in Pest was torn down because the baroque and neoclassical buildings erected between the 1500's and 1700's were considered too small and old-fashioned. They were replaced by buildings whose architecture was influenced by the grandest styles of earlier times. For example, the domed Parliament Building was designed in the neo-Gothic style and based on medieval models.

The Royal Palace is neo-Baroque, while other buildings followed the Italian-inspired styles of the 1400's and 1500's.

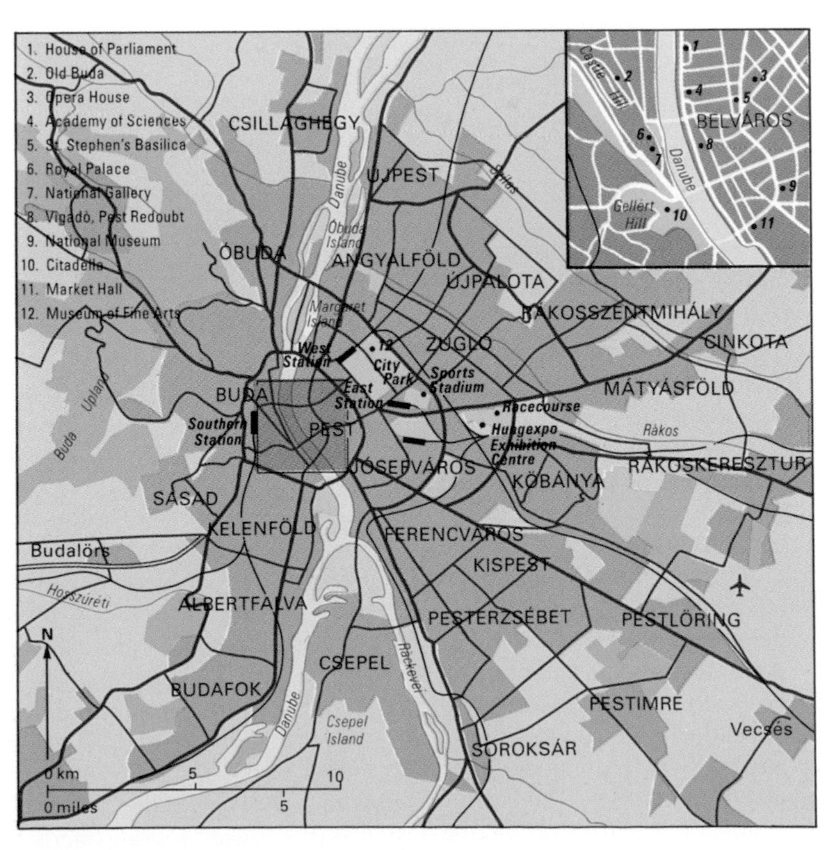

East European Roma

Roma are a nomadic people whose ancestors originally lived in India. They are sometimes called Gypsies, Romanies, or Travellers. About A.D. 1000, they left India and wandered westward through the Middle East, arriving in Europe at the beginning of the 1300's.

Some Roma claimed to have come from a country called Little Egypt. The word *Gypsy* is a shortened form of *Egyptian.* However, most groups prefer the name *Roma* for the people in general rather than *Gypsy,* because *Gypsy* has sometimes been used as an insult. Today, Roma live in all parts of the world, but the largest numbers are found in eastern Europe. According to some estimates, there are about 12 million Roma living throughout the world.

Because Roma have traditionally been a wandering people who chose to live outside the mainstream of society, little is known of their history and culture. As a result, there is an air of romantic mystery about them, and their colorful costumes, lively music, and dancing skills have fascinated people for generations. At the same time, however, they have suffered a great deal of persecution. Even today, Roma experience high rates of poverty, unemployment, and illiteracy.

A history of persecution

At first, the Roma were welcomed in Europe. The European nobility admired—and exploited—their ironworking skills, and the Spanish rulers Ferdinand and Isabella may have used Roma-made weapons to defeat the Moors at Granada in 1492. In Hungary, Roma were employed in the manufacture of instruments of torture as well as the making of weapons.

Although the Spanish Roma were poorly rewarded for their work, they fared much better than the Hungarian Roma, who were slaves to the Magyar kings. The Roma were enslaved by the Romanian nobility, who needed laborers to work their vast estates, and by the Romanian clergy, who believed that Jesus Christ had cursed the Roma. Sold at auction in slave markets, the Roma were forced to work under the most brutal conditions.

These Roma in bright, colorful costumes are taking part in a festival in Hungary.

As the years went by, prejudice against the Roma grew, and they were blamed for a variety of crimes, from theft to kidnapping. In 1782, Hungarian Roma were even accused of cannibalism, and many were driven to the swamps and drowned by Hungarian soldiers.

During World War II, Adolf Hitler condemned the Roma, along with various other religious and ethnic groups, as "racially impure." The Nazis rounded up the Roma and imprisoned them in concentration camps. About 220,000 Roma were murdered in these camps during the war. After the Communists gained control of eastern Europe, they, too, condemned the Roma for their failure to be contributing members of a socialist society.

A love of music and dance

Feelings of hostility toward the Roma have been accompanied by an appreciation of their skills as musicians and dancers. Roma have served as entertainers since the late 1400's, when they were employed as musicians at the court of Matthias Corvinus Hunyadi. Touring bands of Roma with trained

A tame brown bear is caught in the glow of a Romani campfire. Trained bears were once part of the Roma's traveling shows. In addition to providing entertainment, Roma have made their living as fortunetellers and horse traders.

This Romani family in Tulcea, Romania, travels by horse-drawn wagon. Many Romani families consist of a husband and wife, their unmarried children, their married sons, and the sons' wives and children.

A Romanian Romani woman wears the flowery headscarf characteristic of the region. Most Roma speak the language of the people among whom they live, but many also speak their own native tongue, often called *Romany.*

Romani violinists entertain the crowds at a country fair in Hungary, *right.* The violin has a special place in Romani music, and the Roma believe that the wood for the first Romani violin came from the dense forests of Transylvania.

bears once provided entertainment in remote Romanian villages. The Magyar language even has a special word for being entertained by Roma. It is *cigányozni (Gypsying).*

Roma played not only at banquets and special occasions, but also at military events. In the Austro-Hungarian army, recruiting officers had Romani musicians play tunes called *verbunkos* to stir their patriotic spirit, and Roma also provided the music that led troops into battle.

The Roma of southern Spain were the first to perform the *flamenco* style of dance and music that is still popular today. Flamenco dancing, accompanied by guitars and castanets, may include much skillful footwork, finger snapping, and forceful but flowing arm movements.

Romani music played an important role in the works of Franz Liszt, the Hungarian pianist and composer. His *Hungarian Rhapsodies,* in which he used Romani music and folk-dance themes, are among his best-loved compositions. In describing Romani music, Liszt wrote, "The chief characteristic of this music is the freedom, richness, variety, and versatility of its rhythms, found nowhere else in a like degree."